READING IN THE BRAIN

Reading in the Brain

The Science and Evolution
of a Human Invention

STANISLAS DEHAENE

VIKING

VIKING
Published by the Penguin Group
Penguin Group (USA) Inc., 375 Hudson Street, New York, New York 10014, U.S.A.
Penguin Group (Canada), 90 Eglinton Avenue East, Suite 700, Toronto, Ontario,
Canada M4P 2Y3 (a division of Pearson Penguin Canada Inc.)
Penguin Books Ltd, 80 Strand, London WC2R 0RL, England
Penguin Ireland, 25 St. Stephen's Green, Dublin 2, Ireland. (a division of Penguin Books Ltd)
Penguin Books Australia Ltd, 250 Camberwell Road, Camberwell, Victoria 3124, Australia
(a division of Pearson Australia Group Pty Ltd)
Penguin Books India Pvt Ltd, 11 Community Centre, Panchsheel Park, New Delhi – 110 017, India
Penguin Group (NZ), 67 Apollo Drive, Rosedale, North Shore 0632, New Zealand
(a division of Pearson New Zealand Ltd)
Penguin Books (South Africa) (Pty) Ltd, 24 Sturdee Avenue, Rosebank, Johannesburg 2196, South Africa

Penguin Books Ltd, Registered Offices: 80 Strand, London WC2R 0RL, England

First published in 2009 by Viking Penguin, a member of Penguin Group (USA) Inc.

10 9 8 7 6 5 4 3 2

Illustration credits appear on pages 387 and 388.

LIBRARY OF CONGRESS CATALOGING-IN-PUBLICATION DATA
Deheane, Stanislas.
Reading in the brain : the science and evolution of a human invention / Stanislas Deheane.
p. cm.
Includes bibliographical references and index.
ISBN 978-0-670-02110-9
1. Reading, Psychology of.   2.   Reading—Physiological aspects.   I.   Title.
BF456.R2D36 2009
418'.4019—dc22   2009009389

Printed in the United States of America
Set in Warnock Pro
Designed by Amy Hill

For Ghislaine

CONTENTS

CHAPTER 2

The Brain's Letterbox   53

CHAPTER 3

The Reading Ape   121

CHAPTER 4

Inventing Reading   171

CHAPTER 5

Learning to Read   195

Getting oriented in the brain

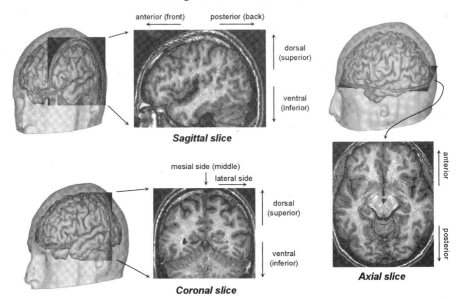

anterior (front) ← → posterior (back)

dorsal (superior)

ventral (inferior)

Sagittal slice

mesial side (middle) → lateral side

dorsal (superior)

ventral (inferior)

Coronal slice

anterior

posterior

Axial slice

The major lobes

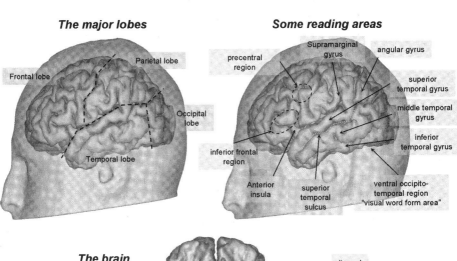

Parietal lobe

Frontal lobe

Occipital lobe

Temporal lobe

Some reading areas

Supramarginal gyrus

angular gyrus

precentral region

superior temporal gyrus

middle temporal gyrus

inferior temporal gyrus

inferior frontal region

Anterior insula

superior temporal sulcus

ventral occipito-temporal region "visual word form area"

The brain seen from underneath

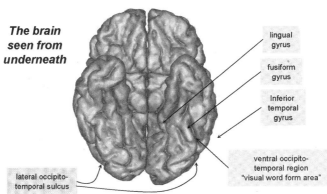

lingual gyrus

fusiform gyrus

Inferior temporal gyrus

ventral occipito-temporal region "visual word form area"

lateral occipito-temporal sulcus

The New Science of Reading

Withdrawn into the peace of this desert,
along with some books, few but wise,
I live in conversation with the deceased,
and listen to the dead with my eyes

—FRANCISCO DE QUEVEDO

*A*t this very moment, your brain is accomplishing an amazing feat—reading. Your eyes scan the page in short spasmodic movements. Four or five times per second, your gaze stops just long enough to recognize one or two words. You are, of course, unaware of this jerky intake of information. Only the sounds and meanings of the words reach your conscious mind. But how can a few black marks on white paper projected onto your retina evoke an entire universe, as Vladimir Nabokov does in the opening lines of *Lolita*:

Lolita, light of my life, fire of my loins. My sin, my soul.
Lo-lee-ta: the tip of the tongue taking a trip of three steps
down the palate to tap, at three, on the teeth. Lo. Lee. Ta.

The reader's brain contains a complicated set of mechanisms admirably attuned to reading. For a great many centuries, this talent remained a mystery. Today, the brain's black box is cracked open and a true science of reading is coming into being. Advances in psychology and neuroscience over the last twenty years have begun to unravel the principles underlying the brain's reading circuits. Modern brain imaging methods now reveal, in just a matter of minutes, the brain areas that activate when we decipher written words. Scientists can track a printed word as it progresses from the retina through a chain of processing stages, each of which is marked by an elementary question: Are these letters? What do they look like? Are they a word? What does it sound like? How is it pronounced? What does it mean?

On this empirical ground, a theory of reading is materializing. It postulates that the brain circuitry inherited from our primate evolution can be co-opted to the task of recognizing printed words. According to this approach, our neuronal networks are literally "recycled" for reading. The insight into how literacy changes the brain is profoundly transforming our vision of education and learning disabilities. New remediation programs are being conceived that should, in time, cope with the debilitating incapacity to decipher words known as dyslexia.

My purpose in this book is to share my knowledge of recent and little-known advances in the science of reading. In the twenty-first century, the average person still has a better idea of how a car works than of the inner functioning of his own brain—a curious and shocking state of affairs. Decision makers in our education systems swing back and forth with the changing winds of pedagogical reform, often blatantly ignoring how the brain actually learns to read. Parents, educators, and politicians often recognize that there is a gap between educational programs and the most up-to-date findings in neuroscience. But too frequently their idea of how this field can contribute to advances in education is only grounded in a few color pictures of the brain at work. Unfortunately, the imaging techniques that allow us to visualize brain activity are subtle and occasionally misleading. The new science of reading is so young and fast-moving that it is still relatively unknown outside the scientific community. My goal is to provide a simple introduction to this exciting field, and to increase awareness of the amazing capacities of our reading brains.

From Neurons to Education

Reading acquisition is a major step in child development. Many children initially struggle with reading, and surveys indicate that about one adult in ten fails to master even the rudiments of text comprehension. Years of hard work are needed before the clockwork-like brain machinery that supports reading runs so smoothly that we forget it exists.

Why is reading so difficult to master? What profound alterations in brain circuitry accompany the acquisition of reading? Are some teaching strategies better adapted to the child's brain than others? What scientific reasons, if any, explain why phonics—the systematic teaching of letter-to-sound correspondences—seems to work better than whole-word teaching? Although much still remains to be discovered, the new science of reading is now providing increasingly precise answers to all these questions. In particular, it underlines why early research on reading erroneously supported the whole-word approach—and how recent research on the brain's reading networks proves it was wrong.

Understanding what goes into reading also sheds light on its pathologies. In our explorations of the reader's mind and brain, you will be introduced to patients who suddenly lost the ability to read following a stroke. I will also analyze the causes of dyslexia, whose cerebral underpinnings are gradually coming to light. It is now clear that the dyslexic brain is subtly different from the brain of a normal reader. Several dyslexia susceptibility genes have been identified. But this is by no means a reason for discouragement or resignation. New intervention therapies are now being defined. Intensive retraining of language and reading circuits has brought about major improvements in children's brains that can readily be tracked with brain imaging.

Putting Neurons into Culture

Our ability to read brings us face-to-face with the singularity of the human brain. Why is *Homo sapiens* the only species that actively teaches itself? Why is he unique in his ability to transmit a sophisticated culture? How does the biological world of synapses and neurons relate to the universe of human cultural inventions? Reading, but also writing, mathematics, art, religion, agriculture, and city life have dramatically increased the native capacities of our primate brains. Our species alone rises above its biological condition, creates an artificial cultural environment for itself, and teaches itself new skills like reading. This uniquely human competence is puzzling and calls for a theoretical explanation.

One of the basic techniques in the neurobiologist's toolkit consists of "putting neurons in culture"—letting neurons grow in a petri dish. In this book,

I call for a different "culture of neurons"—a new way of looking at human cultural activities, based on our understanding of how they map onto the brain networks that support them. Neuroscience's avowed goal is to describe how the elementary components of the nervous system lead to the behavioral regularities that can be observed in children and adults (including advanced cognitive skills). Reading provides one of the most appropriate test beds for this "neurocultural" approach. We are increasingly aware of how writing systems as different as Chinese, Hebrew, or English get inscribed in our brain circuits. In the case of reading, we can clearly draw direct links between our native neuronal architecture and our acquired cultural abilities—but the hope is that this neuroscience approach will extend to other major domains of human cultural expression.

The Mystery of the Reading Ape

If we are to reconsider the relation between brain and culture, we must address an enigma, which I call the *reading paradox*: Why does our primate brain read? Why does it have an inclination for reading although this cultural activity was invented only a few thousand years ago?

There are good reasons why this deceptively simple question deserves to be called a paradox. We have discovered that the literate brain contains specialized cortical mechanisms that are exquisitely attuned to the recognition of written words. Even more surprisingly, the same mechanisms, in all humans, are systematically housed in identical brain regions, as though there were a cerebral organ for reading.

But writing was born only fifty-four hundred years ago in the Fertile Crescent, and the alphabet itself is only thirty-eight hundred years old. These time spans are a mere trifle in evolutionary terms. Evolution thus did not have the time to develop specialized reading circuits in *Homo sapiens*. Our brain is built on the genetic blueprint that allowed our hunter-gatherer ancestors to survive. We take delight in reading Nabokov and Shakespeare using a primate brain originally designed for life in the African savanna. Nothing in our evolution could have prepared us to absorb language through vision. Yet brain imaging demonstrates that the adult brain contains fixed circuitry exquisitely attuned to reading.

The reading paradox is reminiscent of the Reverend William Paley's parable aimed at proving the existence of God. In his *Natural Theology* (1802), he imagined that in a deserted heath, a watch was found on the ground, complete with its intricate inner workings clearly designed to measure time. Wouldn't it provide, he argued, clear proof that there is an intelligent clockmaker, a designer who purposely created the watch? Similarly, Paley maintained that the intricate devices that we find in living organisms, such as the astonishing mechanisms of the eye, prove that nature is the work of a divine watchmaker.

Charles Darwin famously refuted Paley by showing how blind natural selection can produce highly organized structures. Even if biological organisms at first glance seem designed for a specific purpose, closer examination reveals that their organization falls short of the perfection that one would expect from an omnipotent architect. All sorts of imperfections attest that evolution is not guided by an intelligent creator, but follows random paths in the struggle for survival. In the retina, for example, blood vessels and nerve cables are situated *in front of* the photoreceptors, thus partially blocking incoming light and creating a blind spot—very poor design indeed.

Following in Darwin's footsteps, Stephen Jay Gould provided many examples of the imperfect outcome of natural selection, including the panda's thumb.[1] The British evolutionist Richard Dawkins also explained how the delicate mechanisms of the eye or of the wing could only have emerged through natural selection or are the work of a "blind watchmaker."[2] Darwin's evolutionism seems to be the only source of apparent "design" in nature.

When it comes to explaining reading, however, Paley's parable is problematic in a subtly different way. The clockwork-like brain mechanisms that support reading are certainly comparable in complexity and sheer design to those of the watch abandoned on the heath. Their entire organization leans toward the single apparent goal of decoding written words as quickly and accurately as possible. Yet neither the hypothesis of an intelligent creator nor that of slow emergence through natural selection seems to provide a plausible explanation for the origins of reading. Time was simply too short for evolution to design specialized reading circuits. How, then, did our primate brain learn to read? Our cortex is the outcome of millions of years of evolution in a world without writing—why can it adapt to the specific challenges posed by written word recognition?

Biological Unity and Cultural Diversity

In the social sciences, the acquisition of cultural skills such as reading, mathematics, or the fine arts is rarely, if ever, posed in biological terms. Until recently, very few social scientists considered that brain biology and evolutionary theory were even relevant to their fields. Even today, most implicitly subscribe to a naïve model of the brain, tacitly viewing it as an infinitely plastic organ whose learning capacity is so broad that it places no constraints on the scope of human activity. This is not a new idea. It can be traced back to the theories of the British empiricists John Locke, David Hume, and George Berkeley, who claimed that the human brain should be compared to a blank slate that progressively absorbs the imprint of man's natural and cultural environment through the five senses.

This view of mankind, which denies the very existence of a human nature, has often been adopted without question. It belongs to the default "standard social science model"[3] shared by many anthropologists, sociologists, some psychologists, and even a few neuroscientists who view the cortical surface as "largely equipotent and free of domain-specific structure."[4] It holds that human nature is constructed, gradually and flexibly, through cultural impregnation. As a result, children born to the Inuit, to the hunter-gatherers of the Amazon, or to an Upper East Side New York family, according to this view, have little in common. Even color perception, musical appreciation, or the notion of right and wrong should vary from one culture to the next, simply because the human brain has few stable structures other than the capacity to learn.

Empiricists further maintain that the human brain, unhindered by the limitations of biology and unlike that of any other animal species, can absorb any form of culture. From this theoretical perspective, to talk about the cerebral bases of cultural inventions such as reading is thus downright irrelevant—much like analyzing the atomic composition of a Shakespeare play.

In this book, I refute this simplistic view of an infinite adaptability of the brain to culture. New evidence on the cerebral circuits of reading demonstrates that the hypothesis of an equipotent brain is wrong. To be sure, if the brain were not capable of learning, it could not adapt to the specific rules for writing

English, Japanese, or Arabic. This learning, however, is tightly constrained, and its mechanisms themselves rigidly specified by our genes. The brain's architecture is similar in all members of the *Homo sapiens* family, and differs only slightly from that of other primates. All over the world, the same brain regions activate to decode a written word. Whether in French or in Chinese, learning to read necessarily goes through a genetically constrained circuit.

On the basis of these data, I propose a novel theory of neurocultural interactions, radically opposed to cultural relativism, and capable of resolving the reading paradox. I call it the "neuronal recycling" hypothesis. According to this view, human brain architecture obeys strong genetic constraints, but some circuits have evolved to tolerate a fringe of variability. Part of our visual system, for instance, is not hardwired, but remains open to changes in the environment. Within an otherwise well-structured brain, visual plasticity gave the ancient scribes the opportunity to invent reading.

In general, a range of brain circuits, defined by our genes, provides "pre-representations"[5] or hypotheses that our brain can entertain about future developments in its environment. During brain development, learning mechanisms select which pre-representations are best adapted to a given situation. Cultural acquisition rides on this fringe of brain plasticity. Far from being a blank slate that absorbs everything in its surroundings, our brain adapts to a given culture by minimally turning its predispositions to a different use. It is not a *tabula rasa* within which cultural constructions are amassed, but a very carefully structured device that manages to convert some of its parts to a new use. When we learn a new skill, we recycle some of our old primate brain circuits—insofar, of course, as those circuits can tolerate the change.

A Reader's Guide

In forthcoming chapters, I will explain how neuronal recycling can account for literacy, its mechanisms in the brain, and even its history. In the first three chapters, I analyze the mechanisms of reading in expert adults. Chapter 1 sets the stage by looking at reading from a psychological angle: how fast do we read, and what are the main determinants of reading behavior? In chapter 2, I move to the brain areas at work when we read, and how they can be visualized using modern brain imaging techniques. Finally, in chapter 3, I come down to the level of single neurons and their organization into the circuits that recognize letters and words.

I tackle my analysis in a resolutely mechanical way. I propose to expose the cogwheels of the reader's brain in much the same way as the Reverend Paley suggested we dismantle the watch abandoned on the heath. The reader's brain will not, however, reveal any perfect clockwork mechanics designed by a divine watchmaker. Our reading circuits contain more than a few imperfections that betray our brain's compromise between what is needed for reading and the available biological mechanisms. The peculiar characteristics of the primate visual system explain why reading does not operate like a fast and efficient scanner. As we move our eyes across the page, each word is slowly brought into the central region of our retina, only to be exploded into a myriad of fragments that our brain later pieces back together. It is only because these processes have become automatic and unconscious, thanks to years of practice, that we are under the illusion that reading is simple and effortless.

The reading paradox expresses the indisputable fact that our genes have not evolved in order to enable us to read. My reasoning in the face of this enigma is quite simple. If the brain did not evolve for reading, the opposite must be true: writing systems must have evolved within our brain's constraints. Chapter 4 revisits the history of writing in this light, starting with the first prehistoric symbols and ending with the invention of the alphabet. At each step, there is evidence of constant cultural tinkering. Over many millennia, the scribes struggled to design words, signs, and alphabets that could fit the limits of our primate brain. To this day, the world's writing systems still share a number of design features that can ultimately be traced back to the restrictions imposed by our brain circuits.

Continuing on the idea that our brain was not designed for reading, but recycles some of its circuits for this novel cultural activity, chapter 5 examines how children learn to read. Psychological research concludes that there are not many ways to convert a primate brain into that of an expert reader. This chapter explores in some detail the only developmental trajectory that appears to exist. Schools might be well advised to exploit this knowledge to optimize the teaching of reading and mitigate the dramatic effects of illiteracy and dyslexia.

I will also go on to show how a neuroscientific approach can shed light on the more mysterious features of reading acquisition. For instance, why do so many children often write their first words from right to left? Contrary to the accepted idea, these mirror inversion errors are not the first signs of dyslexia, but a natural consequence of the organization of our visual brain. In a majority of children, dyslexia relates to another, quite distinct anomaly in processing speech sounds. The description of the symptoms of dyslexia, their cerebral bases, and the most recent discoveries concerning its genetic foundations are covered in chapter 6, while chapter 7 provides an insight into what mirror errors can tell us about normal visual recognition.

Finally, in chapter 8, I will return to the astonishing fact that only our species is capable of cultural inventions as sophisticated as reading—a unique feat, unmatched by any other primate. In total opposition to the standard social science model, where culture gets a free ride on a blank-slate brain, reading demonstrates how culture and brain organization are inextricably linked. Throughout their long cultural history, human beings progressively discovered that they could reuse their visual systems as surrogate language inputs, thus arriving at reading and writing. I will also briefly discuss how other major human cultural traits could be submitted to a similar analysis. Mathematics, art, music, and religion might also be looked on as evolved devices, shaped by centuries of cultural evolution, that have encroached on our primate brains.

One last enigma remains: if learning exists in all primates, why is *Homo sapiens* the only species with a sophisticated culture? Although the term is sometimes applied to chimpanzees, their "culture" barely goes beyond a few good tricks for splitting nuts, washing potatoes or fishing ants with a stick—nothing comparable to the seemingly endless human production of interlocking conventions and symbols systems, including languages, religions, art forms, sports, mathematics or medicine. Nonhuman primates can slowly learn to recognize novel symbols such as letters and digits—but they never think of inventing them. In my conclusion, I propose some tentative ideas on the singularity of the human brain. The uniqueness of our species may arise from a combination of two factors: a theory of mind (the ability to imagine the mind of others) and a conscious global workspace (an internal buffer where an infinite variety of ideas can be recombined). Both mechanisms, inscribed in our genes, conspire to make us the only cultural species. The seemingly infinite variety of human cultures is only an illusion, caused by the fact that we are locked in a cognitive vicious circle: how could we possibly imagine forms other than those our brains can conceive? Reading, although a recent invention, lay dormant for millennia within the envelope of potentialities inscribed in our brains. Behind the apparent diversity of human writing systems lies a core set of universal neuronal mechanisms that, like a watermark, reveal the constraints of human nature.

CHAPTER 1

How Do We Read?

Written word processing starts in our eyes. Only the center of the retina, called the fovea, has a fine enough resolution to allow for the recognition of small print. Our gaze must therefore move around the page constantly. Whenever our eyes stop, we only recognize one or two words. Each of them is then split up into myriad fragments by retinal neurons and must be put back together before it can be recognized. Our visual system progressively extracts graphemes, syllables, prefixes, suffixes, and word roots. Two major parallel processing routes eventually come into play: the phonological route, which converts letters into speech sounds, and the lexical route, which gives access to a mental dictionary of word meanings.

> The existence of the text is a silent existence, silent until the moment in which a reader reads it. Only when the able eye makes contact with the markings on the tablet does the text come to active life. All writing depends on the generosity of the reader.
>
> —ALBERTO MANGUEL, *THE HISTORY OF READING*

*A*t first sight, reading seems close to magical: our gaze lands on a word, and our brain effortlessly gives us access to its meaning and pronunciation. But in spite of appearances, the process is far from simple. Upon entering the retina, a word is split up into a myriad of fragments, as each part of the visual image is recognized by a distinct photoreceptor. Starting from this input, the real challenge consists in putting the pieces back together in order to decode what letters are present, to figure out the order in which they appear, and finally to identify the word.

Over the past thirty years, cognitive psychology has worked on analyzing the mechanics of reading. Its goal is to crack the "algorithm" of visual word recognition—the series of processing steps that a proficient reader applies to the problem of identifying written words. Psychologists treat reading like a computer science problem. Every reader resembles a robot with two cameras—the two eyes and their retinas. The words we read are painted onto them. They first appear only as splotches of light and dark that are not directly interpretable as linguistic signs. Visual information must be recoded in an understandable format before we can access the appropriate sounds, words, and meanings. Thus we must have a deciphering algorithm, or a processing recipe akin to automatic character recognition software, which takes the pixels on a page as input and produces the identity of the words as output. To accomplish this feat, unbeknownst to us, our brain hosts a sophisticated set of decoding operations whose principles are only beginning to be understood.

The Eye: A Poor Scanner

The tale of reading begins when the retina receives photons reflected off the written page. But the retina is not a homogeneous sensor. Only its central part, called the fovea, is dense in high-resolution cells sensitive to incoming light, while the rest of the retina has a coarser resolution. The fovea, which occupies about 15 degrees of the visual field, is the only part of the retina that is genuinely useful for reading. When foveal information is lacking, whether due to a retinal lesion, to a stroke having destroyed the central part of the visual cortex, or to an experimental trick that selectively blocks visual inputs to the fovea, reading becomes impossible.[6]

The need to bring words into the fovea explains why our eyes are in constant motion when we read. By orienting our gaze, we "scan" text with the most sensitive part of our vision, the only one that has the resolution needed to determine letters. However, our eyes do not travel continuously across the page.[7] Quite the opposite: they move in small steps called saccades. At this very moment, you are making four or five of these jerky movements every second, in order to bring new information to your fovea.

Even within the fovea, visual information is not represented with the same precision at all points. In the retina as well as in the subsequent visual relays of the thalamus and of the cortex, the number of cells allocated to a given portion of the visual scene decreases progressively as one moves away from the center of gaze. This causes a gradual loss of visual precision. Visual accuracy is optimal at the center and smoothly decreases toward the periphery. We have the illusion of seeing the whole scene in front of us with the same fixed accuracy, as if it were filmed by a digital camera with a homogeneous array of pixels. However, unlike the camera, our eye sensor accurately perceives only the precise point where our gaze happens to land. The surroundings are lost in an increasingly hazy blurriness (figure 1.1).[8]

One might think that, under these conditions, it is the absolute size of printed characters that determines the ease with which we can read: small

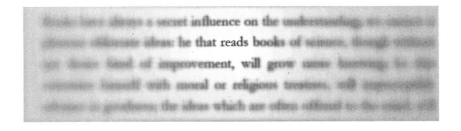

Figure 1.1 The retina stringently filters what we read. In this simulation, a page from Samuel Johnson's *The Adventurer* (1754) was filtered using an algorithm that copies the decreasing acuity of human vision away from the center of the retina. Regardless of size, only letters close to fixation can be identified. This is why we constantly explore pages with jerky eye movements when we read. When our gaze stops, we can only identify one or two words.

letters should be harder to read than larger ones. Oddly enough, however, this is not the case. The reason is that the larger the characters, the more room they use on the retina. When a whole word is printed in larger letters, it moves into the periphery of the retina, where even large letters are hard to discern. The two factors compensate for each other almost exactly, so that an enormous word and a minuscule one are essentially equivalent from the point of view of retinal precision. Of course, this is only true provided that the size of the characters remains larger than an absolute minimum, which corresponds to the maximal precision attained at the center of our fovea. When visual acuity is diminished, for instance in aging patients, it is quite logical to recommend books in large print.

Because our eyes are organized in this way, our perceptual abilities depend exclusively on the number of letters in words, not on the space these words occupy on our retina.[9] Indeed, our saccades when we read vary in absolute size, but are constant when measured in numbers of letters. When the brain prepares to move our eyes, it adapts the distance to be covered to the size of the characters, in order to ensure that our gaze always advances by about seven to nine letters. This value, which is amazingly small, thus corresponds approximately to the information that we can process in the course of a single eye fixation.

To prove that we see only a very small part of each page at a time, George W. McConkie and Keith Rayner developed an experimental method that I like to call the "Cartesian devil." In his *Metaphysical Meditations*, René Descartes imagined that an evil genius was playing with our senses:

> I shall then suppose, not that God who is supremely good and the fountain of truth, but some evil genius not less powerful than deceitful, has employed all his energy to deceive me; I shall consider that the heavens, the earth, colors, figures, sound, and all other external things are naught but the illusions and dreams of which this genius has availed himself in order to lay traps for my credulity. I shall consider myself as having no hands, no eyes, no flesh, no blood, nor any senses, yet falsely believing myself to possess all these things.

Much like the supercomputer in the *Matrix* movies, Descartes' evil genius produces a pseudo-reality by bombarding our senses with signals carefully crafted to create an illusion of real life, a virtual scene whose true side remains forever hidden. More modestly, McConkie and Rayner designed a "moving window" that creates an illusion of text on a computer screen.[10] The method consists in equipping a human volunteer with a special device that tracks eye movements and can change the visual display in real time. The device can be programmed to display only a few characters left and right of the center of gaze, while all of the remaining letters on the page are replaced with strings of x's:

We the pexxxx xx xxx xxxxxx xxxxxx, xx xxxxx xx
↑

As soon as the eyes move, the computer discreetly refreshes the display. Its goal is to show the appropriate letters at the place where the person is looking, and strings of x's everywhere else:

Xx xxx people of txx xxxxxx xxxxxx, xx xxxxx xx
↑

Xx xxx xxxxxx xx xhe United xxxxxx, xx xxxxx xx
↑

Xx xxx xxxxxx xx xxx Xxxxed States, ix xxxxx xx
↑

Xx xxx xxxxxx xx xxx Xxxxxx Xxxxxx, in order to
↑

Using this device, McConkie and Rayner made a remarkable and para-doxical discovery. They found that the participants did not notice the manipulation. As long as enough letters are presented left and right of fixation, a reader fails to detect the trick and believes that he is looking at a perfectly normal page of text.

This surprising blindness occurs because the eye attains its maximum speed at the point when the letter change occurs. This trick makes the letter changes hard to detect, because at this very moment the whole retinal image is blurred by motion. Once gaze lands, everything looks normal: within the fovea, the expected letters are in place, and the rest of the visual field, on the periphery, cannot be read anyway. McConkie and Rayner's experiment thus proves that we consciously process only a very small subset of our visual inputs. If the computer leaves four letters on the left of fixation, and fifteen letters on the right, reading speed remains normal.[11] In brief, we extract very little information at a time from the written page. Descartes' evil genius would only have to display twenty letters per fixation to make us believe that we were reading the Bible or the U.S. Constitution!

Twenty letters is, in fact, an overestimate. We identify only ten or twelve letters per saccade: three or four to the left of fixation, and seven or eight to the right. Beyond this point, we are largely insensitive to letter identity and merely encode the presence of the spaces between words. By providing cues about word length, the spaces allow us to prepare our eye movements and ensure that our gaze lands close to the center of the next word. Experts continue to debate about the extent to which we extract information from an upcoming word—perhaps only the first few letters. Everyone agrees, however, that the direction of reading imposes asymmetry on our span of vision. In the West, visual span is much greater toward the right side, but in readers of Arabic or Hebrew, where gaze scans the page from right to left, this asymmetry is reversed.[12] In other writing systems such as Chinese, where character density is greater, saccades are shorter and visual span is reduced accordingly. Each reader thus adapts his visual exploration strategy to his language and script.

Using the same method, we can also estimate how much time is needed to encode the identity of words. A computer can be programmed so that, after a given duration, all of the letters are replaced by a string of x's, even in the fovea. This experiment reveals that fifty milliseconds of presentation are enough for reading to proceed at an essentially normal pace. This does not mean that all of the mental operations involved in reading are completed in one-twentieth of a second. As we shall see, a whole pipeline of mental processes continues to operate for at least one-half second after the word has been presented. However, the initial intake of visual information can be very brief.

In summary, our eyes impose a lot of constraints on the act of reading. The structure of our visual sensors forces us to scan the page by jerking our eyes around every two or three tenths of a second. Reading is nothing but the word-by-word mental restitution of a text through a series of snapshots. While some small grammatical words like "the," "it," or "is" can sometimes be skipped, almost all content words such as nouns and verbs have to be fixated at least once.

These constraints are an integral part of our visual apparatus and cannot be lifted by training. One can certainly teach people to optimize their eye movement patterns, but most good readers, who read from four hundred to five hundred words per minute, are already close to optimal. Given the retinal sensor at our disposal, it is probably not possible to do much better. A simple demonstration proves that eye movements are the rate-limiting step in reading.[13] If a full sentence is presented, word by word, at the precise point where gaze is focalized, thus avoiding the need for eye movements, a good reader can read at staggering speed—a mean of eleven hundred words per minute, and up to sixteen hundred words per minute for the best readers, which is about one word every forty milliseconds and three to four times faster than normal reading! With this method, called rapid sequential visual presentation, or RSVP, identification and comprehension remain satisfactory, thus suggesting that the duration of those central steps does not impose a strong constraint on normal reading. Perhaps this computerized presentation mode represents the future of reading in a world where screens progressively replace paper.

At any rate, as long as text is presented in pages and lines, acquisition through gaze will slow reading and impose an unavoidable limitation. Thus, fast reading methods that advertise gains in reading speed of up to one thousand words per minute or more must be viewed with skepticism.[14] One can no doubt broaden one's visual span somewhat, in order to reduce the number of saccades per line, and it is also possible to learn to avoid moments of regression, where gaze backtracks to the words it has just read. However, the physical limits of the eyes cannot be overcome, unless one is willing to skip words and thus run the risk of a misunderstanding. Woody Allen described this situation perfectly: "I took a speed-reading course and was able to read *War and Peace* in twenty minutes. It involves Russia."

The Search for Invariants

Can you read, Lubin?

Yes, I can read printed letters, but I was never able to read handwriting.

—MOLIÈRE, *GEORGES DANDIN*

Reading poses a difficult perceptual problem. We must identify words regardless of how they appear, whether in print or handwritten, in upper- or lowercase, and regardless of their size. This is what psychologists call the *invariance problem*: we need to recognize which aspect of a word does not vary—the sequence of letters—in spite of the thousand and one possible shapes that the actual characters can take on.

If perceptual invariance is a problem, it is because words are not always in the same location, in the same font, or in the same size. If they were, just listing which of the cells on the retina are active and which are not would suffice to decode a word, much like a black-and-white computer image is defined by the list of its pixels. In fact, however, hundreds of different retinal images can stand for the same word, depending on the form in which it is

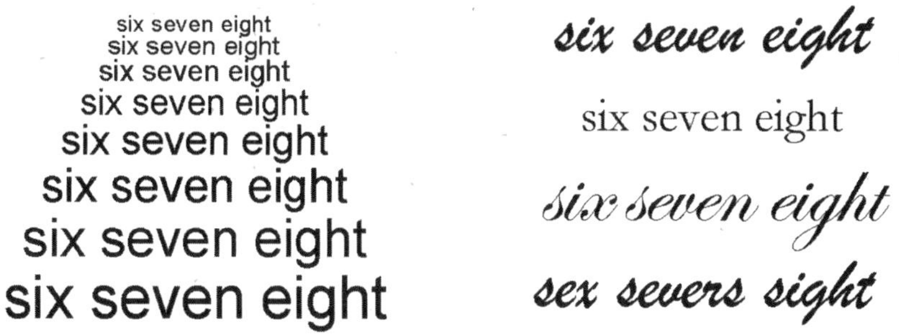

six seven eight
six seven eight
six seven eight
six seven eight
six seven eight
six seven eight
six seven eight
six seven eight

six seven eight

six seven eight

six seven eight

sex severs sight

Figure 1.2 Visual invariance is one of the prime features of the human reading system. Our word recognition device meets two seemingly contradictory requirements: it neglects irrelevant variations in character shape, even if they are huge, but amplifies relevant differences, even if they are tiny. Unbeknownst to us, our visual system automatically compensates for enormous variations in size or font. Yet it also attends to minuscule changes in shape. By turning an "s" into an "e," and therefore "sight" into "eight," a single mark drastically reorients the processing chain toward entirely distinct pronunciations and meanings.

written (figure 1.2). Thus one of the first steps in reading must be to correct for the immense variety of those surface forms.

Several cues suggest that our brain applies an efficient solution to this perceptual invariance problem. When we hold a newspaper at a reasonable distance, we can read both the headlines and the classified ads. Word size can vary fiftyfold without having much impact on our reading speed. This task is not very different from that of recognizing the same face or object from a distance of two feet or thirty yards—our visual system tolerates vast changes in scale.

A second form of invariance lets us disregard the location of words on the page. As our gaze scans a page, the center of our retina usually lands slightly left of the center of words. However, our targeting is far from perfect, and our eyes sometimes reach the first or last letter without this preventing us from recognizing the word. We can even read words presented on the periphery of our visual field, provided that letter size is increased to compensate for the loss of retinal resolution. Thus size constancy goes hand in hand with normalization for spatial location.

Finally, word recognition is also largely invariant for character shape. Now that word processing software is omnipresent, technology that was formerly reserved to a small elite of typographers has become broadly available. Everyone knows that there are many sets of characters called "fonts" (a term left over from the time when each character had to be cast in lead at a type foundry before going to the press). Each font also has two kinds of characters called "cases," the UPPERCASE and the lowercase (originally the case was a flat box divided into many compartments where lead characters were sorted; the "upper case" was reserved for capital letters, and the "lower case" for the rest). Finally, one can choose the "weight" of a font (normal or **bold** characters), its inclination (*italics*, originally invented in Italy), whether it is <u>underlined</u> or not, and any **combination** of *these options*. These well-calibrated variations in fonts, however, are nothing compared to the enormous variety of writing styles. *Manuscript handwriting obviously takes us to another level of variability and ambiguity.*

In the face of all these variations, exactly how our visual system learns to categorize letter shapes remains somewhat mysterious. Part of this invariance problem can be solved using relatively simple means. The vowel "o," for instance, can easily be recognized, regardless of size, case, or font, thanks to its unique closed shape. Thus, building a visual o detector isn't particularly difficult. Other letters, however, pose specific problems. Consider the letter "r," for instance. Although it seems obvious that the shapes r, R, *𝐫* and *𝐫* all represent the same letter, careful examination shows that this association is entirely arbitrary—the shape e, for instance, might serve as well as the lower-case version of the letter "R." Only the accidents of history have left us this cultural oddity. As a result, when we learn to read, we must not only learn that letters map onto the sounds of language, but also that each letter can take on many unrelated shapes. As we shall see, our capacity to do this probably comes from the existence of abstract letter detectors, neurons that can recognize the identity of a letter in its various guises. Experiments show that very little training suffices to DeCoDe, At An EsSeNtIaLly NoRmAl SpEeD, EnTiRe SeNtEnCes WhOsE LeTtErS HaVe BeEn PrInTeD AlTeRnAtElY iN UpPeRcAsE aNd In LoWeRcAsE.[15] In the McConkie and Rayner "evil genius" computer, this letter-case alternation can be changed in between every eye saccade, totally unbeknownst to the reader![16] In our daily reading experience, we never see words presented in alternating case, but our letter normalization processes are so efficient that they easily resist such transformation.

In passing, these experiments demonstrate that global word shape does not play any role in reading. If we can immediately recognize the identity of "words," "WORDS," and "WoRdS," it is because our visual system pays no attention to the contours of words or to the pattern of ascending and descending letters: it is only interested in the letters they contain. Obviously, our capacity to recognize words does not depend on an analysis of their overall shape.

Amplifying Differences

Although our visual system efficiently filters out visual differences that are irrelevant to reading, such as the distinction between "R" and "r," it would be a mistake to think that it always discards information and simplifies shapes. On the contrary, in many cases it must preserve, and even amplify, the minuscule details that distinguish two very similar words from each other. Consider the words "eight" and "sight." We immediately access their very distinct meanings and pronunciations, but it is only when we look more closely at them that we realize that the difference is only a few pixels. Our visual system is exquisitely sensitive to the minuscule difference between "eight" and "sight," and it amplifies it in order to send the input to completely different regions of semantic space. At the same time, it pays very little attention to other much greater differences, such as the distinction between "eight" and "EIGHT."

As with invariance for case, this capacity to attend to relevant details results from years of training. The same reader who immediately spots the difference between letters "e" and "o," and the lack of difference between "a" and "*a*," may not notice that the Hebraic letters ח and ה differ sharply, a fact that seems obvious to any Hebrew reader.

Every Word Is a Tree

Our visual system deals with the problem of invariant word recognition using a well-organized system. As we shall see in detail in chapter 2, the flow of neuronal activity that enters into the visual brain gets progressively sorted into meaningful categories. Shapes that appear very similar, such as "eight" and "sight," are sifted through a series of increasingly refined filters that progressively separate them and attach them to distinct entries in a mental lexicon, a virtual dictionary of all the words we have ever encountered. Conversely, shapes like "eight" and "EIGHT," which are composed of distinct visual features, are initially encoded by different neurons in the primary visual area, but are progressively recoded until they become virtually indistinguishable. Feature detectors recognize the similarity of the letters "i" and "I." Other, slightly more abstract letter detectors classify "e" and "E" as two forms of the same letter. In spite of the initial differences, the reader's visual system eventually encodes the very essence of the letter strings "eight" and "EIGHT," regardless of their exact shape. It gives these two strings the same single mental address, an abstract code capable of orienting the rest of the brain toward the pronunciation and meaning of the word.

What does this address look like? According to some models, the brain uses a sort of unstructured list that merely provides the sequence of letters E-I-G-H-T. In others, it relies on a very abstract and conventional code, akin to a random cipher by which, say, [1296] would be the word "eight" and [3452] the word "sight." Contemporary research, however, supports another hypothesis. Every written word is probably encoded by a hierarchical tree in which letters are grouped into larger-sized units, which are themselves grouped into syllables and words—much as a human body can be represented as an arrangement of legs, arms, torso, and head, each of which can be further broken down into simple parts.

A good example of the mental decomposition of words into relevant units can be found if we dissect the word "unbuttoning." We must first strip off the prefix "un" and the familiar suffix or grammatical ending "ing." Both frame the central element, the word inside the word: the root "button." All three of these components are called "morphemes"—the smallest units that carry some meaning. Each word is characterized, at this level, by how its morphemes are put together. Breaking down a word into its morphemes even allows us to understand words that we have never seen before, such as "reunbutton" or "deglochization" (we understand that this is the undoing of the action of "gloching," whatever that may be). In some languages, such as Turkish or Finnish, morphemes can be assembled into very large words that convey as much information as a full English sentence. In those languages, but also in ours, the decomposition of a word into its morphemes is an essential step on the path that leads from vision to meaning.

A lot of experimental data show that, very quickly and even downright unconsciously, our visual system snips out the morphemes of words. For instance, if I were to flash the word "departure" on a computer screen, you would later say the word "depart" slightly faster when confronted with it. The presentation of "departure" seems to preactivate the morpheme [depart], thus facilitating its access. Psychologists speak of a "priming" effect—the reading of a word primes the recognition of related words, much as one primes a pump. Importantly, this priming effect does not depend solely on visual similarity: words that look quite different but share a morpheme, such as "can" and "could," can prime each other, whereas words that look alike but bear no intimate morphological relation, such as "aspire" and "aspirin," do not. Priming also does not require any resemblance at the level of meaning; words such as "hard" and "hardly," or "depart" and "department," can prime each other, even though their meanings are essentially unrelated.[17] Getting down to the morpheme level seems to be of such importance for our reading system that it is willing to make guesses about the decomposition of words. Our reading apparatus dissects the word "department" into [depart] + [ment] in the hope that this will be useful to the next operators computing its meaning.[18] Never mind that this does not work all the time—a "listless" person is not one who is waiting for a grocery list, nor does sharing an "apartment" imply that you and your partner will soon live apart. Such parsing errors will have to be caught at other stages in the word dissection process.

If we continue to undress the word "unbuttoning," the morpheme [button] itself is not an indivisible whole. It is made up of two syllables, [bʌ] and [ton], each of which can be broken down into individual consonants and vowels: [b] [ʌ] [t] [o] [n]. Here lies another essential unit in our reading system: the grapheme, a letter or series of letters that maps onto a phoneme in the target language. Note that in our example, the two letters "tt" map onto a single sound *t*.[19] Indeed, the mapping of graphemes onto phonemes isn't always a direct process. In many languages, graphemes can be constructed out of a group of letters. English has a particularly extensive collection of complex graphemes such as "ough," "oi," and "au."

Our visual system has learned to treat these groups of letters as bona fide units, to the point where we no longer pay attention to their actual letter content. Let us do a simple experiment to prove this point. Examine the following list of words and mark those that contain the letter "a":

garage

metal

people

coat

please

meat

Did you feel that you had to slow down, ever so slightly, for the last three words, "coat," "please," and "meat"? They all contain the letter "a," but it is embedded in a complex grapheme that is not pronounced like an "a." If we were to rely only on letter detectors in order to detect the letter "a," the parsing of the word into its graphemes would not matter. However, actual measurement of response times clearly shows that our brain does not stop at the single-letter level. Our visual system automatically regroups letters into higher-level graphemes, thus making it harder for us to see that groups of letters such as "ea" actually contain the letter "a."[20]

In turn, graphemes are automatically grouped into syllables. Here is another simple demonstration of this fact. You are going to see five-letter words. Some letters are in **bold** type, others in a normal font. Concentrate solely on the middle letter, and try to decide if it is printed in normal or in bold type:

| List 1 : | HO**RN**Y | **RID**ER | **GRA**VY | **FI**LET |
| List 2 : | **VOD**KA | ME**TRO** | HA**NDY** | **SU**PER |

Did you feel that the first list was slightly more difficult than the second? In the first list, the bold characters do not respect syllable boundaries—in "RIDER," for instance, the "D" is printed in bold type while the rest of the syllable is in normal type. Our mind tends to group together the letters that make up a syllable, thus creating a conflict with the bold type that leads to a measurable slowing down of responses.[21] This effect shows that our visual system cannot avoid automatically carving up words into their elementary constituents even when it would be better not to do so.

The nature of these constituents remains a hot topic in research. It would appear that multiple levels of analysis can coexist: a single letter at the lowest level, then a pair of letters (or "bigram," an important unit to which we will return later), the grapheme, the syllable, the morpheme, and finally the whole word. The final point in visual processing leaves the word parsed out into a hierarchical structure, a tree made up of branches of increasing sizes whose leaves are the letters.

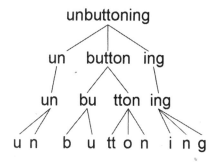

Reduced to a skeleton, stripped of all its irrelevant features like font, case, and size, the letter string is thus broken down into the elementary components that will be used by the rest of the brain to compute sound and meaning.

The Silent Voice

Writing—this ingenious art to paint words and speech for the eyes.

—GEORGES DE BRÉBEUF (FRENCH POET, 1617–1661)

When he paid a visit to Ambrose, then bishop of Milan, Augustine observed a phenomenon that he judged strange enough to be worth noting in his memoirs:

> When [Ambrose] read, his eyes scanned the page and his heart sought out the meaning, but his voice was silent and his tongue was still. Anyone could approach him freely and guests were not commonly announced, so that often, when we came to visit him, we found him reading like this in silence, for he never read aloud.[22]

In the middle of the seventh century, the theologian Isidore of Seville similarly marveled that "letters have the power to convey to us silently the sayings of those who are absent." At the time, it was customary to read Latin aloud. To articulate sounds was a social convention, but also a true necessity: confronted with pages in which the words were glued together, without spaces, in a language that they did not know well, most readers had to mumble through the texts like young children. This is why Ambrose's silent reading was so surprising, even if for us it has become a familiar experience: we can read without articulating sounds.

Whether our mind ever goes straight from the written word to its meaning without accessing pronunciation or whether it unconsciously transforms letters into sound and then sound into meaning has been the topic of considerable discussion The organization of the mental pathways for reading fueled a debate that divided the psychological community for over thirty years. Some thought that the transformation from print to sound was essential—written language, they argued, is just a by-product of spoken language, and we therefore have to sound the words out, through a *phonological route*, before we have any hope of recovering their meaning. For others, however, phonological recoding was just a beginner's trait characteristic of young readers. In more expert readers, reading efficiency was based on a direct *lexical route* straight from the letter string to its meaning.

Nowadays, a consensus has emerged: in adults, both reading routes exist, and both are simultaneously active. We all enjoy direct access to word meaning, which spares us from pronouncing words mentally before we can understand them. Nevertheless, even proficient readers continue to use the sounds of words, even if they are unaware of it. Not that we articulate words covertly—we do not have to move our lips, or even prepare an intention to do so. At a deeper level, however, information about the pronunciation of words is automatically retrieved. Both the lexical and phonological pathways operate in parallel and reinforce each other.

There is abundant proof that we automatically access speech sounds while we read. Imagine, for instance, that you are presented with a list of strings and have to decide whether each one is a real English word or not. Mind you, you only have to decide if the letters spell out an English word. Here you go:

rabbit
bountery
culdolt
money
dimon

karpit

nee

You perhaps hesitated when the letters sounded like a real word—as in "demon," "carpet," or "knee." This interference effect can easily be measured in terms of response times. It implies that each string is converted into a sequence of sounds that is evaluated like a real word, even though the process goes against the requested task.[23]

Mental conversion into sound plays an essential role when we read a word for the first time—say, the string "Kalashnikov." Initially, we cannot possibly access its meaning directly, since we have never seen the word spelled out. All we can do is convert it into sound, find that the sound pattern is intelligible, and, through this indirect route, come to understand the new word. Thus, sounding is often the only solution when we encounter a new word. It is also indispensable when we read misspelled words. Consider the little-known story by Edgar Allan Poe called "The Angel of the Odd." In it, a strange character mysteriously intrudes into the narrator's apartment, "a personage nondescript, although not altogether indescribable," and with a German accent as thick as British fog:

"Who are you, pray?" said I, with much dignity, although somewhat puzzled; "how did you get here? And what is it you are talking about?"

"Az vor ow I com'd ere," replied the figure, "dat iz none of your pizzness; and as vor vat I be talking apout, I be talk apout vat I tink proper; and as vor who I be, vy dat is de very ting I com'd here for to let you zee for yourzelf. . . . Look at me! Zee! I am te Angel ov te Odd."

"And odd enough, too," I ventured to reply; "but I was always under the impression that an angel had wings."

"Te wing!" he cried, highly incensed, "vat I pe do mit te wing? Mein Gott! Do you take me vor a shicken?"

In reading this passage we return to a style we had long forgotten, one that dates back to our childhood: the phonological route, or the slow transformation of totally novel strings of letters into sounds that miraculously become intelligible, as though someone were whispering them at us.

What about everyday words, however, that we have already met a thousand times? We do not get the impression that we slowly decode through mental enunciation. However, clever psychological tests show that we still activate their pronunciation at a nonconscious level. For instance, suppose that you are asked to indicate which of the following words refer to parts of the human body. These are all very familiar words, so you should be able to focus on their meaning and neglect their pronunciation. Try it:

> knee
> leg
> table
> head
> plane
> bucket
> hare

Perhaps you felt the urge to respond to the word "hare," which sounds like a body part. Experiments show that we slow down and make mistakes on words that sound like an item in the target category.[24] It is not clear how we could recognize this homophony if we did not first mentally retrieve the word's pronunciation. Only an internal conversion into speech sounds can explain this type of error. Our brain cannot help but transform the letters "h-a-r-e" into internal speech and then associate it with a meaning—a process that can go wrong in rare cases where the string sounds like another well-known word.

Of course, this imperfect design is also what grants us one of the great pleasures of life: puns, or the "joy of text," as the humorist Richard Lederer puts it. Without the gift of letter-to-sound conversion, we would not be able to enjoy Mae West's wisecrack ("She's the kind of girl who climbed the ladder of success wrong by wrong") or Conan Doyle's brother-in-law's quip ("there's no police like Holmes"). Without Augustine's "silent voice," the pleasure of risqué double entendres would be denied us:

An admirer says to President Lincoln, "Permit me to introduce my family. My wife, Mrs. Bates. My daughter, Miss Bates. My son, Master Bates."

"Oh dear!" replied the president.[25]

Further proof that our brain automatically accesses a word's sound patterns derives from subliminal priming. Suppose that I flash the word "LATE" at you, immediately followed by the word "mate," and ask you to read the second word as fast as you can. The words are shown in a different case in order to avoid any low-level visual resemblance. Nevertheless, when the first word sounds and spells like the second, as in this example, we would observe a massive acceleration of reading time, in comparison with a situation where two words are not particularly related to one another ("BOWL," followed by "mate"). Part of this facilitation clearly arises from similarity at the level of spelling alone. To flash "MATH" eases recognition of "mate," even though the two strings sound quite different. However, crucially, even greater facilitation can be found when two words share the same pronunciation ("LATE" followed by "mate"), and this sound-based priming works even when spelling is completely different ("EIGHT" followed by "mate"). Thus, pronunciation seems to be automatically extracted. As one might expect, however, spelling and sound are not encoded at the same time. It takes only twenty or thirty milliseconds of word viewing for our brain to automatically activate a word's spelling, but an additional forty milliseconds for its transformation into sound, as revealed by the emergence of sound-based priming.[26]

Simple experiments thus lead us to outline a whole stream of successive stages in the reader's brain, from marks on the retina to their conversion into letters and sounds. Any expert reader quickly converts strings into speech sounds effortlessly and unconsciously.

The Limits of Sound

Covert access to the pronunciation of written words is an automatic step in reading, but this conversion may not be indispensable. Speech-to-sound conversion is often slow and inefficient. Our brain thus often tries to retrieve a word's meaning using a parallel and more direct pathway that leads straight from the letter string to the associated entry in the mind's lexicon.

To boost our intuition about the direct lexical route, we have to consider the plight of a make-believe reader who would only be able to mentally enunciate written words. It would be impossible for him to discriminate between homophonic words like "maid" and "made," "raise" and "raze," "board" and "bored," or "muscles" and "mussels." Purely on the basis of sound, he might think that serial killers hate cornfields, and that one-carat diamonds are an odd shade of orange. The very fact that we readily discern the multiple meanings of such homophonic words shows that we are not obliged to pronounce them—another route is available that allows our brain to solve any ambiguity and go straight to their meaning.

A further problem exists for purely sound-based theories of reading: the route from spelling to sound is not a high-speed highway devoid of obstacles. To derive a word's pronunciation from the sequence of its letters is often impossible in the absence of additional help. Consider the word "blood." It seems obvious that it should be pronounced *blud* and that it rhymes with "bud" or "mud." But how do we know this? Why shouldn't "blood" rhyme with "food" or "good"? Why doesn't it sound like "bloom" or "bloomer"? Even the same word root can be pronounced differently, as in "sign" and "signature." Some words are so exceptional that it is hard to see how their pronunciation relates to their component letters ("colonel," "yacht," "though" . . .). In such cases, the word's pronunciation cannot be computed without prior knowledge of the word.

English spelling bristles with irregularities. Indeed, the gap between written and spoken language is centuries old, as attested by William Shakespeare in *Love's Labour's Lost*, where the pedant Holofernes says:

I abhor such fanatical phantasimes, such insociable and point-devise companions; such rackers of orthography, as to speak *dout*, fine, when he should say *doubt*; *det*, when he should pronounce *debt*—d, e, b, t, not d, e, t: he clepeth a calf, *cauf; half, hauf; neighbour* vocatur *nebor; neigh* abbreviated *ne*. This is abhominable—which he would call *abbominable*: it insinuateth me of insanie.

English is an abominably irregular language. George Bernard Shaw pointed out that the word "fish" might be spelled *ghoti*: *gh* as in "enough," *o* as in "women," and *ti* as in "lotion"! Shaw hated the irregularities of English spelling so much that in his will he provided for a contest to design a new and fully rational alphabet called "Shavian." Unfortunately, it never met with much success, probably because it departed too much from all other existent spelling systems.[27]

Of course, Shaw's example is far-fetched: no one would ever read *ghoti* as "fish," because the letter "g," when placed at the beginning of a word, is always pronounced as a hard *g* or a *j*, never as an *f*. Likewise, Shakespeare notwithstanding, in present-day English the letters "alf" at the end of a word are always pronounced *af*, as in "calf" and "half." If letters are taken in context, it is often possible to identify some higher-order regularities that simplify the mapping of letters onto sounds. Even then, however, exceptions remain numerous— "has" and "was," "tough" and "dough," "flour" and "tour," "header" and "reader," "choir" and "chair," "friend" and "fiend." For most irregular words, the recovery of pronunciation, far from being the source of word comprehension, seems to depend on its outcome: it is only after we have recognized the identity of the word "dough" that we can recover its sound pattern.

The Hidden Logic of Our Spelling System

One may wonder why English sticks to such a complicated spelling system. Indeed, Italians do not meet with the same problems. Their spelling is transparent: every letter maps onto a single phoneme, with virtually no exceptions. As a result, it only takes a few months to learn to read. This gives Italians an enormous advantage: their children's reading skills surpass ours by several years, and they do not need to spend hours of schooling a week on dictation and spelling out loud. Furthermore, as we shall discuss later, dyslexia is a much less serious problem for them. Perhaps we should follow Italy's lead, burn all our dictionaries and desine a noo speling sistem dat eeven a θree-yia-old tchaild cood eezilee reed.

There is no doubt that English spelling could be simplified. The weight of history explains a lot of its peculiarities—today's pupils should lament the loss of the battle of Hastings, because the mixture of French and English that ensued is responsible for many of our spelling headaches, such as the use of the letter "c" for the sound *s* (as in "cinder"). Centuries of academic conservatism, sometimes bordering on pedantry, have frozen our dictionary. Well-meaning academics even introduced spelling absurdities such as the "s" in the word "island," a misguided Renaissance attempt to restore the etymology of the Latin word *insula*. Worst of all, English spelling failed to evolve in spite of the natural drift of oral language. The introduction of foreign words and spontaneous shifts in English articulation have created an immense gap between the way we write and the way we speak, which causes years of unnecessary suffering for our children. In brief, reason calls for a radical simplification of English spelling.

Nevertheless, before any revisions can be made, it is essential to fully understand the hidden logic of our spelling system. Spelling irregularities are not just a matter of convention. They also originate in the very structure of our language and of our brains. The two reading routes, either from spelling to sound or from spelling to meaning, place complex and often irreconcilable constraints on any writing system. The linguistic differences between English, Italian, French, and Chinese are such that no single spelling solution could ever suit them all. Thus the abominable irregularity of English spelling appears inevitable. Although spelling reform is badly needed, it will have to struggle with a great many restrictions.

First of all, it is not clear that English spelling, like Italian, could attribute a single letter to each sound, and a fixed sound to each letter. It would not be a simple thing to do because the English language contains many more speech sounds than Italian. The number of English phonemes ranges from forty to forty-five, depending on speakers and counting methods, while Italian has only thirty. Vowels and diphthongs are particularly abundant in English: there are six simple vowels (as in *bat, bet, bit, but, good,* and *pot*), but also five long vowels (as in *beef, boot, bird, bard,* and *boat*) and at least seven diphthongs (as in *bay, boy, toe, buy, cow, beer, bear*). If each of those sounds were granted its own written symbol, we would have to invent new letters, placing an additional burden on our children. We could consider the addition of accents to existing letters, such as ã, õ, or ü. However, it is entirely utopian to imagine a universal alphabet that could transcribe all of the world's languages. Such a spelling system does exist: it is called the International Phonetic Alphabet and it plays an important role in technical publications by phonologists and linguists. However, this writing system is so complex that it would be ineffective in everyday life. The International Phonetic Alphabet has 170 signs, some of which are particularly complex (e ɐ, ɓ, ɕ, ʓ, or ɲ). Even specialists find it very hard to read fluently without the help of a dictionary.

In order to avoid learning an excessive number of symbol shapes, languages with a great many phonemes, such as English and French, all resort to a compromise. They indicate certain vowels or consonants using either special characters such as ü, or groups of letters like "oo" or "oy." Such peculiarities, which are unique to any given language, are far from being gratuitous embellishments: they play an essential role in the "mental economy" of reading, and have to find their place in any kind of spelling reform.

Although we cannot easily assign a single letter shape to each speech sound, we could perhaps try the opposite. Many spelling errors could be avoided if we systematically transcribed each sound with a fixed letter. For instance, if we were to avoid writing the sound f with both the letter "f" and with "ph," life would be much simpler. There is little doubt that we could easily get rid of this and many other useless redundancies whose acquisition eats up many years of childhood. In fact, this is the timid direction that American spelling reform took when it simplified the irregular British spellings of "behaviour" or "analyse" into "behavior" and "analyze." Many more steps could have been taken along the same lines. As expert readers, we cease to be aware of the absurdity of our spelling. Even a letter as simple as "x" is unnecessary, as it stands for two phonemes *ks* that already have their own spelling. In Turkey, one takes a "taksi." That country, which in the space of one year (1928–29) adopted the Roman alphabet, drastically simplified its spelling, and taught three million people how to read, sets a beautiful example of the feasibility of spelling reform.

Yet here again, great caution is needed. I suspect that any radical reform whose aim would be to ensure a clear, one-to-one transcription of English speech would be bound to fail, because the role of spelling is *not* just to provide a faithful transcription of speech sounds. Voltaire was mistaken when he stated, elegantly but erroneously, that "writing is the painting of the voice: the more it bears a resemblance, the better." A written text is not a high-fidelity recording. Its goal is not to reproduce speech as we pronounce it, but rather to code it at a level abstract enough to allow the reader to quickly retrieve its meaning.

For the sake of argument, we can try to imagine what a purely phonetic writing system would look like—one that Voltaire might have considered ideal. When we speak, we alter the pronunciation of words as a function of the sounds that surround them. It would be disastrous if spelling were to reflect the obtuse linguistic phenomena of so-called coarticulation, assimilation, and resyllabification, of which most speakers are usually unaware. A matter of context would end up having the same word spelled differently.

Should we, for instance, use distinct marks for the various pronunciations of plurals? Should we spell "cap driver," under the pretext that the sound *b*, when followed by a *d*, tends to be pronounced like a *p*? At one extreme, should we factor in the speaker's accent ("Do you take me vor a shicken?"). This would be *apsurd* (yes, we do pronounce this word with a *p* sound). The prime goal of writing is to transmit meaning as efficiently as possible. Any servile transcription of sounds would detract from this aim.

English spelling often privileges the transparency of word roots at the expense of the regularity of sounds. The words "insane" and "insanity," for instance, are so deeply related to their meaning that it would be silly to spell them differently because of their slightly different pronunciation. Similarly, it is logical to maintain the silent *n* at the end of "column," "autumn," or "condemn," given that these words give rise to "columnist," "autumnal," or "condemnation."

Transcription of meaning also explains, at least in part, why English spells the same sounds in many different ways. English words tend to be compact and monosyllabic, and as a result, homophony is very frequent (for instance "eye" and "I," "you" and "ewe"). If these words were transcribed phonetically, they could not be distinguished from each other. Spelling conventions have evolved with this constraint in mind. Distinctive spelling for the same sounds complicates dictation, but simplifies the task for the reader, who can quickly grasp the intended meaning. Students who complain about the countless forms of spelling the sound *u* as in "two," "too," "to," or "stew" should understand that these embellishments are essential to the speed at which we read. Without them, any written text would become an opaque rebus. Thanks to spelling conventions, written English points straight at meaning. Any spelling reform would have to maintain this subtle equilibrium between sound and meaning, because this balance reflects a much deeper and more rigid phenomenon: our brain's two reading routes.

The Impossible Dream of Transparent Spelling

Rivalry between reading for sound and reading for meaning is true the world over. All writing systems must somehow manage to address this problem. Which compromise is best depends on the language to be transcribed. Life would certainly be easier if English spelling were as easy to learn as Italian or German. These languages, however, benefit from a number of characteristics that make them easy to transcribe into writing. In Italian as in German, words tend to be long, often made up of several syllables. Grammatical agreement is well marked by resonant vowels. As a result, homonyms are quite rare. Thus, a purely regular transcription of sounds is feasible. Italian and German can afford a fairly transparent spelling system, where almost every letter corresponds to a unique sound.

At the other end of the continuum, there is the case of Mandarin Chinese. The vast majority of Chinese words consist of only one or two syllables, and because there are only 1,239 syllables (410 if one discounts tonal changes), each one can refer to dozens of different concepts (figure 1.3). Thus a purely phonetic writing system would be useless in Chinese—each of the rebuses could be understood in hundreds of different ways! This is why the thousands of characters in Mandarin script mostly transcribe words, or rather their morphemes—the basic elements of word meaning. Chinese writing also relies on several hundred phonetic markers that further specify how a given root should be pronounced and make it easier for the reader to figure out which word is intended. The character 媽, for instance, which means "mother" and is pronounced *ma*, consists of the morpheme 女 = "woman," plus a phonetic marker 馬 = *mă*. Thus, contrary to common belief, even Chinese is not a purely ideographic script (whose symbols represent concepts), nor a logographic one (whose signs refer to single words), but a mixed "morphosyllabic" system where some signs refer to the morphemes of words and others to their pronunciation.[28]

Of course, it is more difficult to learn to read Chinese than to learn to read Italian. Several thousand signs have to be learned, as opposed to only a few tens. These two languages thus fall at the two extremities of a continuous scale of spelling transparency, where English and French occupy intermediate positions.[29] In both English and French, words tend to be short, and homophones are therefore relatively frequent ("right," "write," "rite"). To accommodate these constraints, English and French spelling rules include a mixture of phonetic and lexical transcription—a source of difficulty for the writer, but simplicity for the reader.

In brief, we have only just begun to decipher the many constraints that shape the English spelling system. Will we ever be able to reform it? My own

石室詩士食獅史
石室詩士施氏，嗜獅，誓食十獅。氏時時適市視獅十時，
適十獅適市，是時，適施氏適市。氏視是十獅，恃失勢，
使是十獅逝世。氏拾是十獅屍，市石室。石室濕，氏使侍
拭石室。石室試，氏始試食是十獅屍。食時始識是十獅屍
實十石獅屍。試釋是事。

The story of a poet eating lions in a stone room

A poet named Shi lived in a stone room, and he had the habit of eating lions. He had sworn to eat at least ten. So, from time to time, he visited the lion market around ten. By chance, there were ten of them on the market one day that Shi was there. When he saw those ten lions, Shi killed them for fear of losing them. Then he grabbed their carcasses and stored them in his stone room. Since the room was damp, Shi had it cleaned by a servant. Then Shi started to taste those ten lion carcasses, and while eating them, he realized that those were indeed ten lions. Try to explain it!

Shi, first tone	Shi, fourth tone
詩 poet, poem	室 room
獅 lion	士 person
施 name, apply	嗜 to have the habit of
失 lose	(施)氏 demonstrative
屍 carcass	誓 swear, pledge
濕 damp	適 arrive, adapt, suitable
識 know, knowledge	是 yes, this one
Shi, second tone	視 look, see
石 stone	恃 apprehend, depend
食 eat	勢 situation, power
十 ten	逝 leave, pass
時 hour	世 world
拾 to pick up	市 market
實 true, reality	拭 clean up
Shi, third tone	試 try
史 story	釋 explain
使 order, send	事 thing
始 start	

Figure 1.3 Spelling irregularities are not as irrational as they seem. For instance, although Chinese writing uses up to twenty or thirty different characters for the same syllable, the redundancy is far from pointless. On the contrary, it is very helpful to Chinese readers because the Chinese language includes a great many homophones—words that sound alike but have different meanings, like the English "one" and "won." Here, an entire Chinese story was written with the sound "shi"! Any Chinese reader can understand this text, which would clearly be impossible if it were transcribed phonetically as "shi shi shi . . ." Chinese characters disambiguate sounds by using distinct characters for distinct meanings. Similarly, homophony explains why English sticks to so many different spellings for the same sounds ("I scream for ice cream").

personal standpoint on this score is that drastic simplification is inevitable. We owe it to our children, who waste hundreds of hours at this cruel game. Furthermore, some of them may never recover, handicapped for life by dyslexia or simply because they were raised in underprivileged or multilingual families. These are the real victims of our archaic spelling system. My hope is that the next generation will have become so used to abridged spelling, thanks to cell phones and the Internet, that they will cease to treat this issue as taboo and will muster enough willpower to address it rationally. However, the problem will never be resolved by a simple decree that instates phonological spelling. English will never be as simple as Italian. The dream of regular spelling is something of an illusion, as has been pointed out in a pamphlet that has been circulating in Europe for some time:

The European Union commissioners have announced that agreement has been reached to adopt English as the preferred language for European communications, rather than German, which was the other possibility. As part of the negotiations, the British government conceded that English spelling had some room for improvement and has accepted a five-year phased plan for what will be known as Euro English (Euro for short).

In the first year, "s" will be used instead of the soft "c." Sertainly, sivil servants will resieve this news with joy. Also, the hard "c" will be replaced with "k." Not only will this klear up konfusion, but typewriters kan have one less letter. There will be growing publik enthusiasm in the sekond year, when the troublesome "ph" will be replaced by "f." This will make words like "fotograf" 20 per sent shorter.

In the third year, publik akseptanse of the new spelling kan be expekted to reach the stage where more komplikated changes are possible. Governments will enkorage the removal of double letters, which have always ben a deterent to akurate speling. Also, al wil agre that the horible mes of silent "e"s in the languag is disgrasful, and they would go.

By the fourth year, peopl wil be reseptiv to steps such as replasing "th" by "z" and "w" by "v." During ze fif year, ze unesesary "o" kan be dropd from vords kontaining "ou," and similar changes vud of kors be aplid to ozer kombinations of leters.

After zis fifz yer, ve vil hav a reli sensibl riten styl. Zer vil be no mor trubls or difikultis and evrivun vil find it ezi tu understand ech ozer.

Ze drem vil finali kum tru!

Two Routes for Reading

Before I go any further, I would like to summarize what has been covered so far. All writing systems oscillate between an accurate representation of sound and the fast transmission of meaning. This dilemma is directly reflected in the reader's brain. Two information processing pathways coexist and supplement each other while we read. When words are very regular, rare, or novel, we preferentially process them using a "phonological route," in which we first decipher the letter string, then convert it into pronunciation, and finally attempt to access the meaning of the sound pattern (if any). Conversely, when we are confronted with words that are frequent or whose pronunciation is exceptional, our reading takes a direct route that first recovers the identity and meaning of the word and then uses the lexical information to recover its pronunciation (figure 1.4).

The best support for the existence of these two pathways comes from the study of brain injuries and their psychological consequences. Some patients, after a stroke or a brain lesion, lose the ability to quickly compute the pronunciation of written words.[30] Clearly, their spelling-to-sound conversion route has been severely damaged. Although they read normally before the lesion, their reading now presents all the characteristics of a syndrome called deep dyslexia or phonological dyslexia. They no longer manage to read infrequent words like "sextant" out loud, even when their spelling is very regular. Furthermore, they are helpless when faced with reading neologisms or invented words such as "departition" or "calbonter." Surprisingly, however, they may still be able to understand words used frequently, and generally manage to read aloud even irregular but common words like "eyes," "door," or "women." Occasionally, they confuse one word with another. A deep dyslexic may, for instance, read the written word "ham" as *meat* or the word "painter" as *artist*. The very nature of these errors demonstrates that access to word meaning is largely preserved. If the patient had not understood, at least in part, the word that he attempted to read, he would not even have been able to approach the right meaning. Patients with deep dyslexia appear to recognize written words even though the computation of their pronunciation has become essentially impossible. It is as if one of the reading routes (from vision to sound) was blocked, while information still circulated along the other route (from vision to meaning).

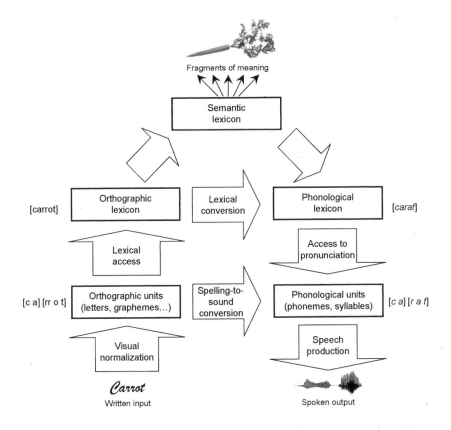

Figure 1.4 Word reading proceeds along several parallel processing routes. To move from the written word (bottom left) to its pronunciation (bottom right), our brain relies on several pathways, indicated here with boxes and arrows. When the word is regular, a surface route directly converts its letters into speech sounds. When the word is irregular, such as "carrot," deeper representations are involved. They can be compared to mental lexicons that attach word form to meaning.

The opposite situation is also on record. Here, patients who suffer from another syndrome, called surface dyslexia, no longer enjoy direct access to word meaning, but have to slowly work their way through a text and sound out the words. In this situation, the limits of the "silent voice" become obvious. Surface dyslexia patients can still read regular words such as "banana," or even neologisms such as "chicopar," but they are virtually helpless with irregular words. They typically standardize them through blind conversion into sounds. One patient, for instance, read "enough" as *inog* and then swore that he had never heard of this bizarre word. Manifestly, the direct route from vision to the mental lexicon had been blocked, while the conversion from print to sound remained in operation.

While the contrast between these two types of patients proves that there are two quite distinct pathways for reading, it also shows that neither of them, alone, allows us to read all words. The direct route, from letters to words and their meaning, can be used to read most words, as long as they are sufficiently frequent, but it does not work for rare or novel words that are not stored in our mental lexicon. Conversely, the indirect route, from letters to sounds and from sounds to meanings, fails with irregular words like "women" and with homophones such as "too," but it plays an irreplaceable role when we learn new words. When we read, both routes are in constant collaboration and each contributes to the specification of word pronunciation. Most of a word's phonemes can be deduced from a letter string through the application of simple spelling-to-sound conversion rules, while occasional ambiguities are resolved by a little nudge from higher lexical and semantic levels. In children, the two routes are often poorly coordinated. Some children rely largely on the direct route. They attempt to second-guess words and often pick a loose synonym for the target (for instance, reading "house" as *home*). Others mumble their way through a sentence and painfully put together an approximate pronunciation based on the words' letters, but cannot go from the loosely assembled sounds to their meaning. Years of practice are needed before the two routes become close enough to give the impression, in expert adults, of a single integrated reading system.

Most of the current models of reading postulate that fluent reading relies on the close-knit coordination between the two reading routes, where each weighs in to a different degree depending on the word to be read (known or unknown, frequent or rare, regular or irregular) as well as the task (reading aloud or text comprehension). In the 1980s and 1990s, some researchers did try to account for these observations with a single-route system. At the time, the emergence of neural network models generated a huge wave of interest. Some researchers saw them as universal learning machines that could acquire any skill without the possession of predefined cognitive architecture. They figured that reading acquisition could be modeled by the connection of letter inputs to phonological outputs, while the intermediate connections were tuned by a powerful learning rule. They hoped to arrive at a single network that could simulate normal reading and its pathologies, but did not have to postulate multiple cortical processing pathways. While these networks, at the time, represented a remarkable advance, particularly in the modeling of spelling-to-sound conversion,[31] most researchers today believe that this approach is inadequate. My own impression is that it is impossible to model reading without a thorough analysis of the brain's architecture, which relies on multiple parallel and partially redundant pathways. Practically all recent models, even though they rely on neural network simulations, incorporate the essential idea of multiple routes for reading.[32] When we tackle the cerebral mechanisms for reading later in this book, we will see that an essential feature of cortical architecture is its organization into multiple parallel paths. Thus, even the two-route model probably underestimates the true complexity of neural reading systems. The separation into two pathways, a spelling-to-sound conversion route and a semantic route, is simply a useful approximation.

Mental Dictionaries

As long as we were dealing with the surface route that converts graphemes into phonemes, it was realistic to imagine that mental reading followed a short list of simple procedures. To store a map of the few hundred English graphemes and their pronunciation would seem to suffice. However, when one considers how the deep route, which recognizes thousands of familiar words, may work, a much more massive store is needed. Cognitive psychologists compare it to a dictionary or "mental lexicon." No doubt we should speak of mental dictionaries in the plural, because our heads, in fact, carry around a number of different types of information about words. As proficient readers, we all possess a lexicon of English spelling that lists the written forms of all the words we know from the past. These orthographic memories are probably stored in the form of hierarchical trees of letters, graphemes, syllables, and morphemes. For instance the entry for the word "carrot" should look like [ca] + [rrot]. But we also maintain a separate "phonological lexicon," a mental dictionary of the pronunciation of words, for instance the fact that "carrot" is pronounced *carat*. We also have a grammatical store that specifies that "carrot" is a noun, that its plural is regular, and so on. Finally, each word is associated with dozens of semantic features that specify its meaning: a carrot is an edible vegetable, elongated in shape, with a characteristic orange color, and so on. These mental dictionaries open up, one after the other, as our brain retrieves the corresponding information. This metaphor holds that our mind houses a reference library in several volumes, from a spelling guide to a pronunciation manual and an encyclopedic dictionary.

The number of entries in our mental dictionaries is gigantic. The extent of human lexical knowledge is often grossly underestimated. I have heard otherwise well-informed people defend the accepted idea that Racine and Corneille wrote their plays using only one or two thousand words. Rumor has it that Basic English, a highly simplified version of the English language that includes only 850 words, should allow us to express ourselves effectively, and some pretend that the vocabulary of certain inner-city adolescents has shrunk to 500 words! All of these ideas are mistaken. We have excellent estimates of the average person's vocabulary, and they easily amount to several tens of thousands of words. A standard dictionary has about 100,000 entries, and sampling procedures suggest that any English speaker knows about 40,000 or 50,000 of them—compound words excluded. Add about the same number of proper nouns, acronyms ("CIA," "FBI"), trademarks ("Nike," "Coca-Cola"), foreign words, and in total, each of our lexicons probably contains between 50,000 and 100,000 entries. These estimates are further proof of the impressive capacities of our brain. Any reader easily retrieves a single meaning out of at least 50,000 candidate words, in the space of a few tenths of a second, based on nothing more than a few strokes of light on the retina.

An Assembly of Daemons

Several models of lexical access manage to imitate the performance of the human reading system under conditions close to those imposed by our nervous system. Almost all of them derive from a set of ideas first defined by Oliver Selfridge in 1959. Selfridge suggested that our lexicon works like a huge assembly of "daemons," or a "pandemonium."[33] This lively metaphor holds that the mental lexicon can be pictured as an immense semicircle where tens of thousands of daemons compete with one another. Each daemon responds to only one word, and makes this known by yelling whenever the word is called and must be defended. When a letter string appears on the retina, all the daemons examine it simultaneously. Those that think that their word is likely to be present yell loudly. Thus when the word "scream" appears, the daemon in charge of the response to it begins to shout, but so does its neighbor who codes for the word "cream." "Scream" or "cream"? After a brief competition, the champion of "cream" has to yield—it is clear that his adversary has had stronger support from the stimulus string "s-c-r-e-a-m." At this point the word is recognized and its identity can be passed on to the rest of the system.

Behind the apparent simplicity of this metaphor lie several key ideas on how the nervous system works during reading:

- MASSIVE PARALLEL PROCESSING: All the daemons work at the same time. There is thus no need to serially examine each of the 50,000 words one by one, a procedure whose duration would be proportional to the size of our mental dictionary. The massive parallelism of the pandemonium thus results in a substantial gain in time.
- SIMPLICITY: Each daemon accomplishes an elementary task by checking to what extent the stimulus letters match its target word. Thus the pandemonium model does not succumb to the pitfall of postulating a homunculus, or the little man who according to folk psychology holds the reins of our brain. (Who controls his brain? Another even tinier homunculus?) In this respect, the pandemonium model can be compared to the philosopher Dan Dennett's motto: "One discharges fancy homunculi from one's scheme by organizing armies of such idiots to do the work."[34]
- COMPETITION AND ROBUSTNESS: Daemons fight for the right to represent the correct word. This competition process yields both flexibility and robustness. The pandemonium automatically adapts to the complexity of the task at hand. When there are no other competitors around, even a rare and misspelled word like "astrqlabe" can be recognized very quickly—the daemon that represents it, even if it initially shouts softly, always ends up beating all the others by a comfortable margin. If, however, the stimulus is a word such as "lead," many daemons will activate (those for "bead," "head," "read," "lean," "leaf," "lend" . . .) and there will be a fierce argument before the "lead" daemon manages to take over.

All of these properties, in simplified form, fit with the main characteristics of our nervous system. Composed of close to one hundred billion (10^{11})

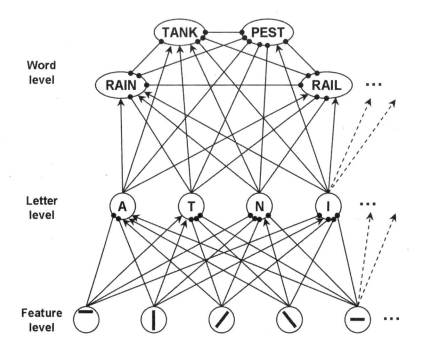

Figure 1.5 The word identification process is similar to a vast assembly where thousands of letter and word units conspire to provide the best possible interpretation of the input string. In McClelland and Rumelhart's model, of which only a fragment is shown here, basic features of the input string activate letter detectors, which in turn preferentially connect to detectors of the words that contain them. The links can be excitatory (arrows) or inhibitory (lines ending with discs). A fierce competition between lexical units finally identifies a dominant word that represents the network's preferred hypothesis about the incoming string.

cells, the human brain is the archetype of a massively parallel system where all neurons compute simultaneously. The connections that link them, called synapses, bring them evidence from the external sensory stimulus. Furthermore, some of these synapses are inhibitory, which means that when the source neuron fires, the firing of other neurons is suppressed. The result has been likened by the Canadian neurophysiologist Donald Hebb to a network of "cell assemblies," coalitions of neurons that constantly compete. It is therefore no surprise that Selfridge's pandemonium has been a source of inspiration for many theoretical models of the nervous system, including the first neural network models of reading. Figure 1.5 shows one of the very first such models, which was introduced by Jay McClelland and David

Rumelhart in 1981.[35] The model includes three hierarchical layers of neu-
ron-like units.

• At the bottom, input units are sensitive to line segments presented on
  the retina.
• In the middle lie letter-detector units that fire whenever a given letter is
  present.
• At the top, units code for entire words.

All of these units are tightly linked by a swarm of connections. This enormous connectivity turns the network dynamics into a complex political game
in which letters and words support, censor, or eliminate each other. If you
study the graph carefully, you will see excitatory connections, represented by
small arrows, as well as inhibitory connections, represented by small circles.
Their role is to propagate the votes of each of the daemons. Each input detector, which codes for a specific feature such as a vertical bar, sends stimulation
to all the letters that contain this particular feature—one might say, for the
sake of simplicity, that each visual neuron "votes" for the presence of these
letters. At the next level, similarly, letter detectors conspire to elect specific
words through stimulation of their corresponding units. The presence of letters "A" and "N," for instance, supports the words "RAIN" and "TANK," but
only partially argues for the word "RAIL" and not at all for "PEST."

Inhibition also contributes toward selection of the best candidate. Thanks
to inhibitory connections, letters can vote against the words that do not contain them. For instance, the unit that codes for letter "N" votes against the
word "RAIL" by inhibiting it. Furthermore, words that compete inhibit each
other. Thus identification of the word "RAIN" is incompatible with the presence of the word "RAIL," and vice versa.

Finally, it is useful to incorporate top-down connections, from words to
their component letters. This process can be compared to a senate where letters are represented by words that, in return, support the letters that voted for
them. Reciprocal connections allow for the creation of stable coalitions that
can resist an occasional missing letter: if one letter "o" is lacking in the word
"crocqdile," for instance, its neighbors will still conspire to elect the word
"crocodile," and in turn the latter will vote for the presence of a middle letter
"o" that is not physically present. In the end, millions of connections are

needed to incorporate the numerous statistical constraints that link the levels of words, letters, and features.

Other subtleties also allow for the whole network to operate smoothly. For instance, word units can have different thresholds for firing. A word encountered frequently has a lower threshold than a rare word, and with an equal amount of bottom-up support has a better chance to win the race. The most recent models also incorporate additions such as a fine-grained coding of letter position. The resulting network has such a complex set of dynamics that it is impossible to fully describe it mathematically. One has to resort to computer simulations in order to determine how long the system takes to converge to the correct word, and how often it misidentifies it.

Parallel Reading

If cognitive scientists burden themselves with all these complex models of reading, it is because their predictions fit remarkably well with empirical measurements. Not only do the models inspired by Selfridge's pandemonium manage to reproduce the results of classic experiments on reading speed and errors, but they often lead to the discovery of subtle novel phenomena that constitute the essential properties of human reading behavior.

If you were asked to design software for written word recognition, it is probable, regardless of the solution you choose, that your program would be increasingly slow as words get longer. It would be natural, for instance, for the software to process letters consecutively, from left to right. Since instructions are processed serially, one would expect a six-letter word to take about twice as long as a three-letter word. In any serial model, recognition time should increase in direct relation to the number of letters.

In light of this example, it is rather remarkable that the relation between reading time and number of letters does not hold for the human brain. In adult expert readers, the time to read a word is essentially independent of its length. As long as the word does not have more than six or seven letters, its recognition takes an approximately constant amount of time, regardless of length.[36] Obviously, this implies that our brain relies on a parallel letter-crunching mechanism that is able to process all letters at once. This empirical result is incompatible with the metaphor of a computer scanner, but it conforms precisely to the pandemonium hypothesis, whereby millions of specialized processors operate simultaneously and in parallel at each of several levels (features, letters, and words).

Active Letter Decoding

Let us pursue this computer metaphor. In a classic computer program, information is typically processed through a series of steps, from the simplest operations to the most abstract ones. It is logical to begin with a subroutine that recognizes individual letters, followed by one that groups them into graphemes, and finally a third that examines potential words. This type of program, however, is typically highly intolerant of errors. Any failure at the first stage usually results in a complete breakdown of the entire recognition process. Indeed, even the best automated character-recognition software, which is delivered with present-day scanners, remains highly sensitive to any decrease in the quality of the image—a few specks of dust on the scanning window are all it takes to turn a page that humans find perfectly readable into garbled text that the computer considers gibberish.

Our visual system, unlike the computer, revels in the resolution of ambiguities. You can experiment on this for yourself by reading the following sentence:

Honey bees savour sweet nectar

Unbeknownst to you, your eye just skimmed effortlessly across a series of difficulties that would have confounded any classic computer program. Did you notice that in the word "bee" the repeated letter is actually a "c"? In fact, the very same letter shapes are used to spell the word "nectar," but your visual system treats the first one as an "e" and the second as a "c." The word "savour" is worse. The letters "a" and "o" also have exactly the same shape—as do the letters "v" and "u"! Such ambiguities are resolved by the context: the string "souour" would not be legal in English, whereas its interpretation as the verb "savour" fits nicely within the rest of the sentence.

In brief, ambiguities that would create a hitch in any current software cannot even be perceived by a human reader. This resistance to degradation, while hardly compatible with a classic computer program, fits nicely into the pandemonium framework, where letters, graphemes, and words support each other through a multitude of redundant connections. The joint conspiracy of letters, words, and context suffices to confer an extraordinary robustness on our mental reading apparatus. Alberto Manguel was right: it is the reader who confers meaning to the written page—his "able eye" gives life to what would otherwise remain a dead letter. Letter and word identification result from an active top-down decoding process whereby the brain adds information to the visual signal.

Psychologists have discovered a remarkable manifestation of this active decoding process with the word superiority effect. In Gerald Reicher's classic experiment,[37] one asks an adult reader to identify which of two possible letters (for instance, D or T) is briefly presented on a screen. The level of difficulty can be adjusted so that the person only occasionally gives the correct answer. In some trials, the letter is presented alone. In others, the same letter (D or T) is presented in the context of other letters that form a word (for instance, HEAD or HEAT). Note that the extra string does not add any useful information. Since the same initial letters HEA_ are present in both cases, the subject must make a decision solely on the basis of the last letter. Nevertheless, quite surprisingly, subjects' performance is much worse when the letter is presented alone than when it is preceded by the redundant letters! Letter identification improves vastly when a letter is presented in the context of a word. It seems that the added votes provided by the word level help to clean up some of the noisiness in the input stimulus. This word superiority effect persists when a letter is inserted in a neologism (GERD or GERT), or even a consonant string that looks like a real word (SPRD or SPRT), but it ceases to work for a string of random letters (GQSD or GQST).[38]

Once again, this phenomenon seems difficult to explain in a strictly linear information processing model, where the identification of individual letters necessarily precedes their combination into larger units. On the contrary, Reicher's word superiority effect underlines the redundancy and parallelism of our procedure for visual word recognition. Even when we focus our attention on a single letter, we automatically benefit from the context in which it is placed. When this context is a word or a word fragment, it gives us access to more levels of coding (graphemes, syllables, and morphemes) whose "votes" add to those of letter units and facilitate their perception. Most models of reading explain the Reicher effect by postulating that letters and words interact within a two-way process: the higher-level grapheme and word detector units conspire to favor the detection of letters that are compatible with their own interpretation of the input string. What we see depends on what we think we're seeing.

Conspiracy and Competition in Reading

The computer metaphor, which compares the human reader to a simple scanner, thus turns out to be inadequate. Word decoding does not proceed in a strictly sequential manner, and the time needed to read a word is not related to the number of letters that it contains. Let us thus return to the pandemonium model, where recognition is achieved by an assembly of daemons. Here the time that it would take for a human assembly to reach an important decision, the "convergence time," would depend not so much on the content of the motion itself as on the disputes that it provokes. When all senators agree, even a complicated law can be adopted without much discussion. Conversely, even a minor aspect of a law, if it touches upon a sensitive issue, can initiate arguments that may go on for a long time before an agreement is reached.

Studies of human reading suggest that the reader's brain behaves much like a mental senate. The recognition of a word requires multiple cerebral systems to agree on an unambiguous interpretation of this visual input. The time that it takes us to read a word thus depends primarily on the conflicts and coalitions that it sets into motion in our cortical architecture.

Over the years, experimental psychologists have discovered that conflicts can occur at all levels of word processing. Within the lexicon, for instance, words have been shown to compete against their "neighbors," or words that are so similar that they have all but one letter in common.[39] The word "hare," for instance, cohabits with its many neighbors "bare," "care," "dare," "fare," "mare," "pare," "rare," "here," "hire," "hale," "hate," "have," "hard," "harm," and "hart," while the word "idea" is a lonely hermit with no neighbors. Experiments show that the number and especially the relative frequency of a word's neighbors play a crucial role in the time it takes us to recognize it.[40]

In many cases, the presence of a number of neighbors helps. The more neighbors a word has, the faster we can tell that it belongs to the English lexicon. The presence of neighbors makes the word "hare" more typical of English spelling than the word "idea." Dense neighborhoods also provide better learning opportunities. We have many more occasions to grasp the pronunciation of words that end in "-are" than of the only one that ends with "-dea." This orphan is therefore less optimally coded at both the visual and phonological levels.

But too many neighbors can also be intrusive and annoying. To understand or name a word often requires unequivocal identification and hence separation from all the neighbors—an operation that can be particularly slow and difficult if the neighbors are more common and thus perform better in the lexical competition. Thus the word "hare" is named relatively slowly because it competes with very frequently used words like "have" and "hard."[41] Our lexicon is an arena where competition is fierce and where frequent words enjoy a significant advantage over their lower-ranking neighbors.

Competition also occurs within the spelling-to-sound conversion route. For example, it takes longer to read the word "beach" out loud than the word "black." In "beach," the input string must be parsed into the complex graphemes "ea" and "ch," whose pronunciation departs radically from that of the single letters "e," "a," "c," or "h." The word "black," on the other hand, is simpler inasmuch as each of its letters maps directly onto its accepted sound. Experimental psychology easily demonstrates this latent conflict between the letter level and the grapheme level. During the naming of words made up of complex graphemes, there is a brief period of unconscious competition that is quickly resolved but leads to a measurable deceleration of responses compared with the naming of more transparent words like "black."[42]

It is noteworthy that most of these conflicts are resolved automatically, without our conscious intervention. When our nervous system is confronted with ambiguity, its fundamental strategy is to leave all possibilities open—something that is only feasible in a massively parallel system where multiple interpretations can be simultaneously entertained. Thanks to this open organization, the subsequent levels of analysis can contribute their own pieces of evidence until a globally satisfactory solution is reached. In some cases, only the context in a sentence allows one to understand the meaning of a word or even its pronunciation—think of a sentence like "the road **winds** through a valley battered by fierce **winds**." In such cases, experiments show that all of the possible interpretations of the word are unconsciously activated, until the context restricts interpretation down to a single meaning.[43] Our reading processes are so efficient that we are hardly ever aware of the ambiguities—unless they are very funny, as in the case of Dorothy Parker, who telegraphed her agent because she had failed to meet a deadline: "Tell the editor I've been too fucking busy—or vice versa."

From Behavior to Brain Mechanisms

Throughout this chapter, we examined how reading can be broken down into a sequence of information processing stages. From image processing in the retina to invariant letter recognition, access to pronunciation, morpheme recognition, all the way up to conflict resolution within the lexicon, the efficiency of the human mechanisms for reading is impressive. In a fraction of a second, with no apparent effort, our brain resolves a visual recognition problem that still defies current computer software. The parallel processing of all a word's characters, the clever resolution of ambiguities, the immediate access to one out of perhaps 50,000 words in the mental lexicon, all point to the remarkable adaptation of our brain to the task of reading.

The amazing efficiency of our reading process only serves to thicken the mystery surrounding its origins. How can our brain be so well adapted to a problem for which it could not possibly have evolved? How can the brain architecture of a strange bipedal primate turned hunter-gatherer have adjusted so perfectly, and in only a few thousand years, to the challenges of visual word recognition? To clarify this problem, we will now turn to the cerebral circuits for reading. An amazing recent discovery shows that there is a specific cortical area for written words, much like the primary auditory area or the motor cortex that exist in all our brains. Even more surprisingly perhaps, this reading area seems to be identical in readers of English, Japanese, and Italian. Does this mean that there are universal brain mechanisms for reading?

CHAPTER 2

The Brain's Letterbox

In 1892, the French neurologist Joseph-Jules Déjerine discovered that a stroke affecting a small sector of the brain's left visual system led to a complete and selective disruption of reading. Modern brain imaging confirms that this region plays such an essential part in reading that it can aptly be called "the brain's letterbox." Located in the same brain area in readers the world over, it responds automatically to written words. In less than one-fifth of a second, a time span too brief for conscious perception, it extracts the identity of a letter string regardless of superficial changes in letter size, shape, or position. It then transmits this information to two major sets of brain areas, distributed in the temporal and frontal lobes, that respectively encode sound pattern and meaning.

We are absurdly accustomed to the miracle of a few written signs being able to contain immortal imagery, involutions of thought, new worlds with live people, speaking, weeping, laughing. . . . What if we awake one day, all of us, and find ourselves utterly unable to read?

—VLADIMIR NABOKOV, *PALE FIRE*

Joseph-Jules Déjerine's Discovery

On a beautiful Paris morning in October 1887, Mr. C, a well-read retired salesman and music lover, was comfortably ensconced in an armchair reading a good book when he suddenly found, to his dismay, that he could no longer read a word! For several days previously, he had occasionally felt weakness or numbness in his right arm or leg and the odd mild difficulty in talking. These brief spells of discomfort, however, had only been very fleeting and had not worried him unduly. Now the problem was far more severe: reading was just impossible! Nonetheless, Mr. C could still speak, recognize objects and the people around him, and even jot down a few notes. It was hard to imagine what could have provoked this frustrating situation.

Convinced that a new pair of eyeglasses would solve his problem, Mr. C consulted the well-known ophthalmologist Edmund Landolt. Unfortunately, Dr. Landolt quickly realized that Mr. C's affliction required more than spectacles. He suspected a neurological problem and decided to enlist the help of the eminent French neurologist Joseph-Jules Déjerine at the Bicêtre Hospital. Mr. C's appointment took place on November 15, 1887. After a remarkably thorough psychological and anatomical examination, Dr. Déjerine was able to make a diagnosis and draw the first scientific conclusions ever made about the cerebral bases of reading.[44] He termed Mr. C's impairment "pure verbal blindness," meaning a selective loss of the visual recognition of letter strings. That such a disease existed implied the presence in the brain of a cortical "visual center for letters" specialized in reading. The first hint of brain specialization for reading had thus been scientifically demonstrated.

In fact, what Déjerine and Landolt established was that their patient no longer recognized individual letters or a written word. Confronted with an array of letters,

> he thinks that he has lost his mind, because he understands that the signs which he fails to recognize are letters, he insists that he can see them perfectly and outlines their shape with his hand but is incapable of naming them. If asked to copy what he sees, he succeeds with great effort by slowly copying each letter one stroke at a time, as though he were working on a technical draft, and by checking each and every curve to ensure the exactness of his drawing. In spite of these efforts, he remains unable to name the letters.

This is the paradox of "verbal blindness": the patient is only blind to letters and words. Visual acuity remains excellent, objects and faces are easily recognized, and the patient can still find his way around in a new environment or even appreciate a painting:

> When shown objects, he names them easily. He can name the parts of all the instruments in an industrial design catalogue. At no point during this examination is his memory at fault; drawings immediately prompt the appropriate word and how to use each object. . . . When handed the newspaper *Le Matin*, which he often reads, the patient says: "It's *Le Matin*, I recognize it by its shape," but he cannot read a single letter in the title.

Nonetheless, clinical examination did reveal some visual deficits. On Mr. C's right, the visual world seemed to be blurred. "Placed in the right halves of the visual fields, objects seem darker, less clear than in the other half" (partial right hemianopsy, in neurological terminology). Furthermore, he no longer distinguished colors on his right (hemi-achromatopsia), and that side of his world appeared to him only in black and white or in shades of gray. However, these deficits cannot explain his disproportionate difficulty with reading. Mr. C was aware of his visual deficit and spontaneously oriented his gaze so that the words he tried to read fell to the left of his fixation point, where his vision was normal. But none of this improved his reading.

Perhaps the best proof that Mr. C suffered from a selective letter-reading deficit came from the fact that he could still recognize numbers. He read strings of Arabic digits with ease and even performed complex calculations without flinching. This essential observation implies that reading digits relies on anatomical pathways partially distinct from those used for reading letters and words. Visually speaking, however, digits and letters have very similar shapes that are arbitrary and interchangeable. In fact, a number of cultures represent numbers using letter shapes—in Arabic, for instance, the shapes V, Λ, ٦, O, and Ɛ, which look like Greek letters, represent the digits 9, 8, 7, 6, and 5. Thus any dissociation between digits and letters cannot be explained by a loss of visual acuity. No pair of glasses, however sophisticated, could conceivably cure Mr. C's reading deficit. His selective loss of reading came from the brain and his impairment obliges us to infer some sort of cortical specialization for letter strings.

Other observations further emphasize the uniqueness of Mr. C's reading problem. Déjerine insists that Mr. C's intellect and language were remarkably preserved. Indeed, the patient continued to express himself with great clarity and used a vocabulary as extensive as the one prior to his stroke. Even more surprisingly, marveled Déjerine, writing was fully spared:

> Spontaneously, the patient writes as well as he speaks. When I compare the numerous specimens of writing that I asked of him, there are no mistakes, no spelling errors, no letter switching. . . . Writing under dictation can still be done easily and fluently, but the patient finds it impossible to read what he has just written down. . . . He gets impatient with these phenomena, writes several letters one after the other and says: "I can still write those letters, so why can't I read them?"

In fact, Déjerine did observe a slight deterioration in the patient's writing, but this was entirely imputable to low-level visual deficits:

> If he is interrupted when he is writing to dictation, he gets confused and no longer knows where he left off; likewise, if he makes an error, he cannot find it. In the past, he could write faster and better, now the characters are larger and traced somewhat hesitantly because, as he says, he can no longer use his eyes to check. Indeed, far from guiding him, the sight of what he is writing seems to trouble him so much that he prefers to write with his eyes closed.

Preservation of writing went hand in hand with muscle memory. The patient could still read letters if he was allowed to trace their contours. "It is therefore his muscular sense that awakens the name of the letter; and the best proof of this is that one can make him say a word, with his eyes closed, by leading his hand in the air to make him trace the contours of its letters." Years later, other neurologists showed that in such cases, tactile reading is also intact: the patient is better at deciphering letters when they are traced on his palm than when they are visually presented! This provides a valuable indication about the level at which the deficit occurs: the motor memory of letter shapes is intact; only their visual recognition is impaired.

Pure Alexia

Mr. C's symptoms are so bizarre that one could be tempted to dismiss them. How was it possible for him to write and even spell the most abstruse words and at the same time remain unable to read letters? Was he a sham or a hysteric? No, said Déjerine, because "we possess several beautiful clinical observations of this variety of verbal blindness." Indeed, modern neurology has since resoundingly confirmed all of Déjerine's initial observations. Today, hundreds of similar cases have been described.[45] Only the terminology has changed: the syndrome is now called "alexia without agraphia" (literally, impaired reading without impaired writing) or "pure alexia."

Why the word "pure"? For at least four reasons, all dutifully noted in Déjerine's original 1892 report, and later fully replicated in:

- Spoken language is intact.
- Writing is preserved.
- Visual recognition of objects, faces, drawings, and even digits can be largely normal.[46]
- Tactile or motor knowledge of letter shapes is preserved.

Although pure alexia has been reported in many patients since Déjerine, modern analyses have revealed that the patients fall into two subcategories.[47] Some, like Mr. C, cannot read even a single letter. They may even have trouble matching upper- and lowercase and cannot see that "A" and "a" are the same letter in different guises.[48] Others, however, can still recognize single letters. This simple fact makes a major difference in everyday life. By slowly identifying all the letters in a word one by one, these patients can typically still decipher words, and pure alexia simply manifests itself as exceedingly slow reading. Contrary to unimpaired readers, who easily read a word at a glance regardless of length, pure alexia patients may require as long as five or ten seconds to read a word, and their reading time increases sharply as the number of letters increases (figure 2.1; for full-color figures, please visit readinginthebrain.com). This characteristic impairment pattern is termed "letter-by-letter reading" and clearly illustrates the nature of the deficit. Pure alexia patients cannot process all of a word's letters simultaneously.[49]

A Lesion Revealed

Mr. C's case suggests that our brain contains specific reading areas that transmit information about the identity of a letter string to the brain's language regions. Déjerine was aware that only autopsy could infallibly pin down the exact location of the areas concerned and grant his observation "the rigorous value of a physiological experiment." Finally, on January 16, 1892, more than four years after his initial stroke, Mr. C died of a second cerebral infarct without having recuperated the ability to read. As the neurologist who followed him regularly noted, "in spite of patient exercises and much effort, he could

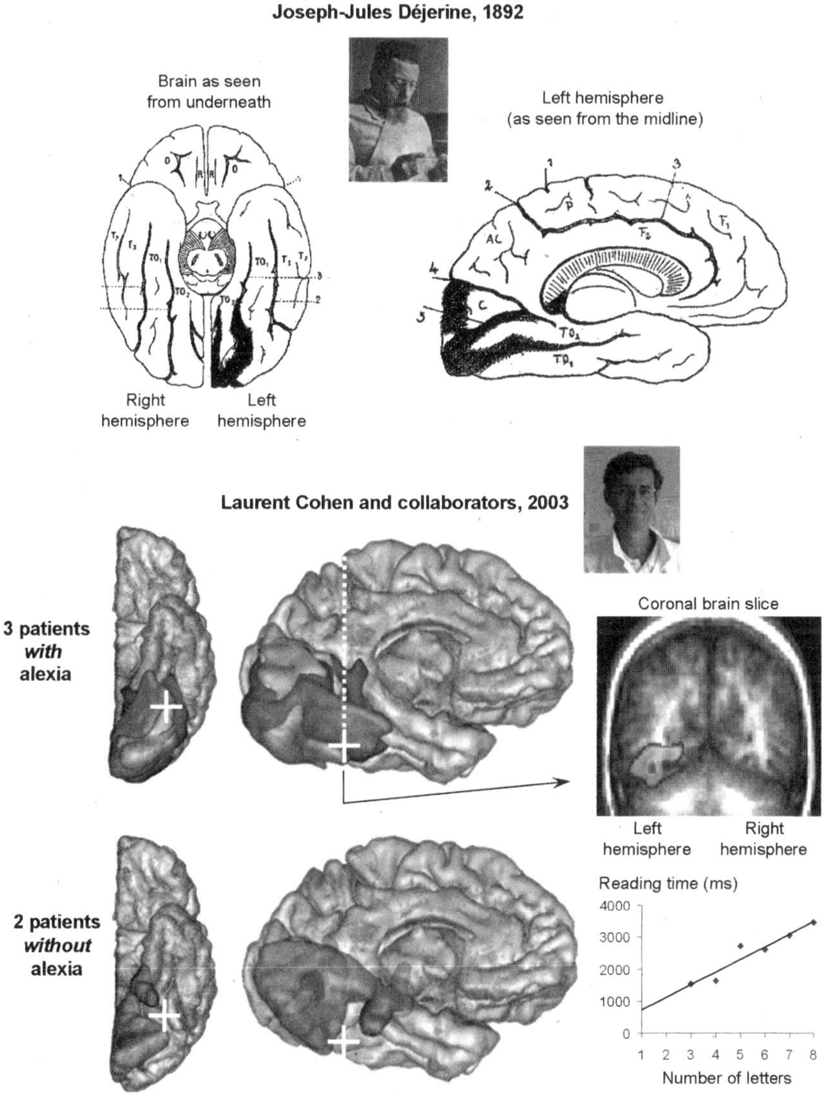

Joseph-Jules Déjerine, 1892

Brain as seen
from underneath

Right   Left
hemisphere   hemisphere

Left hemisphere
(as seen from the midline)

Laurent Cohen and collaborators, 2003

3 patients
with
alexia

2 patients
without
alexia

Coronal brain slice

Left   Right
hemisphere   hemisphere

Reading time (ms)

4000
3000
2000
1000
0

1 2 3 4 5 6 7 8
Number of letters

Figure 2.1 After a stroke, a patient can lose the ability to read and thus become alexic. The autopsy of the first alexic patient, described by Déjerine in 1892 (top), shows lesions similar to those of contemporary patients visualized using magnetic resonance imaging (bottom, after Cohen et al., 2003). In both cases, the inferior and rear part of the left hemisphere is affected. The intersection of several lesions points to a precise cortical site whose lesion systematically affects reading. This area exists in any literate person and always lies in the left ventral occipito-temporal region (white crosshairs). Alexic patients occasionally manage to decipher a word letter after letter. Even then, their reading time remains slow and, unlike with normal subjects, increases with the number of letters. They have lost the ability to recognize letter strings in parallel. *Used with permission of Oxford University Press.*

never regain the meanings of letters and words." Déjerine could now carry out an autopsy. A few weeks later, at the French Société de Biologie, he described what he had found. Postmortem examination had revealed that the patient's right hemisphere was fully intact but that old lesions affected the posterior part of the left hemisphere. They occupied "the occipital lobe, and particularly the circumvolutions of the occipital pole, starting at the base of the cuneus, as well as those of the lingual and fusiform lobules." (Please refer to "Getting oriented in the brain," page xii, for an illustration of these brain regions.) A facsimile of Déjerine's original drawing appears in figure 2.1. The contours of the large "yellow atrophic plaques" appear clearly at the cortical sites where a vascular infarct had rendered Mr. C unable to read four years earlier.

To fully explain how the lesion of some visual areas could selectively alter reading, Déjerine appealed to a concept of "disconnection." In his Société de Biologie report, he emphasized that Mr. C's lesion partially encroached on the "white matter"—the vast fiber tracts that connect distant areas of the brain. Furthermore, the lesion was located in the occipital pole, which is the seat of early visual processing. Finally, it partially affected the left visual cortex, and partially destroyed the long-distance connections of the corpus callosum that bring in visual information from the right-hand vision regions.

With these anatomical clues in mind, Déjerine proposed that Mr. C's brain lesion had disconnected the fibers that fed visual information to a region that he called the "visual center for letters." On the basis of other observations, he confined this hypothetical reading center to the angular gyrus, a fold in the cerebral cortex at the base of the left parietal lobe. Déjerine considered that Mr. C's visual center for letters was intact. This accounted for the patient's ability to still read, spell, and recognize the shape of letters when they were traced on his palm. However, deprived of visual input, this region was literally disconnected and thus unable to apply its knowledge of letters to any visual stimulus. Thus the patient was not blind. He could still see the shapes of letters and process them like any other visual object, but he could no longer recognize them as letters or as words and was suffering from a syndrome aptly termed "verbal blindness."

Modern Lesion Analysis

Over one hundred years after Déjerine's remarkable observation, my neurologist colleague Laurent Cohen and I,[50] together with several other researchers,[51] reopened C's case with new patients. The lesion that Déjerine had described turned out to be remarkably common. Most of our pure alexia patients suffered from lesions in the left occipito-temporal region, in the same general location as Mr. C's (such lesions frequently result from a clot that chokes the left posterior cerebral artery, which supplies blood to this area). Figure 2.1 shows the contours of the lesions we observed in three recent cases. All of them affect the inferior surface of the back of the left hemisphere.

Modern magnetic resonance imaging (MRI) now goes far beyond the tools that were available in Déjerine's time. Not only is it no longer necessary to wait for an autopsy to visualize a patient's lesion, but it is also possible to create computer images of the lesions from several different patients and paste them together in a unified anatomical space that corrects for individual differences in brain size and shape. We are thus able to go further than Déjerine: we can define which parts of the cortex are uniquely responsible for the reading deficit, and separate these from those involved in other associated deficits such as the loss of color vision. The logic is simple. First, take a large number of alexia patients and compute the intersection of the lesions in three-dimensional space. This information averaged across many brains removes the spatial randomness inherent in strokes and isolates the cortical territories most frequently associated with a reading deficit. Second, subtract the lesions of another group of patients who do *not* suffer from alexia. The outcome is a computer image that pinpoints the sites where lesions selectively provoke reading deficits.

This analysis, sketched in figure 2.1, suggests that the most posterior regions of the left hemisphere do not play a selective role in reading. Those occipital regions are frequently impaired in pure alexics, but also in patients who have no reading deficit. Indeed, they are known to be involved in the early stages of visual analysis that are not unique to reading but contribute to the visual recognition of any shape, color, or object. Thus a stroke that affects only these posterior sites provokes general visual impairments. Patients typically experience loss of vision in part of their visual field, and can be totally or partially blind for any stimulus presented to the right of their fixation point. They frequently find it difficult to see the right-hand side of words and program right-sided eye movements. These visual deficits may slow down reading, particularly for sentences that require several eye movements, but they must not be confused with pure alexia in the proper sense.[52]

The critical site for pure alexia lies a few centimeters to the front of the occipital pole, on the bottom side of the left hemisphere. This region, which plays an essential part in this book, is known by several names. Anatomists call it the *left occipito-temporal area*, because it lies at the boundary of the occipital and temporal lobes of the brain, within a groove in the cortical mantle called the lateral occipito-temporal sulcus (see figure 2.2 to understand where these regions are located). Laurent Cohen and I have suggested that it be called the *visual word form area*—an expression that has now become standard in the scientific literature and emphasizes its essential role in the visual analysis of letter and word shape.[53] Both terms are a mouthful, however. In this book, for the sake of simplicity, I will refer to it as "the brain's letterbox," a term that nicely encapsulates what we believe this region does: it takes care of the incoming letters! As we shall see, this brain area indeed plays a crucial role in the fast identification of the letter string and its transmission to the higher areas that compute pronunciation and meaning.

Unlike Déjerine, we now think that visual recognition of letters does not rest primarily on the angular gyrus, on the top rear of the head, but rather in this quite distinct "letterbox" area on the brain's lower side. Déjerine's error can thus probably be attributed to a fairly unusual feature in Mr. C's lesion. Unlike most pure alexia patients, Mr. C's letterbox area was not directly destroyed, but was probably just disconnected, as Déjerine himself suggested. He was simply mistaken about the location of the disconnected region, which actually lay in the ventral visual system rather than in the more dorsal region of the angular gyrus.

As we now know, there are at least three ways for lesions to prevent normal function in the visual word form area.[54] The simplest case is obviously destruction by a direct lesion. However, the region can also be disconnected upstream and thus deprived of visual inputs, as in Mr. C's case, or downstream, so that it cannot send outgoing messages to other brain regions. In any event, the outcome is the same and involves a severe incapacity to recognize the written word.

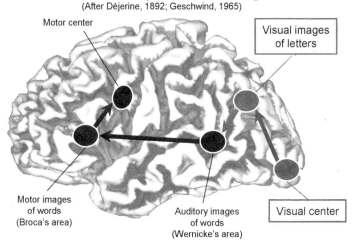

The old neurological model of reading
(After Déjerine, 1892; Geschwind, 1965)

Motor center

Visual images of letters

Motor images of words (Broca's area)

Auditory images of words (Wernicke's area)

Visual center

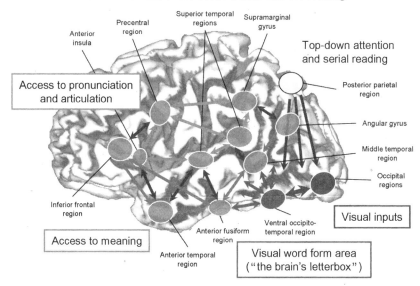

A modern vision of the cortical networks for reading

Anterior insula

Precentral region

Superior temporal regions

Supramarginal gyrus

Top-down attention and serial reading

Access to pronunciation and articulation

Posterior parietal region

Angular gyrus

Middle temporal region

Occipital regions

Inferior frontal region

Access to meaning

Anterior fusiform region

Anterior temporal region

Ventral occipito-temporal region

Visual inputs

Visual word form area ("the brain's letterbox")

Figure 2.2 The classic neurological model of reading (top) is now replaced by a parallel and "bushy" model (bottom). The left occipitotemporal "letterbox" identifies the visual form of letter strings. It then distributes this invariant visual information to numerous regions, spread over the left hemisphere, that encode word meaning, sound pattern, and articulation. All the regions in green and orange are not specific to reading: they primarily contribute to spoken language processing. Learning to read thus consists in developing an efficient interconnection between visual areas and language areas. All connections are bidirectional. Their detailed organization is not yet fully known—in fact, cortical connectivity is probably much richer than suggested in this diagram.

There is a more fundamental issue about which Déjerine was also mistaken. He vastly underestimated the overall complexity of the brain architecture for reading. Déjerine, and many subsequent scientists such as the Harvard neurologist Norman Geschwind,[55] thought in terms of a simple processing chain. They claimed that written words enter the occipital pole in the form of visual patterns and are then sent to the angular gyrus to make contact with the visual images of words. Activation is then propagated to Wernicke's area, the seat of auditory images for words, then on to Broca's area, where articulation patterns are retrieved, and finally to the motor cortex, which controls muscles. The scheme was serial, simple, and linear, much like the production line in a factory. Indeed, it smacks of nineteenth-century mechanical analogies that compared cerebral function to the propagation of electricity or the flow of steam in a locomotive. There was, of course, a filial link between these first neurological schemas of brain functioning and the hydraulic diagrams used by René Descartes to describe reflexes as a circulation of "animal spirits" through pipes in our bodies. The reflex arc was the dominant metaphor for several generations.

Déjerine cannot be blamed for having failed to anticipate a century of research in psychology and neuroscience. Today, a "bushy" vision of the brain, with several functions that operate in parallel, has replaced the early serial model. For one thing, we now realize, after having attempted to program visual shape recognition in computers, that vision is complex and cannot be reduced to a simple chain of cerebral "images." Several intricate operations are needed for the recognition of just one character. Visual analysis is only the first step in reading. Subsequently, a variety of distinct representations must be brought into contact: the roots of words, their meaning, their sound patterns, their motor articulation schemes. Each of these operations typically demands the simultaneous activation of several separate cortical areas whose connections are not organized in linear chains. All the brain regions operate simultaneously and in tandem, and their messages constantly crisscross each other. All the connections are also bidirectional: when a region A connects to a region B, the converse projection from B to A also exists.

These very basic anatomical principles have allowed me to sketch a new version of what a reader's brain looks like (figure 2.2). All regions in this graph are known to contribute to word reading. Nonetheless, my sketch must be viewed as provisional—paradoxically, it is still far too simple and many essential areas and connections are probably missing. Unlike Déjerine's, my diagram does not illustrate a serial vision of brain organization but rather a scheme where several brain areas can be simultaneously activated. As a result, their specific contributions to the reading process continue to be debated. In spite of the battery of imaging tools at our disposal, we are still hard put to assign precise functions to each region in our diagram, because they all operate simultaneously and interact with each other at fast speeds. Déjerine happily pinpointed visual, auditory, and motor images at various places on the cortex, but we cannot. We undoubtedly have a more refined view of cortical functioning, but this complicates things for the neuropsychologist and leads us to wonder if the day will ever come when we will be in a position to achieve an intuitive grasp on so much complex circuitry.

Whether the scientist will ever fully understand the reader's brain is unclear. However, our understanding of the brain circuitry for reading and particularly of the brain's input "letterbox" has certainly progressed. We now know that this visual input area occupies a strategic location and is a funnel through which all visual information about written words seems to flow before it is distributed to a great variety of left-hemispheric areas. This strategic position explains why a lesion in this region leads, as in Mr. C's case, to a total loss of written word processing, as if writing had become a series of meaningless spots. But what are the principles that allow this region to operate?

Decoding the Reading Brain

For more than twenty years, functional brain imaging techniques have revolutionized the study of the human brain by literally allowing us to "read in the brain." Their enormous potential lies in their capacity to visualize brain activity directly, at the very moment when a human volunteer performs a mental operation like reading. In many ways, the use of imaging experiments is far more straightforward than the study of lesion data. Who can tell to what extent the brain reorganizes itself after an injury? The effects of a lesion can be felt far from the affected area, because the damage can deprive other regions of input by disconnecting them. In the weeks following a brain lesion it is not unusual to observe a considerable reorganization of brain activity as the patient regains lost functions via circuits radically different from those used by a healthy brain. Finally, lesions are unpredictable. They can be extensive and their location biased by the anatomical distribution of arteries, so that they cannot be relied upon as precise indicators of functional organization of the healthy brain. If we did not have the means to visualize the normal human brain, our task would be much like that of an apprentice who attempts to learn clockmaking by studying broken clocks.

Brain imaging rests upon a simple principle, anticipated as early as the eighteenth century by the French chemist Antoine-Laurent Lavoisier. Like any other organ of the body, the brain expends more energy when it is at work than when it is at rest. In his "first memoir on animal breathing" (1789), Lavoisier foresaw how this simple idea might eventually be applied to the measurement of brain activity:

> This type of observation leads to the possibility of compar-
> ing the application of forces that would not appear to be
> related. One can evaluate, for instance, how many pounds of
> weight correspond to the efforts of a man who recites a
> speech or a musician who plays an instrument. One might
> even assess the mechanical content in the work of the phi-
> losopher as he thinks, the man of letters as he writes, the
> musician as he composes. These effects, considered as purely
> mental, have something physical and material which allows
> them, in this respect, to be compared to those of a laborer.

Two hundred years would elapse before this simple idea was put into prac-tice. In 1988, Steve Petersen, Michael Posner, Marcus Raichle, and their col-leagues were the first to visualize which brain areas consume energy when we read.[56] They used positron emission tomography (PET) to reveal the func-tional organization of the language areas of the brain. Lavoisier could not have foreseen the peculiar combination of psychology and nuclear physics that this technique would require. Volunteers are injected with a small amount of radioactive water in which the standard oxygen atom (^{16}O) is replaced by oxy-gen 15. This liquid quickly blends into the bloodstream and spreads through-out the body. In the brain, in particular, radioactivity accumulates in regions where blood flow is the fastest and which happen to be those where brain activity is most intense. As a result, the local peaks of radioactivity directly reflect the brain's hot spots.

After a few dozens of seconds, the oxygen 15 atoms inside the volunteer's brain spontaneously return to their stable state (oxygen 16) by emitting a positron—an antimatter particle that is the exact mirror image of the electron. When a positron collides with an electron, both annihilate each other and emit two high-energy photons. The scanner detects these particles of light as they escape from the scalp in opposite directions. Arrays of crystals placed all around the subject's head feed this information into a powerful computer that in turn reconstructs the location where the disintegration initially occurred. Finally, a 3-D picture is obtained that reveals, slice by slice, the precise distribution of the energy consumed by the brain—hence the term "tomography," taken from the Greek words *tomos* ("slice") and *graphein* ("drawing").

On the psychological side, Petersen and his collaborators also innovated by asking their volunteers to perform a series of increasingly complex tasks. They first took images of the brain at rest, when the person was not stimulated and was not thinking of anything in particular. They then measured brain activity when the volunteer viewed and then repeated either written or spoken words out loud. Finally, they made one last measurement while the same person generated verbal associations by thinking of a verb that could plausibly be connected to each of the words presented (e.g., "eat" if the stimulus was "cake"). The authors hoped that by subtracting two consecutive conditions, they would be able to identify the brain regions successively involved in visual and auditory word recognition, articulation, and the mental manipulation of word meaning.

The images from Petersen's experiment were so spectacular that they appeared on the front pages of newspapers all over the world and have now become an integral part of brain science. For the first time in history, the areas responsible for language had been photographed in the living human brain (figure 2.3). To be sure, the pictures would have to be perfected later. The key findings, however, would all be replicated. Whenever subjects looked at written words, the region dedicated to vision, situated at the back of the head, was activated. Another small region of the left hemisphere, right at the border between the occipital and temporal lobes, also showed up—it coincided very nicely with the area that we earlier termed "the brain's letterbox." Listening to a spoken word did not put any of these regions into motion but did trigger

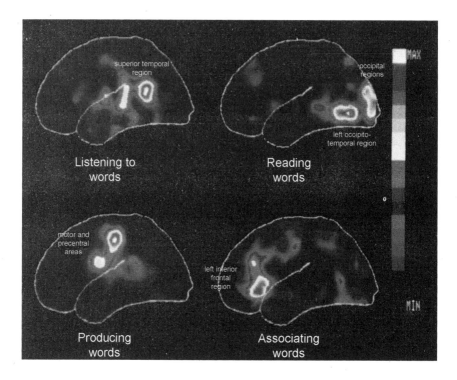

Figure 2.3 A historic image: the brain areas for language, as first revealed by PET scanning (data from Petersen et al., 1989; image courtesy of Marcus Raichle). Relative to the fixation of a small dot, silent reading (top right) activates processes of visual word recognition located in the rear part of the left hemisphere. Depending on the task, information is then transmitted to regions coding for speech sounds (top left), speech production (bottom left), or the manipulation of word meanings (bottom right). *Used with permission of Mike Posner and Marc Raichle.*

activity in a distinct set of sites in the superior and middle temporal cortex, corresponding to processing of hearing and speech perception. The production of a spoken word, on the other hand, stimulated an anterior region of the left hemisphere, close to the area first linked to speech production by the French neurologist Paul Broca in the nineteenth century, as well as the left and right motor areas. Finally, semantic associations mobilized yet another region, the left interior prefrontal cortex often associated with creative thought.

Within this set of areas, only one, the left occipito-temporal region, appeared to play a central and specific role in reading. Strikingly, its localization coincided neatly with the site identified thanks to lesion analysis as the seat of pure alexia. Petersen and his colleagues noted that this region was stimulated only by written words and was not part of the low-level visual areas that are aroused by a visual pattern like a checkerboard. They proposed that this region serves as a link between early visual analysis and the rest of the language system. It is the filter through which visual information must flow in order to enter the language system. It selectively analyzes incoming images for the presence of letters and forwards them to other brain areas that subsequently transform them into sound and meaning.

Reading Is Universal

The key role played by the visual word form area in reading has now been corroborated by another imaging technique that has largely supplanted the PET scan. It is called functional magnetic resonance imaging, or fMRI for short. This extraordinarily flexible method presents two key advantages over PET: it is available at any hospital, and is totally harmless since it does not require the injection of radioactive substances. In functional MRI, fluctuations in the blood's oxygen content modulate the magnetic resonance signal, which is detected by the fMRI machine. Blood cells contain a high concentration of hemoglobin, the molecule responsible for carrying oxygen. Roughly speaking, if a hemoglobin molecule does not transport oxygen, it behaves like a small magnet and disrupts the local magnetic field, thus reducing the signal received by the fMRI machine. When a hemoglobin molecule does contain oxygen, it immediately becomes transparent in the magnetic field—a change that is seen by the machine as a small but measurable increase in the resonance signal.

These detectable fluctuations in the oxygen content of the blood allow us to measure brain activity. Blood vessels and neurons are closely linked. Whenever a brain region is activated, its blood vessels dilate and in one to five seconds the area in question receives a fresh influx of oxygenated blood. By measuring blood oxygenation, fMRI indicates how neural activity increased in the preceding few seconds. Furthermore, it provides a three-dimensional image of any region of the brain. My French colleague Denis Le Bihan, one of the world's most eminent magnetic resonancing experts, compares this method to spying on a gardener to foresee where seeds will sprout. Without actually knowing where they were sown, we can find them by seeing where the gardener takes his watering can. To track blood flow is a comparably indirect way of monitoring neurons at work—a roundabout but in fact highly sensitive method.

Over and above its other advantages, fMRI is very fast. A single fMRI examination can provide several thousand consecutive images of the entire brain with a spatial resolution of a few millimeters. The method yields a whole-brain image every two or three seconds, compared to one every twelve minutes with PET scanning. fMRI is also remarkably sensitive. There is no need to average out findings over long periods of time and across dozens of subjects. A single written word is enough to trigger a transient activation blip in the letterbox area. A few minutes of data suffice to map the intense brain activity evoked by written words in the left ventral visual cortex of any reader.

Thanks to this amazing imaging tool, we can now measure the extent to which the cerebral network for reading varies from one person to the next. Based on all the human subjects we have ever scanned, we find that the visual word form area always lies at the same position in the ventral visual brain. The exact location, of course, varies marginally from one person to another, because the details of individual cortical folds are unique—much as two crumpled sheets of paper never have exactly the same shape. Nonetheless, we are all equipped with a letterbox area that is invariably found in approximately the same position.

Scientists rely on two systems to map cortical locations. In the first of these, developed thanks to two centuries of postmortem analysis, major landmarks of the brain have been identified and labeled, much like the moon's craters. In these terms, the letterbox area is always said to fall in the lateral occipito-temporal sulcus, a crack that runs along the fusiform region of the left hemisphere (figure 2.4).[57] A second system, less dependent on abstruse anatomical labels, was invented by the French surgeon Jean Talairach and later refined by researchers at the Montreal Neurological Institute. It is a geometrical positioning system, much like the earth's GPS, that consists of three perpendicular axes normalized for brain size. In this system, the location of the letterbox area turns out to be remarkably similar, both across individuals and even across experimental laboratories, with differences of no more than roughly five millimeters.[58] Even the brains of Chinese and Japanese readers, as we shall see later, are equipped with a brain area specific to reading, and its position is approximately the same as ours.[59] Note that the direction in which we read (from left to right or right to left) does not seem to affect its place in

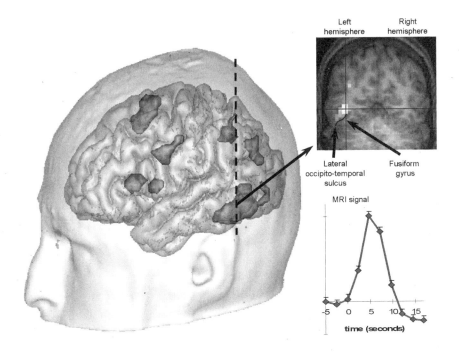

Figure 2.4 Functional magnetic resonance imaging (fMRI) can locate the brain areas involved in reading in just a few minutes. Participants read words presented at random intervals. After each word, reading areas show a characteristic increase in MRI signal, which reaches a peak about five seconds later. The active network varies depending on the exact task and the nature of the control state. However, it always includes the visual word form area, "the brain's letterbox." This region is systematically located deep in the left lateral occipito-temporal sulcus, next to the fusiform gyrus.

the left hemisphere. The location of the region remains unchanged, on the edge of the left occipito-temporal fissure[60] even in readers who read from right to left.

That the location of this brain activation is identical in every human brain may seem uncanny. Reading is a cognitive, social, and cultural activity that dates back five thousand years and whose surface forms are different from one culture to another. Furthermore, our individual command of reading varies greatly from person to person, depending on how we learned to read. Some of us simply mumbled our way through hours of phonics drill where we were asked to connect sounds to letters; others, who followed the Montessori system, traced sandpaper letters with our fingers; while yet others were subjected to the miseries of the so-called whole-word method. What is amazing is that in spite of these vast differences in the way we learned to read, we all call on the same area of the brain to recognize the written word.

A Patchwork of Visual Preferences

The functional properties of the letterbox region are also similar from one person to the next. In figure 2.5, the seven brains scanned show selective activation to written words, with no trace of activation for the oral presentation of the same words.[61] This very stable finding, already observed by Steve Petersen and his colleagues in their 1998 study, demonstrates clearly that in most cases speech does not stimulate this region, which thus appears exclusively reserved for the written word. There is one exception, however, which occurs when people are encouraged to spell a word in their mind's eye. For instance, they could be asked whether the spoken word contains a "descending" letter such as "p," "q," or "j," whose strokes fall below the line. In such cases, the letterbox area is mildly activated by spoken words—only because participants imagine the written word.[62] This also takes place when Japanese subjects have to think of writing a word.[63] Finally, when people are asked to focus on the fine difference between two speech sounds, like *dip* and *tip*, the letterbox area again lights up, probably because access to the written letters "d" and "t" helps sound discrimination.[64] What these examples have in common is that they appear to call for a top-down access from speech sounds to letters. In such cases, a signal heading backwards probably travels along the reading routes in the opposite direction to the one usually used for reading. But apart from these peculiar cases, the letterbox region generally only reacts to written words and ignores spoken words.

All evidence thus points toward the fact that this region is dedicated to visual analysis. The question is whether it is an all-purpose area that processes any visual object, or whether it is specialized for reading. The answer, once again, is surprising: our brain divides visual labor into categories, each of which is processed by a given patch of the cortex. Thus a piece of the area prefers writing to a broad variety of other visual stimuli, and this preference is universally present, in all individuals, at the same location. Recognition of houses and landscapes preferentially calls upon the regions closest to the

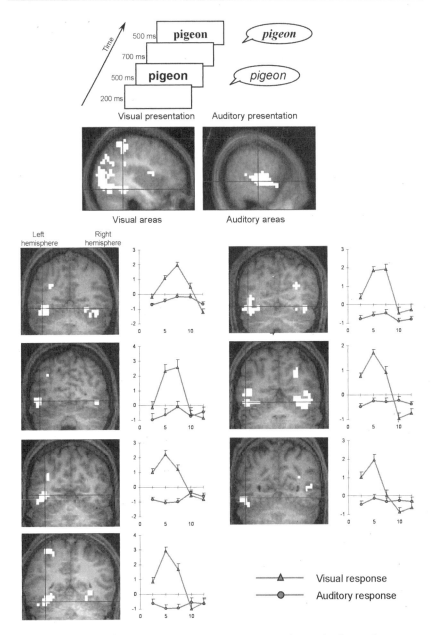

Figure 2.5 Activation of the visual word form area can be easily observed in any literate person. In this experiment, the participants either saw or heard a pair of words and had to judge whether they were identical or different. In seven different readers, written words activated the left occipito-temporal region, "the brain's letterbox," at an astonishingly reproducible location in spite of variability in the folding patterns of cortex. Note that spoken words did not activate this region (after Dehaene et al., 2002). *Used with permission of Wolters Kluwer/Lippincott, Williams & Wilkins.*

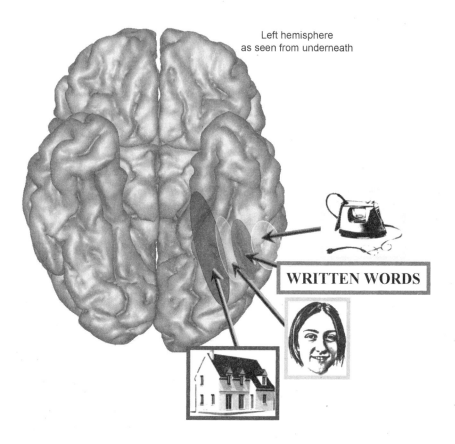

Left hemisphere
as seen from underneath

WRITTEN WORDS

Figure 2.6 A patchwork of specialized visual detectors covers the underside of the brain. Each cortical region preferentially responds to a certain category of objects. This preference pattern occurs in the same order in all individuals, from houses to faces, words, and then objects. Reading always activates an area located between the peak responses to faces and to objects (after Ishai et al., 2000; Puce et al., 1996). *Adapted with permission of Alumit Ishai.*

brain's midline. As one moves laterally toward the sides of the brain, other regions prefer faces to written words. Finally, on the edge of the brain, an entire sector responds to objects and tools (figure 2.6).[65]

The neuroradiologist Aina Puce was the first scientist to make use of the sensitivity and high spatial resolution of fMRI to explore this visual patchwork in a number of individual brains. She flashed faces or meaningless letter strings such as "XGFST" at volunteers, who were asked to examine them carefully. In every case, two small specialized regions were activated at the same

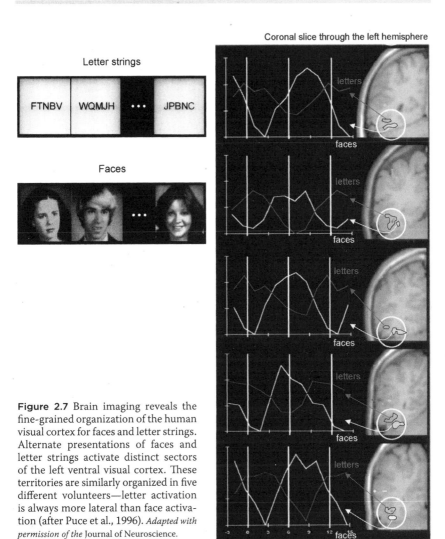

Figure 2.7 Brain imaging reveals the fine-grained organization of the human visual cortex for faces and letter strings. Alternate presentations of faces and letter strings activate distinct sectors of the left ventral visual cortex. These territories are similarly organized in five different volunteers—letter activation is always more lateral than face activation (after Puce et al., 1996). *Adapted with permission of the* Journal of Neuroscience.

relative locations: faces preferentially stimulated an inferior cortical site in the fusiform gyrus, while words excited a more lateral region just next to it in the occipito-temporal fissure (figure 2.7).[66] This result is surprising. Even if brain regions finally specialize in one of these categories, we would have expected the division of labor to be a haphazard procedure that varies randomly from one person to the next. Not so—reading acquisition seems to be a highly constrained process that systematically channels information to the same hot spots in the brain.

We should be cautious, however, about overemphasizing the amount of specialization in the visual cortex. On the rough spatial scale available with fMRI, portions of the visual cortex often show a preference for certain categories of stimuli, but not a total dedication to any one of them. As was first shown by Jim Haxby and his collaborators at the National Institutes of Health, our cortex does not appear to be strictly subdivided into discrete territories with sharp boundaries. Rather, visual preferences overlap.[67] Thus, even within the regions that respond the most to faces, the sight of other categories of objects such as words, tools, or animals provokes substantial activity. But we must bear in mind that each of the elementary cubes or voxels visualized with fMRI measures up to two or three millimeters per side. This is large by cellular standards, and may comprise as many as a million neurons or more. No wonder, then, that these cells do not all exhibit the same preferences. When spatial resolution is increased to resolve details as small as one millimeter, category selectivity becomes spectacular, with some patches of the cortex that respond to only one category, for instance faces, and nothing else.[68] In animals, direct recordings from single neurons within such a patch show that the vast majority of them prefer faces.[69]

What emerges from these studies of the visual cortex is a patchwork of neurons each specialized for a certain category of shapes. Neurons with similar preferences tend to cluster together although they are often somewhat intermingled and create zones of partial preference. Distinct zones prefer faces, objects, digits, or letters. In order to identify them, scientists label them the "face area" or the "word form area." These labels are useful and indicative, but we should remember that all we can actually see are peak responses to faces or to words and not the underlying neurons. The peaks whose positioning is fairly regular say one thing—but in the background there are other less prominent summits that are perhaps as essential to reading or to face recognition. In the final analysis, each visual category encoded is a complex landscape of valleys and hills that covers the inferior surface of the two hemispheres, and testifies to the incredible subtlety of vision.

How Fast Do We Read?

Quite apart from cortical topography, words and faces also have different preferred hemispheres. When we recognize a word, the left hemisphere plays the dominant role. For faces, the right hemisphere is essential. Although both hemispheres are initially equally stimulated, words quickly get funneled to the left and faces to the right, in only a few dozens of milliseconds. This lateralization is another invariant and essential feature of reading.

Had we been obliged to rely on PET and fMRI alone, we would never have discovered the speed of the hemispheric sorting process. Both methods, which measure blood flow and oxygenation, are too slow for the real-time visualization of brain activity. When a cortical region is activated, several seconds elapse before blood flow increases. Like astronomers who pick up light emitted decades earlier by distant stars, fMRI only allows us to see brain activity that occurred several seconds earlier. Such delays and smearing severely limit our observation of brain activity.

Fortunately, other imaging techniques, electro- and magnetoencephalography, can monitor brain activity in real time. They operate on the principle that neuronal activity generates electrical and magnetic responses that can be instantly detected at a distance. Neurons, thanks to their dendrites, collect the voltage changes received from thousands of other neurons. When many neurons, all neatly aligned in a formation perpendicular to the cortical surface, simultaneously receive electrical inputs, the total amount of current available is sufficient to allow for it to be measured from outside the skull. The resulting measurement constitutes the well-known electroencephalogram, or EEG for short.

Electroencephalography is based on an old idea, first applied to the human brain by Hans Berger in 1924. It essentially consists in the use of a voltmeter to measure the differences in voltage generated on the scalp's surface by neural currents. Because these voltages are very small, in the range of a millionth of a volt, a sensitive amplifier is required. In 1968, David Cohen and his colleagues at MIT designed another more sophisticated method called magnetoencephalography, or MEG. This technique detects minuscule changes in magnetic fields produced by neural currents. The brain's magnetic signals are extremely weak, on the order of a few femtoteslas, or a billion times smaller than the earth's magnetic field, and the method had therefore to be particularly sensitive. Although it is a costly technique, its spatial precision, which far exceeds that of EEG, makes it invaluable for the visualization of real-time brain activity.

Both EEG and MEG allow for excellent temporal resolution, because the transfer of the brain's electromagnetic signals to the sensors used to measure them is virtually instantaneous. One can thus obtain a series of snapshots of the brain in action. When applied to reading, these methods have revealed the extraordinary speed at which the brain sorts visual images. Antti Tarkiainen and his colleagues at the University of Helsinki used MEG to measure magnetic activity in the brain while volunteers viewed words and faces (figure 2.8).[70] Their results reveal two distinct stages of visual processing in the cortex. In the first, observed approximately 100 milliseconds after the image first appeared on the retina, the two types of images could not be told apart: words and faces activated similar regions of the occipital pole, at the back of the head. These regions performed a first-pass analysis and probably extracted some elementary lines, curves, and surfaces from the input image. At this stage, the brain did not recognize the types of stimulus with which it was confronted. Only 50 milliseconds later, however, sorting of the input images began. Words now evoked a significant response strongly lateralized in the left hemisphere. For faces, the exact opposite occurred, with the magnetic potential clearly predominant on the right side of the head.

EEG also easily records these early visual events. They are seen as negative voltages that suddenly appear on the back of the head in roughly 170 milliseconds. For words, their amplitude is much greater in the left hemisphere than in the right. A computer reconstruction of the point where these electromagnetic waves originate in the cortex places them toward the posterior part of the occipito-temporal fissure—exactly where we found the letterbox area with functional MRI. These imaging techniques thus confirm our belief that this region plays an early and specific role in visual word recognition. Once the first few steps in vision have occurred, specialized systems come into play. Presumably, reading and face recognition put such different demands on our visual system that it cannot content itself with generic image processing algorithms.

Electrodes in the Brain

Measurement of brain activity from outside the skull is fine, but the neuroscientist's dream has always been to delve into the human brain itself. Is such an investigation ever ethically possible? A more invasive technique, which involves direct electrical contact with the brain's surface, does exist, and provides a unique window into brain specialization for reading. With this tech-

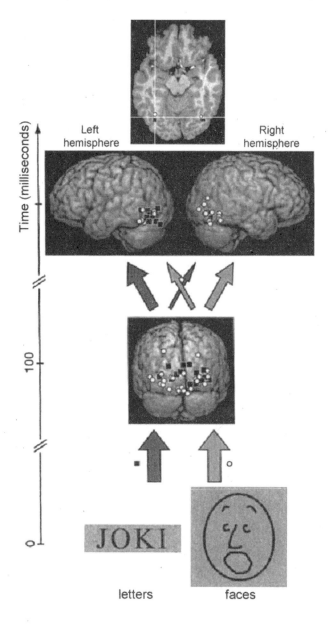

Figure 2.8 Magnetoencephalography is a technique that tracks, millisecond by millisecond, the time course of brain activity during the recognition of faces and letter strings. Initially, and up to about 100 milliseconds after image onset, faces and strings evoke very similar patterns of occipital activation. By 150 milliseconds, however, strings are channeled to "the brain's letterbox" in the left hemisphere, while faces activate a symmetrical region in the right hemisphere (after Tarkiainen, Cornelissen, & Salmelin, 2002). *Adapted with permission of Oxford University Press.*

nique electrodes are placed directly on the cortex or even deep into the brain tissue. Such a method, of course, can only be applied to certain specific patients. It is never attempted without an excellent clinical motive, and requires informed consent by the patient. Its main purpose is to study epilepsy. Some patients suffer from frequent seizures and fail to respond to anti-epileptic drugs. Their only hope for relief is surgery to remove the damaged region where the fits originate, and in most cases surgery leads to a happy, seizure-free recovery. However, the region to be excised must be carefully pinpointed. The challenge is to completely remove the affected tissue but spare the healthy brain areas that may be immediately adjacent and whose removal could have disastrous consequences.

Successful surgery for epilepsy depends on the accurate identification of the areas of the brain where seizures originate. For one or two weeks prior to surgery, a few dozen electrodes are inserted around the region where damage is suspected in order to ensure accurate identification. Direct contact with the cortex allows for precise and highly sensitive monitoring of the surrounding electrical signals. In modern epilepsy clinics, these signals are digitized continuously while patients are filmed day and night. Thus even the slightest hint of an impending brain storm can be reconstructed and traced back to the area where it originated. Between two seizures, the electrodes also record brain signals that, for the most part, arise in healthy regions. With the patient's consent and collaboration, these intracranial signals can be exploited to study, with unprecedented precision, how the brain reacts to external inputs such as words or faces.

In the 1990s, the neurologists Truett Allison, Gregory McCarthy, and their colleagues at Yale University launched an impressive research program that led them, in due course, to record intracranial signals from over one hundred patients.[71] Their surgical technique consisted in wrapping the temporal and occipital lobes with strips of electrodes placed under the brain's protective membrane (the dura) in direct contact with the cortical surface (figure 2.9). Placed every five or ten millimeters, these sensors provided a window into the successive stages of reading and confirmed the incredible speed of our visual system. About 180 milliseconds after an image appeared on the retina, high-amplitude negative waveforms appeared on some electrodes facing the ventral surface of the occipital and temporal lobes. As expected, those signals were mostly found in the left hemisphere for words and in the right hemisphere for faces.

What was puzzling about this study was the extreme spatial selectivity of

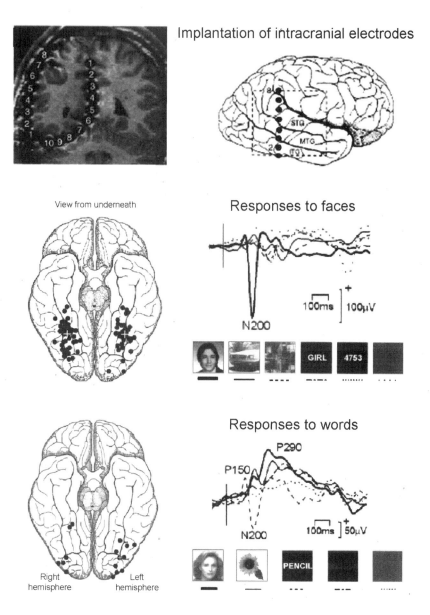

Implantation of intracranial electrodes

View from underneath

Responses to faces

N200

100ms | 100µV

GIRL   4753

Responses to words

P290

P150

N200

100ms] 50µV

PENCIL

Right
hemisphere

Left
hemisphere

Figure 2.9 In epileptic patients, electrodes placed directly on the surface of the cortex expose brain specialization with exceptional spatial and temporal accuracy. On some electrodes, when a certain category of images is presented, the electrical signal suddenly deviates between 150 and 200 milliseconds. Some sites prefer faces, others favor written words. When the data from multiple patients are placed in a standard anatomical space, faces appear to preferentially engage the right hemisphere, while word responses predominate in the left (after Allison et al., 1999). *Adapted with permission of Oxford University Press and* Cerebral Cortex.

the responses. It was not unusual for a single electrode to react massively to words while its neighbors, only a few millimeters away, manifested no response at all. Even more surprising, an electrode might respond quite vigorously to words, but demonstrate no hint of a rejoinder to other categories such as faces, objects, or meaningless shapes. This observation implied the existence of microterritories, some of which were dedicated to words.

In summary, the evidence from direct recordings inside the skull converges quite nicely with all of the other data at our disposal, regardless of whether it comes from patients like Mr. C, from PET scans, from functional MRI, or from electromagnetic imaging techniques. The ventral surface of every human brain contains a systematic arrangement of visual recognition devices, all tuned to different categories of images, and all systematically organized with the same overall layout. In particular, the letterbox area is systematically sandwiched between other areas responsive to faces and objects. This methodical arrangement of the visual brain's patchwork calls for an explanation.

In the case of faces, Nancy Kanwisher, professor of cognitive neuroscience at MIT, has championed a simple evolutionary hypothesis. In primates, where social life is particularly developed, a specialized cortical area devoted to face recognition has most probably evolved over time.[72] However, while this Darwinian approach is not altogether implausible for faces, it seems that no such evolutionary pressure can explain the existence of patches of the cortex earmarked for letters and words. What allowed the primate's brain to anticipate the invention of writing and foresee a specific region for the written word?

Before we lift part of the veil on this paradox, we should take a closer look at what the letterbox area actually accomplishes. Does its activity reflect a genuine specialization for words, or does the area simply provide a generalized response to the lines and curves that constitute letters? The mere comparison of words and faces cannot solve this problem. What is needed is a more careful scrutiny of how the brain solves the problem of invariant word recognition.

Position Invariance

As mentioned earlier, the visual system's capacity for spatial invariance is impressive. A good reader can recognize words regardless of how they are positioned (assuming, of course, that they do not exceed our retina's limited resolution). But is the letterbox area responsible for this form of perceptual invariance? Experiments that I have carried out with Laurent Cohen suggest that this is so.[73] We asked volunteers to gaze continuously at a small crosshair on a screen and then silently read the words that appeared to the left or right of this point. As is well known, the retina transmits crossed visual information to the brain: words that appear on the left side of a screen project onto the right half of the retina and are then transmitted to the visual centers of the right hemisphere. Likewise, words presented on the right side of a screen end up in the left hemisphere. Brain imaging indeed confirmed that our experiment segregated the input information in one or the other hemisphere. In the occipital regions, and specifically in a region known as V4, functional MRI showed activations strictly confined to

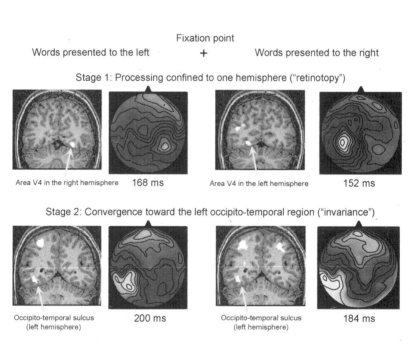

Figure 2.10 Regardless of where they appear on the retina, words are channeled toward "the brain's letterbox" in the left occipito-temporal cortex. In this experiment, participants gazed at a central crosshair while words were presented to the left or to the right. Around 150–170 milliseconds after word onset, a first negative wave appeared on the side of the head opposite to the stimulus. It was associated with the activation of a visual area called V4, located in the back of the brain. At this stage, visual information remained confined to one hemisphere. By 180–200 milliseconds, however, a second negativity appeared on the left side of the head, regardless of where the stimulus word had appeared. MRI confirmed that activation converged to the left visual system (after Cohen et al., 2000). *Adapted with permission of Oxford University Press.*

the hemisphere opposite to the one where the words were presented. This finding was confirmed with electroencephalography: up to about 160 milliseconds, the brain potentials evoked by the stimulus word were confined to the side of the scalp opposite to the location that was stimulated (figure 2.10).

At this point, however, the course of brain activity changed abruptly. In less than 40 milliseconds, all activation switched to the left hemisphere, with the most spectacular transformation occurring for words presented on the left side of gaze. These words, whose initial contact had been made with the right hemisphere, were suddenly transferred to the left and in approximately 200 milliseconds were processed like words presented on the right.

Functional MRI helped us to pinpoint the site where visual information was transferred from one hemisphere to the other. What we found was that the signals from the left and right halves of the retina converged onto the left occipito-temporal area, to the very same letterbox area that was damaged in patients with impaired reading. Activation of this region in the left hemisphere had the same spatial contours and intensity regardless of whether words were presented to the left or right of the fovea. In other words, spatial invariance begins in the letterbox area. Other experiments confirmed that this very same region is also the first point at which repetition of the written word, first on one side of the screen, then on the other, is recognized—a clear indication of spatial invariance.[74]

To achieve invariance, information that comes from two halves of the retina must, in due course, be channeled to the left hemisphere. This is achieved by nerve connections sent from the visual regions of the right hemisphere to the left letterbox area. The vast majority of these connections travel through the corpus callosum, a large bundle of nerve fibers that links the two hemispheres. This architecture leads to a rather strange prediction: if an injury or stroke damages transmission across the corpus callosum, there should be a reading impairment, but it will be restricted to the left half of the visual field. Words presented on the left will continue to activate the visual areas in the right hemisphere, but the information will not be able to find its way to the left-hemisphere language areas—it will remain blocked on the right. Thus a patient who sees words on both sides of a screen will be unable to read those that appear on the left.

This curious syndrome of "hemi-alexia" is not just something Laurent Cohen and I pulled out of a hat. He and I observed this peculiar phenomenon in two patients in whom the posterior part of the corpus callosum had been damaged.[75] Figure 2.11 shows the brain activity for one of them (Patient AC)

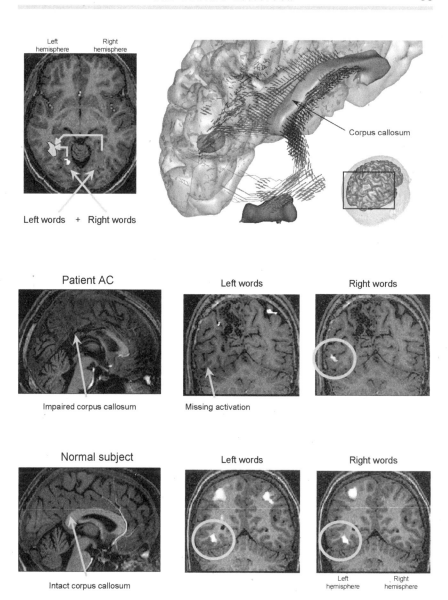

Figure 2.11 Visual invariance rests in part upon the corpus callosum, a vast bundle of fibers interconnecting the two cerebral hemispheres. When a word is presented to the left of the fixation crosshair, it is initially processed by visual areas in the right hemisphere, and must then be transmitted to the left. An array of fibers, which can be visualized with diffusion MRI, links those regions across the corpus callosum (top right). In one of our patients (AC), this fiber tract was lesioned. As a result, interhemispheric transmission was interrupted, and this person became unable to read words presented on the left (after Molko et al., 2002). *Adapted with permission of Oxford University Press and* The Journal of Cognitive Neuroscience.

measured using functional MRI. When words appeared on the right side of the screen, this young man could easily read them. However, when they appeared on the left, he only managed very laboriously to name them. He reported that he could not see the word itself, but only an indistinct shape that it took him more than two seconds to identify. Indeed, MRI showed that words on the left did not activate his visual letterbox area. They did, however, induce enhanced activity in other areas such as the prefrontal cortex, and presumably reflected AC's arduous search for the right word.

In brief, AC had lost an anatomical pathway essential to reading. This route, which travels through the corpus callosum, helps letters that enter the right hemisphere find their way into the left-hemisphere letterbox region. The route is tortuous, however, and has a price. Letters that appear on the right side of our gaze are at a clear advantage: they go straight into the left hemisphere and do not have to travel any distance to reach the letterbox area. Letters that appear on the left, however, first reach the right hemisphere and must then move across the two hemispheres through several centimeters of callosal cable. As a result, even in normal readers, reading is always a bit slower and more error-prone when words appear on the left side of fixation rather than on the right. The increased length of the transfer and, perhaps most crucially, the reduced flow of information transmitted through the corpus callosum are costly for word recognition. Thus, in the human brain, positional invariance is incomplete: not all zones of the retina are equally efficient at reading, and, like AC, we all see words somewhat better on our right.[76]

One more imaging technique, called diffusion magnetic resonance imaging, allowed us to visualize the impaired fiber tract that played havoc with AC's capacity to read. We used a modified form of magnetic resonance imaging. It can measure the direction of nerve fibers in the living human brain and works in the following way. As everyone knows, water molecules are in constant random "Brownian" motion. In a hot cup of coffee, Brownian motion is what makes a drop of milk melt and spread into the whole drink. In our bodies, water molecules also drift slowly. A clever magnetic resonance trick allows us to measure this diffusion. Roughly speaking, it consists of magnetizing the brain twice in opposite directions. For motionless molecules, the magnetization cancels out and the net effect is nil. Molecules in motion, however, create a measurable signal, proportionate to the amount of motion in the direction measured.

What happens when diffusion is measured in many different directions? In liquids, it is irrelevant—a drop of milk in coffee is dispersed equally fast in all directions. In biological tissues, however, cellular membranes restrict motion. The brain's white matter, in particular, consists mostly of large bundles of nerve fibers that act as tubes: water molecules can move freely along the main axis of the fiber tracts, but cannot easily move crosswise. At each point in the brain, the direction of maximal water diffusion is like an arrow that indicates the axis of the main fiber tracts. If all the little local arrows are connected with the help of powerful computer tracking software, a 3-D image of nerve fiber tracts can be obtained. The technique is comparable to determining the direction of a highway by photographing car taillights with a very slow shutter. The taillights always point in the direction in which the road is heading.

Diffusion MRI provides superb information about the connectivity of the human brain, and the evidence gathered is very novel. Before the invention of diffusion MRI, even with dissection it was extremely difficult to track connections in the human brain. In fact, the only plausible chart at our disposal dated back one hundred years to . . . Joseph-Jules Déjerine himself. Only he was sufficiently skilled at dissection and perseverant enough to gather information by performing autopsies.

The diffusion image taken of AC's brain immediately revealed an obvious anomaly.[77] The rear part of his corpus callosum and a large chunk of the neighboring white matter failed to demonstrate the standard directionality of water diffusion. The signal had lost the directional selectivity that is habitually found at this particular location in the brain and that indicates the presence of a major fiber tract. The fibers connecting the two hemispheres had obviously been damaged and the water molecules were diffusing more freely. This abnormal diffusion allowed us to track the course of the impaired fiber tract even in regions where the standard anatomical MRI looked normal. We then mapped its path in a normal brain and were able to come up with an image of AC's impaired fiber bundle (figure 2.11). As we expected, damage ran from the visual regions of the right occipital cortex to the neighborhood of the left letterbox region, thus neatly explaining AC's hesitant reading of words on the left.[78]

All these examples serve to demonstrate the power of modern brain imaging. In the laboratory, we routinely visualize not only the anatomy of the brain, but also activity in given regions, their time course, and even the direction of the nerve fibers that connect them. Diffusion MRI has become an indispensable tool in clinical neurology. It is routinely used to diagnose strokes and a number of other white matter pathologies such as multiple sclerosis. The information it provides, however, is strictly anatomical. Even if we can see a connection, brain imaging does not give us any idea of how or when it is used. But this situation should not last. New and promising research indicates that information about brain activity is also present in the diffusion signal.[79] Research in this area makes such regular and gigantic strides that we can count on new discoveries every year. We can be optimistic that in the near future, advances in brain imaging will make it possible to map anyone's brain circuits for reading in a matter of minutes.

Subliminal Reading

As we have just seen, the letterbox area recognizes words regardless of where they appear. When we read, this is the first brain region where the location of words is irrelevant. This invariance for spatial location, however, is only one of the fundamental properties of efficient word recognition. Character shape also varies widely. A proficient reader can recognize that an "A" and an "a" are the same letter in different case or understand that a MiXtUrE oF uPpEr- aNd LoWeRcAsE LeTtErS does not alter word meaning. What we must ask ourselves is how case invariance is implemented in the brain. Does it call on the same regions as those that implement spatial invariance? Do the two occur at the same time, or do they require a series of successive operations that call on distinct cortical mechanisms?

To address these issues, Thad Polk and Martha Farah performed a simple experiment.[80] They used functional MRI to measure brain activity while volunteers read words in mixed case, such as "HoTeL." Although unfamiliar, these stimuli triggered almost the same amount of brain activity as words written in homogeneous same case. In particular, the activity profile in the letterbox area, even when stimulated with such bizarre stimuli as "ElEpHaNt," was entirely normal. Thus Polk and Farah suggested that this region houses an abstract representation of letters and words where neurons no longer take any notice of case.

It is not, however, obvious that these results are entirely conclusive. The relatively poor spatial resolution of brain imaging does not allow us to see if the same neurons respond to "HoTeL," "HOTEL," and "hotel." The total activity evoked by these strings may be identical even if different neurons are involved. Thus a mere activation overlap does not confirm that case invariance is handled at this site in the brain.

This problem could appear unsolvable, were it not for an elegant technique that measures brain responses to pairs of consecutive words.[81] On each trial, volunteers are presented with two stimuli at very close intervals. The idea is to compare the total activity evoked when both of these stimuli represent the same word, possibly in upper- and lowercase ("hotel" followed by "HOTEL"), and when they represent different words ("radio" followed by "HOTEL"). We know from experiments with animals that neurons are very sensitive to repetition. They quickly adapt to repeated stimulation and discharge less the second time an image is presented. A high level of response reappears, however, when a novel image is introduced. When the MRI signal adapts and rebounds in this way, it tells us indirectly that the neurons in a given area have spotted the repetition of a stimulus and its subsequent change. By changing the stimulus shape, we can now ask what counts as a repetition for that brain area—does it treat, for instance, "hotel" and "HOTEL" as two distinct shapes or as the same word?

There is one last refinement to this procedure. Ideally, volunteers should not know that words will reappear. If they can see that words are repeated, their level of attention drops. This provokes a dip in the MRI signal that spreads to the attention-related areas of the brain, and it is no longer possible to tell whether the signal reflects specific, local, perceptual invariance or a simple global change in the overall state of attention.

In my experimental work using this method, I have taken great care to prevent conscious perception of repeated words. The first word in each experimental pair is always presented very briefly, for only twenty-nine milliseconds, or a duration equivalent to less than a single movie frame. These are the standard conditions for subliminal perception. The first word presented is invisible because it is sandwiched between two meaningless arrays of geometrical shapes. Instead of the word, participants in this experiment perceive only a brief flicker followed by a second word that they believe is the only word to see. Control experiments confirm that they have no information at all about the hidden word.

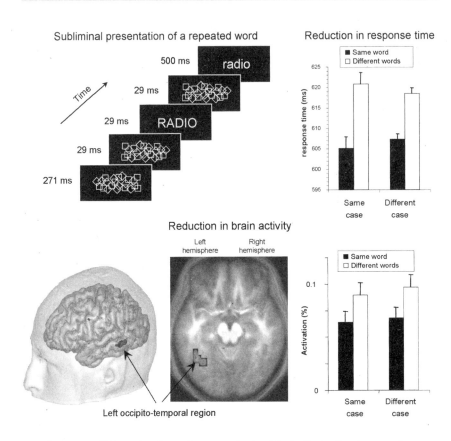

Subliminal presentation of a repeated word

Reduction in response time

Reduction in brain activity

Left occipito-temporal region

Figure 2.12 Written words can be recognized subliminally, in the absence of any awareness. In this experiment, a word was presented for a very short duration (29 milliseconds). Masking shapes that appeared just before and just after rendered it invisible. Nevertheless, this nonconscious presentation accelerated the participants' responses when the same word later reappeared as a consciously visible target (top right). Brain imaging showed that the left occipito-temporal area was responsible for this subliminal priming effect: its activity diminished when the word was repeated, even when its graphic form changed radically from uppercase to lowercase (after Dehaene et al., 2001). *Adapted with permission of Nature Neuroscience.*

What is surprising under such conditions of subliminal presentation is that the masked word, although invisible, still activates part of the reading circuitry in the brain. In particular, the letterbox area reacts to word repetitions—its activation is reduced when the same word is presented twice in a row (figure 2.12). Crucially, it reacts in the same way regardless of whether the words appear in same case ("HOTEL"-"HOTEL") or in different case ("hotel"-"HOTEL"). This is a specific effect that is not seen in other regions of the brain. For instance, regions of the occipital lobe that are involved in lower-level visual processing only reduce their activity when the visual object presented twice is exactly the same ("HOTEL"-"HOTEL"), but cease to do so as soon as case changes ("hotel"-"HOTEL"). They probably only encode elementary features on the retina, and therefore interpret any visual change as a new stimulus—even when a word is the same in different guises, as in "hotel" and "HOTEL." In contrast, the letterbox area carries out a more abstract operation: it recognizes that "a" and "A" represent the same stimulus. Brain imaging therefore demonstrates that this brain region simultaneously computes both space and case invariance.[82]

In more recent experiments, scientists have begun to sort out the many steps involved in letter and word recognition. Consider the words "anger" and "range." The two are anagrams of a very peculiar type. Not only are they made up of the same letters, but one word can be converted into the other by moving just one letter from the first position to the last. Now suppose that these two words are consecutively flashed at you, with a change in case in order to avoid any low-level visual similarity ("ANGER"-"range"), but also with a slight horizontal shift so that all of their shared central letters, "a," "n," "g," and "e," appear at exactly the same point on the screen. With this simple trick, almost all of the letters in the words can be repeated without the whole word ever reappearing. This method leads us to question, once again, what counts as a repetition for the letterbox area. Does letter repetition provoke an adaptation effect, or must the entire word be repeated? In other words, is this brain region only interested in single letters, or does it encode larger-sized units, perhaps pairs of letters or even an entire word?

The answer is—all of the above! In fact, the adaptation process in the letterbox area reveals several successive levels of letter coding, organized hierarchically from the back toward the front of the occipito-temporal cortex, with an increasing degree of abstraction.[83] The lowest coding level deals with single letters. It is found in the rearmost sector of the occipito-temporal cortex, in both hemispheres. It is concerned only with the repetition of individual letters. Whenever letters reappear, there is a drop in the fMRI signal regardless of whether they belong in a different word or not. Furthermore, what is essential is that the letters be repeated at the same location on the screen. When a word is shifted to the left or to the right by just one letter, the adaptation effect disappears in the letterbox area. This suggests that its neurons no longer recognize that the same letters are still present. In other words, position invariance does not exist at this level—even if this region does resist a change from lower- to uppercase. We therefore think that the letterbox area contains a bank of abstract letter detectors that can each note the presence of a given letter at a certain location, regardless of case.

If you now move forward one centimeter on the left occipito-temporal cortex, word coding suddenly becomes more invariant. Here the cortex is sensitive to the similarity of words such as "ANGER" and "range," even when their letters are not exactly aligned. Thus this region still codes for units below the whole-word level (perhaps single letters or small letter chunks), but it does so with some tolerance for changes in spatial location. It recognizes that the spelling of the two words is similar, even if their position is different. It is therefore probably at this level that our brain begins to code for morphological relations between words, like, for instance "depart" and "departure."[84] Word roots, however, have not yet been encoded for meaning. This specific brain region could thus wrongly attribute a shared root to the words "depart" and "department." In fact, all that our visual system does at this stage is to quickly parse the word input into a hierarchical tree that identifies the most likely letters, graphemes, syllables, and morphemes present in the string.

If you again move forward by another centimeter, you reach a still more selective neural code. Here, in the most anterior part of the letterbox area, the cortex begins to respond to the entire word. At this point, activation tends to be reduced when the same word is repeated ("ANGER" followed by "anger"), but not when the same letters form a different word ("ANGER" followed by "range"). A string of letters is coded as an integrated whole—or at the very least, the brain undoubtedly recognizes large chunks of letters such as "rang" that distinguish the two words from each other in my example.

What we learn from these studies is clear. First, reading is a sophisticated construction game—a complex cortical assembly line is needed to progressively put together a unique neural code for each written word. Second, conscious reflection is blind to the true complexity of word recognition. Reading is not a direct and effortless process. Rather, it relies on an entire series of unconscious operations. In the cases I have just described, all of the successive stages of brain activity occurred while the participants did not even suspect that they were being presented with a hidden word. Clearly, subliminal strings of letters can be unconsciously "bound" together—the letters do not float independently but can form stable combinations that unconsciously encode the difference between anagrams like "recent" and "center." The entire visual word recognition process, from retinal processing to the highest level of abstraction and invariance, thus unfolds automatically, in less than one-fifth of a second, without any conscious examination.[85]

How Culture Fashions the Brain

It could be argued that the visual operations I have just described are not specific to reading. When our brain recognizes the similarity between the words "RADIO" and "radio," perhaps it simply adjusts for letter size. A generic visual process of size invariance may identify the letters "O" and "o" in the same way that we recognize, say, a fork regardless of its size, position, or the angle from which it is viewed. What proof do we have that the letterbox region actually implements operations unique to reading?

In fact, case invariance itself provides strong evidence that our visual system is attuned to reading. We are so accustomed to the associations between uppercase and lowercase letters that we no longer see how arbitrary they are. Only a few letters look alike in upper- and lowercase ("o" and "O," "u" and "U"). However, many others seem to be paired almost at random. Nothing predisposes the shape "a" to serve as the lowercase version of the letter "A"—we could easily imagine an alphabet where "e" might be the lowercase of "A," and "g" the lowercase of "R." The association between lower- and uppercase letters is a convention that we adopt when we learn to read. Interestingly, brain imaging can be used to demonstrate that the letterbox area has adapted to this convention.[86] In one of my experiments, I showed volunteers a list of words whose letter shapes, with the exception of size, were very similar in upper- and lowercase: "COUP"-"coup," "PUCK"-"puck," "ZOO"-"zoo," and so on. I then added another list of words made up of letters whose lowercase and uppercase forms bore no resemblance to each other and were simply a matter of convention: "GET"-"get," "EAGER"-"eager," "READ"-"read," and so on. Needless to say, the participants easily recognized these words in spite of their different shapes. Neuroimaging revealed which brain area encodes this type of cultural invariance. The results again supported the central role played by the left occipito-temporal letterbox region. This brain area was the only one in which activation declined whenever words were repeated, irrespective of whether the relation between the two words depended on cultural practice or on visual resemblance (figure 2.13). For the letterbox area, therefore, the letters "G" and "g" are as similar as "O" and "o"—a clear proof that this region has adapted to the conventions of our alphabet. That neurons respond in the same way to the shapes "g" and "G" cannot be attributed to an innate organization

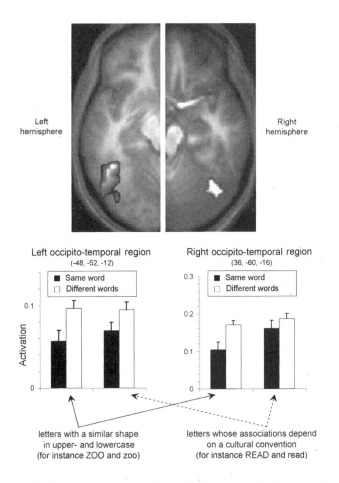

Figure 2.13 The left occipito-temporal "letterbox" area contains the cultural conventions of the alphabet. Its activity diminishes whenever the same word is presented twice, thus indicating that it recognizes this word even when case changes. Crucially, it still recognizes words when their lowercase and uppercase letters look radically different (e.g., "E" and "e") and are only related due to cultural convention. The symmetrical area of the right hemisphere, in contrast, seems to respond only to visual similarity, not to cultural conventions (after Dehaene et al., 2004). *Adapted with permission of* Psychological Science.

of vision. It necessarily results from a learning process that has incorporated cultural practices into the appropriate brain networks. In this respect, it is interesting to contrast the left letterbox area with the symmetrical visual region in the right hemisphere. This second region recognizes words that look alike, such as "ZOO" and "zoo," but it fails miserably on conventional associations like "GET" and "get." It would appear that the right hemisphere, in most right-handers, only applies generic visual mechanisms of size and position invariance to the written word. Only the left letterbox area has internalized the cultural practices unique to reading.

Much additional evidence exists concerning cultural specialization in the letterbox area. Using brain imaging, my colleagues and I have shown that it does not merely respond passively and innately to anything that merely resembles a letter or a word. Rather, it adapts actively to the task of reading by compiling statistics about letters that belong together. As a result, strings of letters do not always stimulate it equally well—spelling in the language familiar to the reader must be respected. For example, this region responds far better to strings that form an existing or plausible word, such as "CABINET" or "PILAVER," than to strings that violate spelling rules, like a string of consonants such as "CQBPRGT" (figure 2.14).[87] It also prefers customary letter combinations, like "WH" or "ING," to rare or impossible ones such as "HW" or "QNF."[88] Even valid letter strings can fail to activate the letterbox area if the person who is scanned hasn't learned to read them—thus Hebrew characters provoke strong occipito-temporal activation in Hebrew readers, but not in English readers.[89]

Mr. C, Déjerine's first patient, can be considered a fine example of what cultural training can do. In spite of his serious problem with letter strings, he could still read Arabic numbers. Visually speaking, of course, these stimuli are very similar. The shapes of letters are arbitrary, and one could easily imagine an alternative alphabet in which "52314" or "CQBPRGT" were words, while the string "CABINET" would be meaningless. Yet the letterbox area prefers strings of letters to strings of digits such as "52314."[90] Thus the letterbox area is not merely determined by visual stimuli, but also by the cultural history of the reader's brain. When we learn to read, a subset of our visual neurons adapts to the letters and languages that we are taught. Digits are no doubt processed elsewhere, probably in a more posterior area of the left and right hemispheres. Electrical recordings of brain potentials show that this cultural

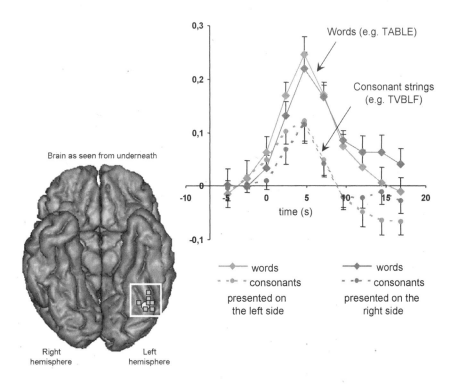

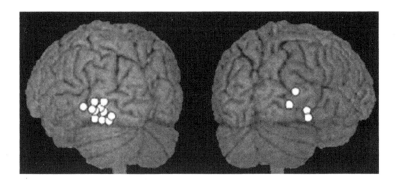

Figure 2.14 The left occipito-temporal "letterbox" area incorporates the regularities of the spelling system in the reader's language. Functional MRI (top row) shows that it responds better to real words than to consonant strings that violate the spelling rules of the participants' language—a preference for regular stimuli that is independent of the retinal location where the stimuli appear (after Cohen et al., 2002). Another technique with a higher temporal resolution, magnetoencephalography (bottom image), shows that this preferential response to words occurs about 150 milliseconds after the appearance of the word on the retina (after Tarkiainen et al., 1999). *Adapted with permission of Oxford University Press.*

sorting process occurs remarkably quickly. It takes only about 150 to 190 milliseconds for the left occipito-temporal region to display a preference for well-formed letter strings over strings of random consonants or digits.[91]

The Brains of Chinese Readers

There are so many different writing systems that one might expect enormous discrepancies in the brain organization and activation patterns of readers from different countries—especially those that use alphabetic rather than non-alphabetic writing systems. Surprisingly, however, this is not the case. In all cultures, in spite of large differences in surface form, written words are always processed by very similar brain circuits. In particular, the left occipito-temporal letterbox area plays a prominent role in all readers, with only minimal differences linked to the shape and internal structure of characters.

Consider the case of Chinese and Japanese. It was long speculated that the recognition of Chinese characters was a more global process than our decoding of our alphabet. Chinese readers were thought to rely largely on their right-hemisphere visual system, which is considered more "holistic." However, modern brain imaging has clearly refuted this hypothesis. When Chinese readers are scanned, the *left* occipito-temporal region activates, with a significant lateralization toward the left hemisphere, at a spot essentially identical to the letterbox area.[92] It is amazing to think that in spite of geographical distance, different imaging methods, brain morphologies, education strategies, and writing systems, brain activations related to word recognition in Chinese readers lie only a few millimeters away from those of English readers.

Fluent readers of both Roman letters and Chinese characters provide a unique opportunity to study the universality of visual word recognition in the same brain. Since the 1970s, China has implemented an official transcription system for writing Chinese. This system is called Pinyin—a word that means "assembling sounds." It relies on the twenty-six letters of our Latin alphabet, plus a number of graphemes such as "zh," "ch," and "ang." The word for "bank," for instance, is transcribed as "yínháng." Chinese children often learn Pinyin before they are taught the traditional Chinese characters, and thus the younger generation is frequently bilingual, or rather, "bigraphal"—children have two distinct visual entry points into the same spoken language. Nevertheless, brain imaging shows that the regions activated by these two forms of writing overlap almost completely in the left ventral temporal cortex.[93]

The letterbox area also demonstrates very similar functional properties for readers of Chinese and the Roman alphabet. In particular, it prefers proper Chinese characters to visually similar meaningless shapes.[94] This property is strictly analogous to the greater response to words rather than consonant strings observed with alphabet readers. It demonstrates that this region has adapted itself to the constraints of Chinese writing. Rather than letters, a Chinese reader's letterbox area must contain a hierarchy of detectors tuned to the semantic and phonetic markers that make up the internal structure of Chinese characters.

In fact, I know of little data to support the idea of a "holistic" recognition of Chinese characters. The experimental results that are available suggest rather that Chinese characters, much like letter strings, are coded as a hierarchical pyramid of increasingly large bundles of visual features. Thus a Chinese character can be primed by one of its fragments, in much the same way as an English word can be primed by one of its morphemes ("hunt" followed by "hunter").[95]

Japanese and Its Two Scripts

My colleague Kimihiro Nakamura and I have also studied Japanese reading.[96] If you open any Japanese newspaper, you will see a funny mixture of characters (figure 2.15). Many are in the Kanji notation, a vast set of more than three thousand characters borrowed from Chinese that represent words and their roots. However, others are written in Kana, another set of forty-six symbols that represent syllables and can transcribe any sequence of Japanese speech sounds. Many nouns are written as frequently in Kanji as in Kana, but grammatical words, verb endings, adjectives, and adverbs are often written only in Kana. Thus it is not unusual for a verb to be transcribed using a mixture of Kanji and Kana.

In spite of these complex rules, so distinctly different from those of English, brain scans of Japanese readers show that visual recognition of Kanji and Kana also relies on the left occipito-temporal area (figure 2.15). In both cases, activation is lateralized to the left hemisphere, and its peak almost coincides with the

Kanji characters

神	/kami/	God
神社	/jiN-ja/	Temple
神経	/shiN-kei/	Nerve
精神	/sei-shiN/	Mind
神主	/kaN-nushi/	Priest

Kana characters

か	/ka/	
かみ	/ka-mi/	Paper
かさ	/ka-sa/	Umbrella
あか	/a-ka/	Red
たから	/ta-ka-ra/	Treasure

Overall active network

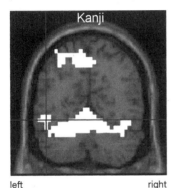
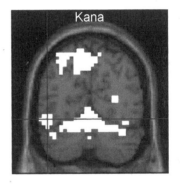

left
hemisphere

right
hemisphere

Differences between the two notations

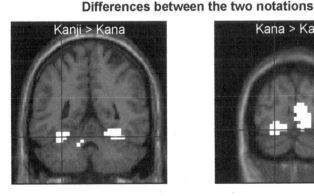
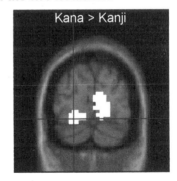

Figure 2.15 Reading Japanese characters activates a region essentially identical to that found in English readers. Japanese relies on two distinct writing systems. Kanji characters, which number in the thousands, denote word meanings and word roots (left column), while the forty-six Kana characters encode syllables (right column). In spite of these differences, both notations call upon the same left occipito-temporal region, at a location virtually identical to that found in readers of the Roman alphabet. Only subtle differences are seen: Kanji characters encroach more on the midline temporal regions of both hemispheres, while Kana characters cause a slightly greater activation of the occipital pole (after Nakamura et al., 2005).

letterbox location activated in English readers. Furthermore, damage to this area frequently causes pure alexia for both Kanji and Kana in Japanese patients.[97]

Although the brain reacts in much the same way to the two notations, careful scrutiny reveals a few differences. In the early visual regions of the occipital pole, words presented in Kana prompt slightly more activation than those in Kanji—perhaps because many of them are composed of several characters and therefore extend a bit farther on the retina. Conversely, words written in Kanji produce more activation in a small bilateral region of the ventral temporal lobe—perhaps because they call for a slightly more global or "holistic" perception mode. Lesion studies also confirm that the brain's reading networks are not identical for Kanji and Kana. For instance, in rare cases a stroke patient may retain the ability to read in one script but not in the other.[98] The two notations thus probably depend on different microterritories in the cortex, although all are located within the same general area.

On a more general scale, visual word recognition clearly rests on universal cerebral mechanisms. In all cultures, the same area in the left occipito-temporal region is in charge of written word recognition, exquisitely adapting its hierarchical architecture to the specific requirements of each writing system.

Beyond the Letterbox

Any word we read is initially funneled through the letterbox area, which plays a dominant and universal role in the recognition of writing. However, what happens beyond this initial stage, once the shape of a word has been analyzed into letters, graphemes, and morphemes? Where does this information go? What cortical routes allow us to retrieve the pronunciation and meaning of words?

Anatomy can be used to track the principal nerve fibers that transmit information from the visual system toward the rest of the brain. Diffusion MRI has begun to reveal that large bundles of connections link the brain's cortical areas. In particular, a major pathway, the inferior longitudinal bundle, which runs across the temporal lobe from its rearmost part to its anterior pole, has recently been mapped.[99] Its anatomy suggests that it collects the information from all the brain areas devoted to vision, including the letterbox area, and shuttles it farther along the temporal lobe. In addition to this major cortical highway, local U-shaped fibers connect successive temporal regions, thus establishing a feed-forward chain (figure 2.16). All of these nerve cables, sheathed in a biological insulator called myelin, allow for the very fast transmission of written words to the entire temporal lobe.

It should be borne in mind, however, that brain imaging only allows us to see the main structures of the human brain, without going into its finer details.

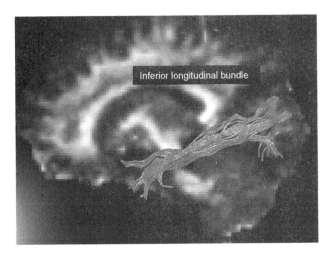

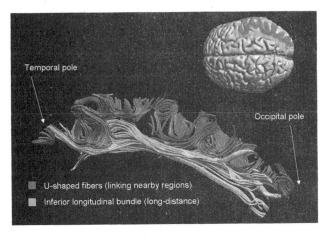

Figure 2.16 The left occipito-temporal region, at the back of the brain, analyzes written words and transmits the results to other distant areas through vast fiber bundles that can now be visualized with diffusion MRI. The image shows the reconstruction of a major anatomical pathway, the inferior longitudinal fasciculus, which projects toward the front of the temporal lobe. All along the cortical surface, numerous U-shaped fibers also connect nearby regions step by step (after Catani et al., 2003). *Used with permission of Oxford University Press.*

It is likely that some small connections, which may be equally important from a functional standpoint, are not nicely packaged together into macroscopic bundles and cannot be picked up by our current imaging methods. Only a microscope can reveal the brain's wiring in full detail. A postmortem study carried out by anatomists from Lausanne, Switzerland, has shown that, in the human brain, neurons from the ventral temporal lobe broadcast information to brain regions that are much more distant and dispersed than what research on monkeys led us to believe.[100] Because these rare conductors are essentially invisible with magnetic resonance imaging, the scientists who carried out this study were obliged to wait for a very special set of circumstances. By a pure coincidence, one of their patients, who had initially suffered from a stroke in the right occipito-temporal region, died from an unrelated condition three weeks after his attack. The anatomists processed his brain with a chemical solution that marked the nerve endings that had recently died. This process revealed all of the direct connections from the damaged area. The results were astounding. It turned out that this small right visual area projected toward a remarkably diverse set of cortical regions and particularly across the brain's midline into the left hemisphere. Connections were found almost everywhere, but their density peaked in the language areas, particularly Broca's area (inferior frontal cortex) and Wernicke's area (posterior part of the superior temporal cortex).

If we assume that brain organization is symmetrical, it is safe to conclude that the left letterbox area also sends a comparable number of projections to the same regions in both hemispheres. This brief survey of brain connectivity therefore suggests that this occipito-temporal region acts as an essential switchboard for the reading circuit. From this central spot, connections fan out in all directions and broadcast a word's identity to a great many other cortical areas in parallel.

Direct proof of the broadcasting phenomenon was obtained by Ksenija Marinkovic, Anders Dale, Eric Halgren, and their colleagues at MIT. For the first time, they managed to track the propagation of a written word inside the brain in real time.[101] Thanks to a modern magnetoencephalography machine with 204 channels, they measured the minuscule magnetic fields generated in the brain when many of its neurons are active in synchrony. They also developed new computer software that turned these signals into a movie of brain activation as it unfolded on the surface of the cortex. Their film revealed the time course of the brain mechanisms associated with reading (figure 2.17). Of

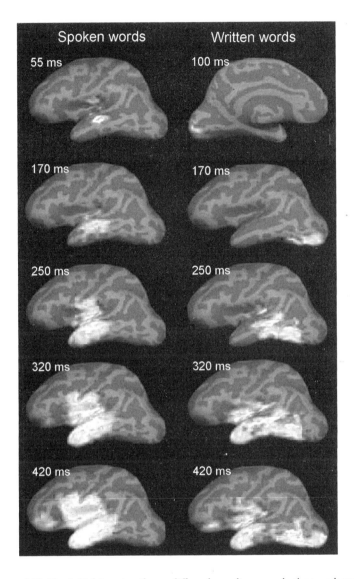

Figure 2.17 The initial input pathway differs depending on whether we hear or read a word, but cortical activity later converges onto the same set of language areas. Each image, obtained with magnetoencephalography, represents a snapshot of activity on the cortical surface at a different moment after the onset of a word. To facilitate visualization, the folds of human cortex were smoothed by a computer algorithm. During reading (right column), activation starts in the occipital pole around 100 milliseconds. By 170 milliseconds, it extends into the left occipito-temporal region, "the brain's letterbox" that extracts the visual word form. Immediately afterwards, an explosion of activity occurs in multiple temporal and frontal regions of the left hemisphere, most of which are shared with spoken word processing (left column) (after Marinkovic et al., 2003). *Used with permission of* Neuron.

course it was only a reconstruction, not a direct view of the cortex, but such a realistic picture of the way a word progresses through the brain had never been obtained before.

As expected, the film showed that activation started at the occipital pole, which is associated with early visual processing. Roughly 170 milliseconds later, it became strongly lateralized to the left hemisphere and focused precisely on the letterbox area. Then, all of a sudden, only 250 milliseconds after the word first appeared, there was a major burst of activity. Stimulation spread extensively into the temporal lobes and activated the superior, middle, and inferior temporal regions of both hemispheres. Finally, after 300 milliseconds, it edged even more forcefully into the left hemisphere, with an extension into the temporal pole, the anterior insula, and Broca's area. This intense activation pattern continued for several hundred milliseconds, with a further foray into frontal regions and a surprising return to posterior visual regions.

What I find most interesting about this experiment is that, after about 250 milliseconds, the activated areas ceased to be exclusively restricted to vision— all of them could be activated equally well by a spoken word. Thus, within the cortical chain reaction that accompanies reading, the letterbox area is the last region to be reserved exclusively for visual word processing. Beyond this point, the vision areas of the brain begin to work in tight conjunction with the spoken language networks.

Sound and Meaning

In chapter 1, we saw that psychologists postulated a dual-route model for reading. They maintained that the brain comprises two parallel paths that can be used simultaneously, depending on what is to be read:

- Infrequently used words and neologisms move along a phonological route that converts letter strings into speech sounds.
- Frequently used words, and those whose spelling does not correspond to their pronunciation, are recognized via a mental lexicon that allows us to access their identity and meaning.

A whole series of brain imaging experiments has relied on this theory to decide which areas of the brain concentrate most on pronunciation and which compute the identity and meaning of words.

Two different strategies can be used to test the dual-route model at the brain level. The simplest one consists of the measurement of brain activity while a volunteer reads different kinds of letter strings. Although the task is fixed, the stimuli vary, and they are designed to optimally activate one of the two brain routes. For example, some experiments have contrasted the reading of real words such as "house" or "pants" to that of pseudo-words such as "houts" or "panse."[102] Regions that demonstrate the greatest response to words are considered to be part of the lexical route, while regions that show more activation for pseudo-words are thought to be involved in the direct phonological route. Indeed, since there is no entry for "houts" in the mental lexicon, such a string can only be read using a spelling-to-sound conversion.

More subtle contrasts have been used. For instance, does the frequency of words modulate brain activation, in such a way that infrequently used words like "sextant" or "chalice" behave much like pseudo-words and follow the spelling-to-sound path? Do brain regions differentiate between words whose pronunciation corresponds to their spelling, such as "bloom," and words like "blood" where this is not the case? The psychological dual-route model maintains that the former tend to take the spelling-to-sound route, whereas the latter can only be read correctly if they are first recognized by the mental lexicon.

A second method, which complements the previous one, consists of the presentation of a single list of words with varying tasks,[103] picked because they selectively emphasize information pertaining to one of the cerebral routes. For example, in one block of trials, participants are asked to judge whether the words rhyme—a phonological task that emphasizes the conversion from graphemes to phonemes and the mental representation of speech sounds. In the other, they are asked to evaluate whether the words are synonymous or not—a semantic task that requires access to the mental lexicon and draws attention to meaning. Finally, there is a spelling task where the same participants are asked to assess whether two words end with the same letters, regardless of how they sound (e.g., "napkin" and "mountain")—an orthographic judgment that might amplify spelling information in the brain. In each case, it is expected that brain activity will increase in the areas that encode information in the format needed to solve the task.

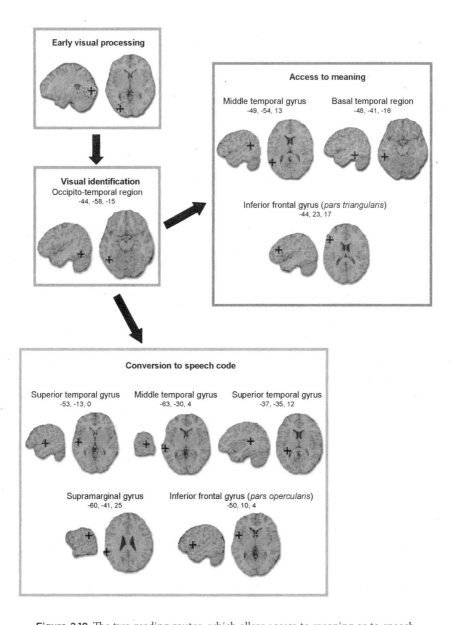

Figure 2.18 The two reading routes, which allow access to meaning or to speech sounds, involve distinct sets of brain areas. This reanalysis of dozens of experiments shows that after a shared period of visual processing (top left), two major sets of cortical circuits distinguish access to word meaning (top right) from the decoding of letters into speech sounds (bottom). The two circuits are intertwined and occasionally abut in the same anatomical region. In the inferior frontal cortex, for instance, distinct nearby regions for meaning and for articulation seem to cohabit (after Jobard, Crivello, & Tzourio-Mazoyer, 2003). *Used with permission of* Neuroimage.

By and large, these two methods arrive at the same results.[104] The two reading routes do indeed correspond to two distinct networks of brain areas (figure 2.18). It is therefore correct to maintain that separate networks are dedicated to sound and meaning—the psychological model appears largely supported by brain imaging results.

From Spelling to Sound

The brain route employed to decode letters into sounds essentially involves the superior regions of the left temporal lobe, which are well known for their role in speech sound analysis. The left inferior prefrontal and precentral cortices, which contribute to articulation, are also engaged—but it is in the superior temporal lobe that visual letters and spoken sounds first meet. This is clearly shown in a simple experiment illustrated in figure 2.19.[105] During functional MRI, participants were either made to see letters alone, or hear speech sounds alone, or do both simultaneously. In the case where they were given both together, the letter and the sound were either congruent (for instance letter "o" and sound *o*) or incongruent (letter "a" and sound *o*). A vast portion of the superior temporal lobe responded to the sight of the letter—but only a small subregion called the planum temporale reacted to compatible letters and sounds. At this cortical location, a sound that fits with a letter increases brain activity, but a conflict between spelling and sound reduces it. Thanks to magnetoencephalography, this effect can be precisely timed. The conversion of a letter into a sound starts only 225 milliseconds after the letter first appears on the retina, and its compatibility with a spoken sound is recognized after about 400 milliseconds.[106]

Several cues suggest that the planum temporale is one of the most crucial regions for speech decoding. It is asymmetrical, with a larger surface in the left hemisphere than in the right. Many researchers see this anatomical asymmetry as one of the major probable causes of the lateralization of language to the left hemisphere. Not only is the left planum temporale already bigger than the right prior to birth, but the brains of infants are already powerfully and asymmetrically activated when they listen to speech in the first few months of life.[107] In childhood, this region quickly learns to process relevant speech sounds in the native language and to ignore unnecessary ones.[108] A native

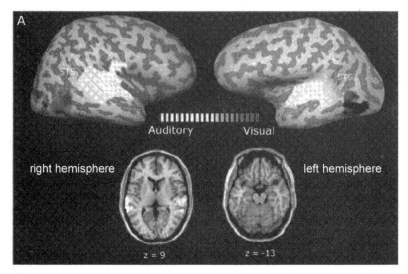

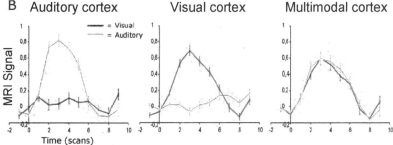

Figure 2.19 The superior temporal region contributes to the conversion of letters into sounds. Brain imaging easily separates visual areas, which are activated by the shapes of letters, from auditory areas, which are activated by speech sounds. Part of the superior temporal cortex, in white, is multimodal, which means that it activates to both sound and sight. Within these regions, smaller sectors are sensitive to the match or mismatch between a letter and a simultaneously heard sound (after van Atteveldt et al., 2004). *Adapted with permission of* Neuron.

speaker of Japanese, for instance, as we all know, cannot hear the difference between an *r* and an *l*, since these sounds are not used to discriminate words in that language. This peculiar form of "linguistic deafness" arises in the left planum temporale and nearby regions—toward the end of the first year of life, these areas lose their ability to distinguish an *r* from an *l*.

Because it is a meeting point for visual and auditory inputs, the planum temporale no doubt plays an important role when young children learn to read. Indeed, it is ideally equipped to learn the correspondence between letters and sounds. When a beginner learns to decode the letters "b" and "a" into the sound *ba*, his planum temporale receives both the letters and the speech inputs and can thus establish the relationship between them. Later on, the links between graphemes and phonemes become automatic. Letter-to-sound conversion becomes so automatic that the letter "A" paired with the wrong sound *o* makes our superior temporal cortex emit a transient mismatch signal.

The precise pathways used for the conversion from letters to sounds are not yet fully understood. In the case of a single letter, the superior temporal region probably receives information straight from early visual areas. However, when input is a whole string of letters, complex processing is needed to parse it into graphemes and syllables. This process is serial in nature, and probably employs the inferior parietal region that lies just above the planum temporale.[109] That region, together with part of Broca's area (the opercular region), creates a phonological circuit that is engaged every time we recite something mentally. It belongs to an articulatory or phonological loop that we use when, for instance, we store a phone number in verbal memory.[110] This buffer, or storage space, is probably essential for the pronunciation of long written words—for example, to decipher a pharmaceutical formula like *acetylsalicylic acid* (aspirin).

Avenues to Meaning

The brain network that analyzes word meaning is quite distinct from that which converts letters into sounds. Semantics mobilizes a widespread array of regions, some of which are shown in figure 2.18. Crucially, not one of them is exclusive to the written word. Rather, they all activate as soon as we think about concepts conveyed by spoken words[111] or even images.[112] In fact, they can easily be located with a classic image association test, in which a person is asked to judge which of three pictures belong together—a pyramid, a palm tree, and a hammer.

A peculiar property, inherent to some of these semantic regions, is that they can be active even before a volunteer is asked to perform a specific task. The posterior temporo-parietal region, in particular, is active even when we are at rest in the dark. Typically, it does not activate more when a volunteer hears or sees words—but it "deactivates" to a point below the initial baseline level whenever meaningless pseudo-words are presented.[113] Why is this so? The moment we wake up, our brain is instantly active and aware of what is going on around it. This mental reflection process spontaneously calls upon a semantic network that manipulates words and images—and thus many of the concerned areas are active even when we are at rest. Because of this constant buzz of brain activity, functional MRI does not have an absolute zero that would allow for the easy identification of any additional brain activation brought on by external stimulation. In some regions such as the angular gyrus, reading meaningful words causes as much activity as the ongoing buzz and thus fails to induce any additional change—as if the area were not concerned by reading at all. When we receive a stimulus that does *not* make any sense, however, like the pseudo-word "larpic," spontaneous activity briefly lapses and causes a temporary "de-activation"—an apparent decrease in activation which suggests that this area was in fact actively engaged in reading words.

Because their activation fluctuates constantly, even in the resting state, the brain areas that code for meaning cannot be easily visualized directly—they must be studied with refined experimental methods. One of these, introduced earlier in this book, is the priming method. Consider the words "sofa" and "couch," whose meanings are essentially identical even if their letters are different. Joe Devlin and his collaborators[114] presented volunteers with synonyms like these in pairs, where the first word was flashed subliminally, and only the second word was actually seen. They found that only one area of the brain, the left middle temporal region, recognized the similarity between them in spite of their superficial difference. This area was activated less by two synonyms than by two words with unrelated meanings, such as "honey" and "couch," suggesting that it was active for longer when it had to encode two different meanings, but synonyms only activated it once. With a number of elegant manipulations, Devlin and his colleagues demonstrated a major difference in the way words are encoded by the middle and ventral regions of the left temporal lobe. As I have already explained at length, the ventral letterbox region only deals with the visual coding of letters. Devlin found that its activity decreased whenever words looked alike visually ("hunt" and "hunter"), even if their meanings had nothing to do with each other ("horn" and "horny"). Furthermore, this region failed to pick up the fact that "sofa" and "couch" meant roughly the same thing—exactly what we would expect if this region focused entirely on the spelling of these words. In contrast, the left middle temporal cortex manifested a clear sensitivity to word pairs that were semantically related, such as "sofa"/"couch" or "hunt"/"hunter." It was not fooled by superficially similar pairs like "horn"/"horny."

Other studies confirm that this second region is dedicated to the meaning of individual words. In Japanese readers, it is the only region that recognizes that a word is same regardless of whether it is written in Kanji or Kana, two scripts whose written forms are visually unrelated.[115] It is also sensitive to the semantic relations between words that are spoken rather than written.[116] Thus this region seems to code for the fragments of a word's meaning, regardless of how one arrives at the meaning, be it in a picture, a spoken word, or a written string.

The brain's networks for meaning, however, are not limited to simply processing single words. Regions near the front of the temporal lobe appear to concentrate on the combinations of meanings that words achieve when they are assembled into sentences.[117] Another region, farther forward in the inferior frontal cortex, seems to select one meaning out of many—it is triggered, together with temporal regions, whenever we hear sentences that contain many ambiguous words, such as "the shell was fired toward the tank."[118] Think of how desperately these brain areas must search for a stable state as the following ambiguous headlines are read:

Some Pieces of Rock Hudson Sold at Auction
Deaf Mute Gets New Hearing in Killing
Two Convicts Evade Noose, Jury Hung
Milk Drinkers Are Turning to Powder
Dealers Will Hear Car Talk at Noon
Sex Education Delayed, Teachers Request Training
Include Your Children When Baking Cookies[119]

Overall, although researchers have managed to map several of the relevant brain areas, how meaning is actually coded in the cortex remains a frustrating issue. The process that allows our neuronal networks to snap together and "make sense" remains utterly mysterious. We do know, however, that meaning cannot be confined to only a few brain regions and probably depends on vast arrays of neurons distributed throughout the cortex, of which the frontal and temporal regions in figure 2.18 are just the tip of the iceberg. Although these regions activate when we access the gist of a word, they probably do not store the meaning itself, but merely facilitate access to semantic information spread out elsewhere in the cortex. According to the neurologist Antonio Damasio,[120] they may well operate as "convergence zones" that gather and send signals to many other regions. They thus, in all likelihood, collect dispersed fragments of meaning and bundle them together into articulated sets of neurons that constitute the genuine neuronal substrate of word meaning.

Take the verb "bite." As you remember what it means, your mind briskly evokes the body parts involved: the mouth and teeth, their movements, and perhaps also the pain associated with being bitten. All of these fragments of gesture, motion, and sensation are bound together under the heading "bite." This link works in both directions: we pronounce the word whenever we talk about this peculiar series of events, but to hear or read the word brings on a swarm of meanings.

The lateral temporal region seems to play an essential role in the mediation between the shapes of words and the elements that constitute their meaning. This region can be subdivided into subregions that specialize in different categories of words. Faces, people, animals, tools, vegetables . . . all of these categories seem to map onto partially distinct regions, which each collect fragments of meaning from different sources: the parietal lobe for numbers and body parts, the occipital area V4 for colors, the area V5 for motion, the precentral and anterior parietal areas for action and gestures, the Brodmann area 10 for intentions and beliefs, the temporal pole for proper names . . .

Here again, brain lesions have provided unique and peculiar information. It is not unusual for a stroke to selectively deprive a patient of a narrow domain of knowledge. Some patients, for instance, lose all spoken and written knowledge of animals and cannot even recognize them in pictures.[121] They cannot answer simple questions such as "Where does a lion live?" or "What color is an elephant?" But they do retain normal knowledge about other domains such as verbs, tools, or famous people. In other patients, animals are preserved, but another category is lost. Distinct lesion sites, spread within the left temporal and frontal lobes, tend to be associated with these category-specific impairments.[122] Functional imaging has also helped to sharpen knowledge about these localizations: whenever we think of nouns or verbs, animals or tools, distinct cortical regions light up.[123]

Finer semantic distinctions remain to be explored. Recently, Friedemann Pulvermuller and his colleagues compared the brain response to words that all belonged to the same category of action verbs, but had different detailed contents. Some verbs such as "bite" or "kiss" evoked mouth movements, others hand actions ("write," "throw"), and yet others actions performed by legs and feet ("kick," "walk").[124] A remarkable fragmentation of meaning was observed: each of the verb types activated a distinct sector of the premotor cortex, at the location represented in the "motor homunculus," or map of our body. Thus word meaning seems to be literally embodied in our brain networks. A string of letters only makes sense if it evokes, in a few hundred milliseconds, myriad features dispersed in the sensory, motor, and abstract brain maps for location, number, intention . . .

By tying each word form to its semantic features, the hierarchical connections of the temporal lobe solve what has been termed the grounding problem for meaning. As stressed by Ferdinand de Saussure, forms and meanings are arbitrarily related: any string of letters can represent any concept. When we learn a language, however, this arbitrariness ceases. We cannot help seeing an animal behind the word "lion," and find it impossible to read the word "giraffe" as a verb—each of these words has become solidly attached, by temporal lobe connections, to the many dispersed neurons that give it a meaning.

A Cerebral Tidal Bore

Perhaps the easiest way to describe how activation spreads through the dozens of fragments of meaning dispersed in the brain is to compare it to a tidal bore. Some rivers are subject, twice a day, at high tide, to a peculiar phenomenon whereby the leading edge of a massive wave reaches deep into their estuaries. If conditions are right, the wave can travel dozens of miles upstream. No salt water ever reaches this far inland—the tidal bore simply relays a distant rise in water level that spreads in synchrony into the river's entire system. Only an airplane or satellite can get the true measure of this beautiful natural phenomenon. For a few minutes, a whole network of streams is simultaneously swollen by a powerful surge of water, simply because they all flow into the same sea.

A written or spoken word probably activates fragments of meaning in the brain in much the same way that a tidal bore invades a whole riverbed. If you compare a word like "cheese" with a non-word like "croil," the only difference lies in the size of the cortical tidal wave that they can bring on. A known word resonates in the temporal lobe networks and produces a massive wave of synchronized oscillations that rolls through millions of neurons. This tidal bore goes even as far as the more distant regions of the cortex as it successively contacts the many assemblies of neurons that each encode a fragment of the word's meaning. An unknown word, however, even if it gets through the first stages of visual analysis, finds no echo in the cortex and the wave it triggers is quickly broken down into inarticulate cerebral foam.

Can the cerebral tidal bore be tracked in real time? In 1995, I gave a simple demonstration of the remarkable speed at which lexical access operates during reading.[125] My experiment consisted of recording the electroencephalogram from sixty-four electrodes while volunteers examined letter strings. Some of these, such as "PGQLST," consisted solely of consonants and thus violated the basic spelling rules of English. Others built words belonging to four categories: numbers, proper names, animals, and action verbs. By tracking the voltage changes on the scalp's surface one millisecond at a time, I was able to determine the duration of two essential stages in reading: the orthographic filter, which accepts legal letter strings, and the semantic filter, which sorts words according to meaning. Only 180 milliseconds were needed for consonant strings to be separated from real words. The topography of the test suggested that this early filtering took place in the left occipito-temporal letterbox region. Crucially, only 80 additional milliseconds were needed for the four categories of words to begin to activate distinct sectors of the cortex. Thus only a quarter of a second elapsed before the tidal bore reached the shores of meaning.

Nowadays, more-advanced magnetoencephalography techniques allow us to track the wave of cerebral activation not just on the scalp, but also on the cortical surface itself. With this technology, movies can be made, in slow motion, to reveal the succession of areas contacted by a single word. Panagiotis Simos and his collaborators at the University of Texas have used these tools to trace the cortical trajectory of words as they took different cerebral paths. They simply asked participants to name letter strings that belonged to three categories:

- Pseudo-words or meaningless strings of letters like "trid" or "plosh" that respect the spelling rules of English. Since they do not have an entry in the mental lexicon, these strings can only be named through the spelling-to-sound conversion route, which applies pronunciation rules with no regard for meaning. It was therefore expected that these strings would activate the superior temporal areas associated with speech sounds, but not the middle temporal areas associated with meaning.

- "Pseudo-homophones" or strings of letters such as "hed" or "wimen" that sound like words, but are misspelled. Again, these stimuli can only be pronounced via the spelling-to-sound route. However, once their pronunciation is retrieved, they evoke a meaning. One would therefore expect the brain activation to start in the superior temporal cortex, and then expand into the middle temporal cortex.

- Finally, real words such as "have" or "eye" that are frequent in English but whose pronunciation is irregular. With such strings, it was expected that brain activation would flow in the opposite direction. In order for pronunciation to be recovered, the word would first have to be found in the mental lexicon. Thus the tidal bore would first reach the lexical and semantic areas, and only later those encoding speech sounds.

The results conformed to this simple picture. At up to 150 milliseconds, brain activity remained confined to the visual occipital cortex. Around 200 milliseconds, it spread into the left letterbox area. At this stage, there were no detectable differences between the three categories of stimuli. Immediately afterward, however, they began to prompt different reactions. Pseudo-words and pseudo-homophones activated the superior temporal region, particularly the left planum temporale, which houses a representation of the pronunciation of speech sounds. Irregular words activated the left middle temporal region, which contains a semantic "convergence zone," and only later caused a secondary wave of stimulation into the auditory areas. Finally, the semantic areas were never strongly activated by pseudo-words devoid of meaning, but did fire, albeit late, in response to homophones such as "hed."

These brain imaging results dovetail nicely with conclusions from a great many psychological studies of reading. Do we have to pronounce words mentally before we understand them? Or can we move straight from a letter string to its meaning and skip the pronunciation stage? Both may happen, but it all depends on the type of word. Common words, as well as words whose spelling is irregular, head straight for the lexical areas of the middle temporal lobe—in Simos's experiment, the time it took participants to name these was well predicted by the speed at which this region lit up. Other words, whether they are regular, rare, or unknown, are first pronounced mentally using the auditory areas of the superior temporal lobe before a meaning is attached to them—in Simos's experiment, their naming speed was predicted by the latency with which the auditory areas first became active.

Brain Limits on Cultural Diversity

Having established this schematic picture of the brain's two reading routes, we can now return to the issue of how it copes with different scripts. The world's many writing systems enjoy very different degrees of spelling transparency (figure 2.20).[126] Some languages, like Italian, have a transparent writing system, in which the spelling-to-sound rules are simple and almost without exceptions: if you can pronounce the letters, you can read any Italian word or pseudo-word. Other languages are more opaque. In English, the same letter string "ough" reads differently in "tough" and "though." In this type of language, conversion rules do not suffice. In order to read written words aloud, an English reader must possess a vast lexicon if he is to retrieve a word's pronunciation. Spelling-to-sound rules do exist, but the innumerable exceptions mean that lexical memory is constantly solicited.

Chinese and Japanese obviously occupy one end of the orthographic transparency scale. In these languages, readers must commit thousands of characters to memory (about 7,000 Chinese characters are indispensable for everyday reading, but up to 40,000 are used in literature). Each of them must be mapped onto the corresponding spoken word. These mappings are not entirely arbitrary—indeed, many Chinese characters have a phonetic marker whose spelling is relatively consistent: one shape spells *ma*, another *ti*, and so on. Nonetheless, there are so many markers, and so many exceptions to their regular pronunciation, that the concept of a spelling-to-sound "rule" barely applies here. Chinese writing relies primarily on probabilistic associations between characters and speech sounds that must be learned one at a time because no abstract rule governs their structure.

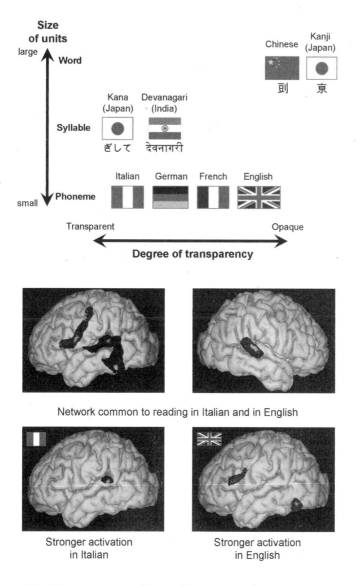

Network common to reading in Italian and in English

Stronger activation
in Italian

Stronger activation
in English

Figure 2.20 The transparency of the spelling system influences the organization of the reader's brain. Writing systems differ in the size of the units they denote (phonemes, syllables, or whole words) and in their degree of transparency (the regularity of the relation between symbols and speech sounds). Italian spelling is very regular, while English spelling bristles with irregularities and exception words. When contrasting brain activations in Italian and English, small modulations are seen within an overall shared network. Italian causes stronger activation in auditory areas of the superior temporal lobe, while English puts greater emphasis on the occipito-temporal area as well as on the left inferior frontal region (after Paulesu et al., 2000).

What is the impact of cultural variations on the brain's reading circuits? Once again, cross-cultural generalization would appear to prevail. Although brain imaging has revealed a few discrepancies between Italian, English, Chinese, and Japanese readers, these are limited to local modulations of the universal circuits (figure 2.20). The two reading routes described earlier exist in all cultures and reside in similar areas of the brain. The only difference consists in the way that each language makes use of the routes. Chinese characters or Kanji that represent whole words tend to provoke greater activation in the areas involved with the mental lexicon (particularly the left middle temporal region). Writing systems that express speech sounds using relatively simple rules, like alphabetic writing, Chinese Pinyin, or the Japanese Kana script, tend to activate the auditory areas directly via the spelling-to-sound conversion route (particularly the left superior temporal region and the angular gyrus).[127]

Similar cultural distinctions can also be found for alphabetic writing systems.[128] Italian, which is a very transparent language, produces somewhat greater activation in the auditory areas of the temporal lobe than English, probably because of its simpler spelling-to-sound rules. English, whose system is opaque, leads to greater activation in the left inferior frontal region involved in semantic and lexical selection and to modestly increased activity in the left occipito-temporal region. These patterns can probably be attributed to the fact that the computation of the visual word form requires more work in English. While an Italian reader can be content with identifying a series of letters before converting them into sound, an English reader must pay attention to the context in which letters occurs if he is to extract complex graphemes such as "ou," "gh," or "ing"—the only units that allow for a certain amount of spelling-to-sound regularity.

In some ways, the differences between languages are comparable to those that exist within languages and separate regular from irregular words. When we read an irregular word like "colonel," our brain activity resembles that of a Chinese reader as we must rely largely on a lexico-semantic route that parses this string as a known "character" with its own pronunciation. In contrast, when we read a new but regular string such as Nabokov's "Lo-lee-ta," we, like an Italian reader, rely on a direct spelling-to-sound route that goes through the superior temporal lobe.

In the final analysis, all these findings lean toward a fundamental univer-

sality of reading circuits. In spite of the diversity of writing systems and transcription rules, people the world over, by and large, solicit the same brain
areas when they read. Chinese characters, alphabetic strings, Hebraic letters,
and Japanese Kanji all take very similar cortical processing routes. Furthermore, when they enter the visual cortex, all written stimuli are channeled to
the left letterbox region, where they are recognized regardless of their exact
shape, size, and location. This package of visual information is then shuttled
on one of two main routes: one that converts it into sound, the other into
meaning. Both routes operate simultaneously and in parallel—one or the
other gets the upper hand, depending on the word's regularity.

Reading and Evolution

The remarkably universal characteristics of the brain areas for reading bring
us face-to-face once again with the enigma surrounding their evolution. How
has our brain come to possess cerebral circuits specialized for reading? Why
does visual word recognition always engage the same region of the brain in all
readers, at positions that are always within a few millimeters of each other? By
what impossible coincidence is this region equipped with all the features
needed for efficient reading, including a capacity for spatial invariance, an
ability to learn the abstract shapes of letters, and adequate connections to
other language areas? Wouldn't these properties appear to imply that our
brain is predisposed to reading? Not at all—irrefutable logic dictates that this
is impossible. The invention of reading is far too recent for our genome to
have adapted to it. During the hundreds of thousands of years that saw the
slow emergence of our species, specific devices for language and socializing
may have evolved in our brain—but this cannot apply to reading, which dates
back only a few thousand years.

   If it is then true that we use the visual system of a primate for reading, we
must now concentrate on the organization of other primates' brains. What
does a macaque do with the brain areas that we now devote to reading? How
have these areas been diverted to a new cultural activity as novel as the recognition of written words?

CHAPTER 3

The Reading Ape

Reading rests upon primitive neuronal mechanisms of primate vision that have been preserved over the course of evolution. Animal studies show that the monkey's brain houses a hierarchy of neurons that respond to fragments of visual scenes. Collectively, these neurons contain a stock of elementary shapes whose combinations can encode any visual object. Some macaque monkey neurons even respond to line junctions resembling our letter shapes (e.g., T, Y, and L). Those shapes constitute useful invariants for recognizing objects. According to the "neuronal recycling" hypothesis, when we learn to read, part of this neuronal hierarchy converts to the new task of recognizing letters and words.

Man in his arrogance thinks himself a great work, worthy the interposition of a deity. More humble and I believe true to consider him created from animals.

—CHARLES DARWIN, *NOTEBOOKS*, 1838

I cannot swallow Man making a division as distinct from a chimpanzee as an *Ornithorhynchus* [platypus] from a horse; I wonder what a chimpanzee would say to this?

—CHARLES DARWIN, *LETTER TO HUXLEY*, 1863

*N*o one defended the evolutionary origins of human cognitive competence more forcefully than Charles Darwin. In *The Descent of Man* (1871) and *The Expression of Emotions in Man and Animals* (1872), he collected hundreds of observations that pointed to the animal origins of human behavior. Even language, he claimed, did not escape this rule: "Man has an instinctive tendency to speak, as we see in the babble of our young children." Darwin was the first scientist to draw a systematic comparison between language learning and the acquisition of singing behavior in birds.

Even in its attempt to understand the higher intellectual faculties, social emotions, moral consciousness, and cultural inventions that mark our species, Darwin's message was clear: chimpanzees and other primates hold the keys to our origins.

Darwin's theories apply also to reading. The neural circuits used for reading will only be fully understood if we probe similar circuits used for vision in other primates, which are not fundamentally different in humans. A hundred years of experiments have now begun to unravel the neuronal mechanisms that allow all primates, including humans, to see. Do these mechanisms also hold the secret to our uniquely human ability to read? Are our literate brains endowed with neurons that specialize in letters, syllables, and whole words? What type of neuronal code is engraved in our brains when we learn to read?

Of Monkeys and Men

In Chicago in the 1930s, Heinrich Klüver, a behavioral scientist, and Paul Bucy, a neurosurgeon, joined forces to explore brain lesions and their peculiar effects on the brains of macaque monkeys.[129] After a bilateral lesion of the temporal lobes, these primates displayed a surprising set of symptoms, now known as the Klüver-Bucy syndrome. Massive changes occurred in their visual, eating, social, and sexual behaviors. The impaired animals behaved as if they no longer recognized their fellow apes, objects, or food items. They explored the world around them haphazardly using their mouths and attempted to copulate with or ingest very inappropriate objects—a form of sightlessness that Klüver and Bucy, like the German neurologist Heinrich Lissauer (1890), called "psychic blindness." The animals were not genuinely blind to light, but no longer recognized shapes simply by seeing them.

A couple decades later, neurophysiologists Karl Pribram and Mortimer Mishkin pinpointed the brain regions responsible for the monkey's visual impairment.[130] They discovered that "psychic blindness" came from lesions to the ventral temporal lobe in both hemispheres (figure 3.1). Monkeys with such lesions can no longer discriminate objects. Moreover, they become incapable of learning that a piece of food is being systematically hidden under a red cube, and not under a striped green sphere. The researchers found that even after hundreds of training trials, the monkeys continued to pick one of the two objects at random. They further noted that the animals were not blind: they moved around normally and located or grasped the objects around them with ease.

As early as 1890, Heinrich Lissauer had described a similar phenomenon in humans. He observed that after a stroke in the left temporal region, some patients could no longer recognize objects visually, even though they could still handle

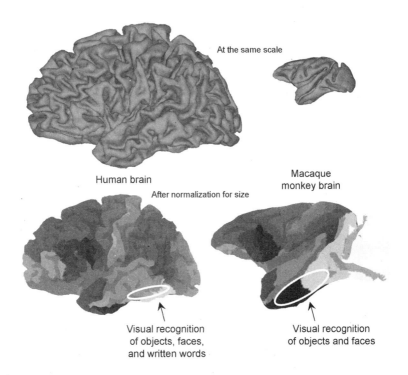

At the same scale

Human brain
After normalization for size

Macaque
monkey brain

Visual recognition
of objects, faces,
and written words

Visual recognition
of objects and faces

Figure 3.1 The occipito-temporal region associated with reading in humans plays a more general role in object and face recognition in all primates. In spite of massive differences in cortical surface, a great many similarities in the micro-organization of the cellular layers of cortex point to a systematic correspondence between regions in the human and monkey brains.

them properly and their visual acuity was otherwise normal.[131] This impairment, which Lissauer called "psychic blindness" or "visual agnosia," provided the first documented indication that visual object recognition in humans and primates relies on specific brain pathways.

In humans, the temporal regions responsible for object recognition occupy the lower side of the brain. In the macaque monkey, they tend to be more lateral and to extend toward the front tip of the temporal lobe (figure 3.1). This difference in shape can probably be attributed to an enlargement of the associative regions of the human cortex that have to do with language processing and semantic networks, particularly in the anterior and lateral temporal lobes.

Expansion of the associative regions would appear to have driven visual recognition toward the lower side of the brain, but a clear homology or similarity due to shared ancestry still relates our brain to that of other primates. In 1910, after a careful examination of the micro-organization of the cortex, and specifically of how cortical neurons neatly cluster into layers, the anatomist Korbinian Brodmann proposed a subdivision of the human cortical areas that is still largely accepted today. Anatomy shows that the same brain areas can be observed, albeit with some differences, in both macaques and humans. In particular, the areas that Brodmann labeled 20 and 37 are present in both species and both deal with visual recognition. Crucially, the "letterbox" area, which is activated when we read, falls squarely within Brodmann's area 37. Thus reading is not handled by a new and uniquely human brain area. We recognize the written word using a region that has evolved over time and whose specialty, for the past ten million years or more, has been the visual identification of objects. The brain's letterbox overlaps nicely with the general site where a lesion induces the "psychic blindness" discovered by Lissauer, Klüver, and Bucy.

Neurons for Objects

Any object projects a very different image on the retina when lit from two different angles or when presented either on the left or on the right of visual fixation. This variability poses a severe challenge to the neural circuits that construct the categories with which we recognize the world. When we need to identify an object, it is irrelevant whether it is near or far, to our left or to our right, upright or horizontal, in light or in shade. We can easily recognize it regardless of illumination, inclination, distance, and location. In the course of evolution, the brain was undoubtedly subjected to intense selective pressures that obliged it to arrive at an invariant perception of the world. We now know that invariance is an essential characteristic of the inferior temporal lobe. Monkeys with ventral temporal lesions can no longer recognize objects invariantly. Unlike undamaged monkeys, they fail to extend learning to new conditions of lighting, size, or location of a learned shape.[132] In consequence, brain lesion experiments show that the inferior temporal cortex is a key player in the collection of invariant visual information—an innate competence that

Figure 3.2 Neurons in the monkey's inferior temporal cortex can be remarkably selective to the sight of particular objects, faces, or scenes. Among the one hundred images that were successively presented to this neuron, only the sight of a chair strongly enhanced its firing (after Tamura & Tanaka, 2001). *Used with permission of Oxford University Press and* Cerebral Cortex.

is probably related to our ability to recognize that "RADIO," "radio," and even "RaDiO" represent the same word.

Recently, recordings of single neurons have begun to reveal a fine-grained neuronal code for visual objects in macaque monkeys. As early as the late 1960s, David Hubel and Torsten Wiesel recorded neuronal activity in the primary visual area of the cat, and found that neurons discharged in response to simple bars of light—paving the way to work that was to be awarded the Nobel Prize in 1982. In the 1970s and 1980s, several pioneering neuroscientists (Robert Desimone, Charles Gross, David Perrett, Keiji Tanaka) followed in their footsteps and moved their electrodes forward in the macaque brain.[133] They found that to elicit a discharge in the inferior temporal neurons, they had to present monkeys with more complex stimuli than simple lines. They put together a great variety of images, shapes, faces, and objects and presented them visually, one at a time, while an electrode recorded the neuronal discharges.

The selectivity of the neurons' discharges immediately struck them. Frequently a neuron responded to a given face alone, or to only one particular object out of dozens of others (figure 3.2). This selectivity was all the more remarkable because it was accompanied by a great capacity for constancy in the face of massive changes in the details of the input image. In the primary visual cortex, neural responses are conditioned by a very narrow input window on the retina (the "receptive field"). Neurons in the inferior temporal cortex, however, are completely different. Their receptive fields are vast and they frequently respond to objects appearing almost anywhere in the visual scene. At any given point in this vast reception zone, each neuron maintains a preference for a favorite specific object. This bias is true even when the image slips by several degrees, when its size is multiplied or divided by two,[134] or when the light and resulting shadows change.[135]

What happens when the object turns around? Our visual system displays significant difficulties with invariance for rotation. In the earliest stages of visual recognition, the consecutive positions of a rotating face are not coded by the same neurons. The right, front, and left profiles activate neighboring cortical patches that partially overlap with one another (figure 3.3).[136] When an object rotates on the retina, most of the inferior temporal neurons respond only to a specific view. As the object rotates away from this preferred view, the neurons tolerate about 40 degrees of rotation and then cease

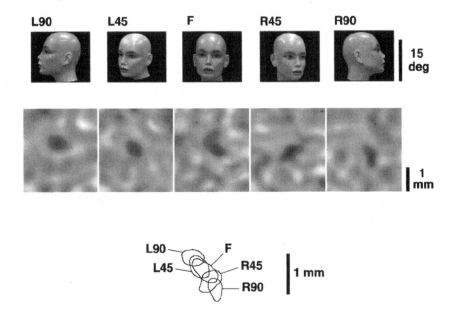

Figure 3.3 Small patches of the monkey's inferior temporal cortex respond to faces. The presentation of a face activates a cortical column, about 500 microns wide, that can be visualized with optical imaging as a dark spot on the surface of the cortex. As the face rotates, the successive views activate nearby and partially overlapping sectors, thus contributing to the elaboration of a neural code invariant for rotation (after Tanaka, Tomonaga, & Matsuzawa, 2003). *Used with permission of* Cerebral Cortex.

to respond. A few neurons, however, are more abstract and react to an object regardless of its position in space (figure 3.4).[137] Simply put, these invariant neurons seem to pool input from many view-specific neurons, each roughly adjusted to a given angle. They essentially detect the presence of a certain object by pooling across all of the possible viewpoints from which it might be seen.

In brief, the visual invariance problem appears to be solved through a series of successive processing stages, each implemented within the inferior temporal cortex. At the top level of this visual hierarchy, activity in groups of neurons remains constant even when an object moves around, recedes, turns, or casts new shadows. This mechanism predates the acquisition of reading by millions of years—but its existence plays a key role in our ability to recognize words anywhere on a page and in any font and size.

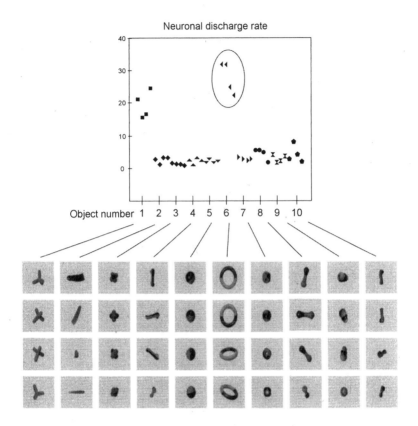

Figure 3.4 Some inferior temporal neurons respond to shapes in a selective and invariant way. This neuron vigorously discharges to the presentation of a ring and, to a lesser extent, to a "tripod" (bottom left). The response is largely independent of the object's orientation in space (after Booth & Rolls, 1998). *Used with permission of Cerebral Cortex.*

Grandmother Cells

The physiological observation that a single neuron can respond to one image out of a thousand is stupefying. Is our cortex really covered by millions of ultra-specialized neurons? The physiologist Horace Barlow once proposed, tongue in cheek, that the brain contains "grandmother cells," or cells that only respond to a single familiar person. Although Barlow's statement had all the appearances of a joke, he was in fact right, or at least close to the truth. The brains of monkeys, like those of humans, contain neurons that are so specialized they appear to be dedicated to a single person, image, or concept. For instance, a neuron that responded exclusively to Hollywood superstar Jennifer Aniston was once recorded in the anterior temporal region of a human epilepsy patient.[138] It did not appear to matter whether the stimulus was a color photograph, a close-up of her face, a caricature, or even her name in writing—only Jennifer seemed to excite this neuron!

The concept of a grandmother neuron, however, must be mitigated by several observations. Even when this amazing selectivity is uncovered in a single neuron, it must result from computation by a much larger network. The experiments to which I refer involved the insertion of an electrode at random in the visual brain. If one can find a specialized neuron in this haphazard way, there are no doubt millions of others waiting to be discovered. Their specificity, moreover, necessarily results from the collective operation of many cells. In the final analysis, the selective response by a single cell is like the tip of an iceberg: we can only see it because of the underlying mass of cells that conspire to create a hierarchy of detectors. For all we know, a single neuron, on its own, can only perform a relatively elementary computation on input. On the output side, furthermore, a single neuron alone does not have much clout: only a coalition of a few hundreds of cells can influence other groups of cells. Each visual event or face we recognize must therefore be encoded by several clusters of selective cells, or what is known as a "sparse" coding scheme.

Picturing the entire process that leads from the retina, where millions of photoreceptors only react to blotches of light, to the neurons that detect the presence of Jennifer Aniston, is a mind-boggling feat. The detailed neuronal organization of visual recognition is only just beginning to be uncovered. Anatomically, we know that the macaque monkey's inferior temporal cortex is organized like a pyramid. The visual image enters at the base of the pyramid, and myriad consecutive connections convey it from the primary visual cortex, at the back of the head, to the front end of the temporal pole (figure 3.5).[139] This anatomical progression is accompanied by an increase in the complexity of the images that make a neuron fire. At each stage, the recombination of responses by neurons from the lower level allows new neurons to respond to increasingly complex portions of the image. Our visual system is very precisely wired to reassemble the giant jigsaw puzzle created by the retina when it explodes incoming images into a million pixels.

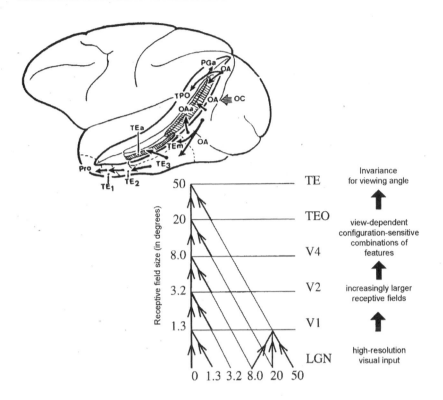

Figure 3.5 The occipital and inferior temporal areas are organized as a hierarchy of increasing invariance. They are successively interconnected according to a "synaptic pyramid." At each stage, the size of the receptive field, which is the region of the retina to which the neurons respond, increases by a factor of two or three. In parallel, the visual preferences of the neurons become increasingly complex and invariant (after Rolls, 2000). *Used with permission of* Neuron.

If we could climb the neuronal pyramid step by step, synapse by synapse, and make recordings from single neurons encountered along the way, from the primary visual cortex to the inferior temporal lobe, we would see three types of changes:

- First, the preferred images that make the neuron fire would become increasingly complex. A small, inclined bar is enough to bring on a significant discharge in the primary visual cortex. More complex curves, shapes, fragments of objects, or even entire objects or faces are, however, needed to trigger neurons at the higher levels.

- Second, neurons would begin to respond to increasingly broader portions of the retina. Each neuron is defined in terms of its receptive field, or the place on the retina to which it responds. The receptive fields broaden by a factor of two or three at each step. This means that the part of the retina to which the preferred object must be presented for the neuron to fire doubles or triples in diameter at each step.
- Finally, an increasing degree of invariance is present. Early on, neurons are sensitive to changes in location, size, or lighting of the incoming picture. In higher-level areas, in the move up the hierarchy, neurons tolerate increasingly significant shifts and distortions of the input image.

Functional brain imaging on human volunteers shows that hierarchical organization and increasing invariance also hold for our visual cortex.[140] In humans as in other primates, the concept of neuronal hierarchy provides a simple though still hypothetical solution to the issue of visual invariance. When our cortex is called on to identify an object, it must learn what it looks like from different angles. Learning mechanisms allocate a set of neurons to each view of the object. They then wire the illustrations together so that they collectively excite the same neurons at the next level further up the pyramid. The net result is an invariant neural circuit that tolerates considerable changes in viewing position. This simple idea can be replicated at each step. The neurons responsible for recognizing Jennifer Aniston's profile collect input from the neurons at lower levels that identify a fragment of her face. These neurons are able to recognize an eye or a nose because the preceding level has already detected the patterns of light and dark that are compatible with the presence of these features at a given location on the retina.

In brief, the keys to the primate's visual system are the notions of hierarchy and parallel functioning. The mental image that is first split on the retina into myriad "pixels" is progressively recomposed by a pyramid of neurons, all operating simultaneously. This approach might at first seem inefficient, because millions of neurons must be dedicated to each of the possible fragments that make up a single visual scene. The burden on the nervous system, however, is relatively modest because the scene can be distributed across a gigantic array of simple parallel processors. Much as a colony of ants is more "intelligent" than a single ant, collective action by millions of neurons accomplishes operations far more complex than what could be achieved by a single neuron. The vast number of computing units, in fact, leads to considerable savings in processing time. Single neurons are slow computers. They receive and transmit information in about ten milliseconds, which is a million times slower than the speed of an electronic microprocessor. Yet by combining the activity of millions of neurons, our visual system becomes the most efficient of computers: it only takes one-sixth of a second to spot a face, regardless of its identity or location.[141]

The architecture of the brain has inspired many programmers. Several computer models of the visual hierarchy I have described are now available.[142] The best are now close to human performance, both in terms of speed and the extent of image distortion they can tolerate. Thanks to these artificial neuronal networks, automatic face recognition no longer needs to be viewed as something out of science fiction. It is part of real life—even the simplest digital camera now includes face and smile recognition.

An Alphabet in the Monkey Brain

The pyramid model maintains that the neural code for a given visual object consists of a hierarchy of neurons, each of which detects the presence of a fragment of this object in the input image. Most object-sensitive neurons probably react to simplified and restricted views of objects or their parts. With this hypothesis as a starting point, the Japanese neuroscientist Keiji Tanaka made a remarkable discovery: the monkey brain contains a patchwork of neurons dedicated to fragments of shape. Collectively, these primitive shapes constitute a sort of "neuronal alphabet" whose combinations can describe any complex form.

To study the neuronal code for objects, Tanaka and his colleagues came up with a procedure that starts out with a complex scene and then progressively simplifies it (figure 3.6). They looked for a picture that made a neuron fire vigorously and then whittled it down to the simplest possible shape that still triggered firing. Consider, for instance, a neuron that initially responded to the sight of a cat. Tanaka found that the neuron still fired as vigorously to the sight of two superimposed discs. Another neuron liked the shape of an apple, but in fact discharged equally well for a black disc with a little stalk. A third neuron, triggered by the sight of a cube, actually seemed to detect only the

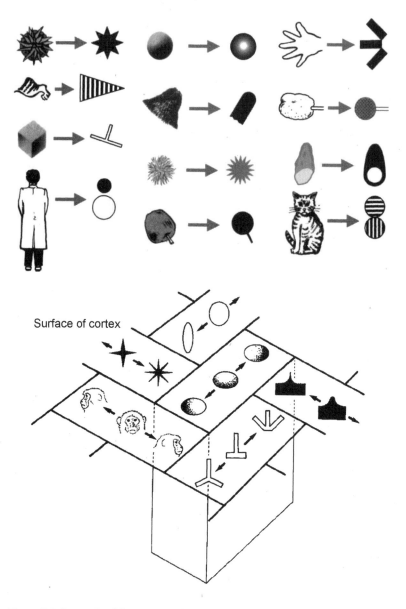

Figure 3.6 Image simplification reveals a microtopography in the temporal cortex. Starting from a stimulus that strongly activates a neuron, the image is progressively streamlined until a much simpler one is found that still evokes as strong a response (top). Some of these minimal shapes are reminiscent of letters (O, T, Y, E, etc.). The neurons that respond to them are organized in vertical columns transversal to the cortex. As one moves in a perpendicular direction along the cortical surface, the preferred shapes vary continuously, defining local sectors of shape preference (after Tanaka, 2003). *Used with permission of* Cerebral Cortex.

Y-junction where its central edges met. Most neurons in the inferior temporal cortex fired regardless of such dramatic simplifications of an image.[143]

After hundreds of recordings, Tanaka was able to reconstruct the approximate patchwork of object neurons on the monkey's cortical surface. He discovered a systematic layout. The preferences of different neurons varied smoothly across the cortex, with neighboring neurons tending to code for similar shapes (figure 3.6). For instance, there was a patch of cortex dedicated to variants of Y and T. Other cortical sectors specialized in shapes that looked like stars, the simplified profile of a face, or the figure 8. An entire cortical catalog of elementary shapes emerged. At a given cortical location, all the neurons within a vertical column of cortex were interested in a somewhat similar shape. But each was also sensitive to minute metric deviations from the basic prototype.

As a whole, these neurons provide an alphabet of shape variations that make a rough categorization of an image (is it a face or not?) and single out its details. (Is it Jennifer Aniston's face? How old is she? Is she happy?) By combining the responses from millions of such detectors, the brain encodes each of the billions of pictures we might ever meet in nature.

One of Tanaka's former students, Manabu Tanifuji, followed up on the preceding experiments and clarified how multiple columns of neurons encode arbitrary objects. He used an optical recording technique that allowed him to see a large chunk of the cortical code associated with a visual object.[144] Whenever the brain activates, the amount of light that it reflects is reduced via mechanisms that are only partially understood. Tanifuji employed a technique that measures the neuronal activity of an entire brain area using a sensitive video camera that detects small changes in the reflection of light on the cortical surface. With the help of a powerful zoom lens, the camera tracks active structures only a few tenths of a millimeter apart—roughly the size of Tanaka's shape-coding columns.

In one experiment, Tanifuji presented a monkey with the picture of a fire extinguisher—the idea was to stimulate the cortex with objects for which evolution could not possibly have played any part. The cortical patchwork that coded for this object immediately appeared in the form of a sparse array of dark spots representing neural activity. This "cortical signature" of the fire extinguisher appeared each time it was presented. Next, Tanifuji removed the handle and hose. At the sight of the red tank alone, some spots vanished, others remained as active, and, perhaps more surprisingly, others even lit up at the sight of the "naked" extinguisher. In other words, the assembly was reconfigured, and a unique neural code was immediately attributed to every new picture, in spite of the fact that only part of it was familiar.

By recording from single neurons in each of the spots, Tanifuji was able to shed some light on the underlying combinatorial code. On one spot, the sight of the handle was enough to trigger the neurons—as did any V-like shape composed of several spikes, like the contours of a hand or of a cat's ears. At another spot, the neurons responded to the handle alone—but a straight line didn't work, the handle had to be curved like a J. The response of a third group of neurons was even more subtle. It reacted to the extinguisher's tank, as well as to any other elongated and somewhat rectangular shape, but all firing ceased whenever features like the handle or the hose were added.

Tanifuji's experiments contribute substantially toward clarifying how our temporal cortex encodes any visual image, even the first time it is seen. The trick is to exploit an alphabet of elementary shapes, each coded by specific populations of neurons, and their millions of potential combinations. Not only do different neurons react to the different parts of an object, but they are also concerned with how the parts are arranged in space. A similar combinatorial code exists, furthermore, at each of several hierarchical levels in the visual system—only the scale and degree of complexity of the preferred shapes increase in the move up the hierarchy. The primary visual area (V1) appears to be largely dedicated to the detection of thin lines and object contours. By the secondary visual area (V2), neurons are already sensitive to combinations of lines with well-defined inclinations or curves.[145] Farther up, in the posterior part of the inferior temporal cortex (an area called TEO), they are tuned to simple combinations of curves.[146] Neurons' selective response to, say, the shape of an F can be reduced to the detection of a simple conjunction of elementary curves, each placed at a relatively fixed location: the top bar, the upper left angle, the middle bar, and so on. In brief, the same scheme seems to be repeated at each hierarchical level: neuronal selectivity results from a conjunction of the more elementary features coded by neurons at lower levels.

Proto-Letters

Perhaps the most striking feature of the inferior temporal neurons is that many of their preferred shapes closely resemble our letters, symbols, or elementary Chinese characters (figures 3.4 and 3.6). Some neurons respond to two superimposed circles forming a figure eight, others react to the conjunction of two bars to form a T, and others prefer an asterisk, a circle, a J, a Y . . . For this reason, I like to call them "proto-letters." That these shapes are so deeply embedded in the preferences of neurons in the brains of macaque monkeys is quite amazing. By what extraordinary coincidence is this cortical stock so similar to the alphabet that we inherited from the Hebrews, Greeks, and Romans? The reading paradox reaches its apex with this mysterious resemblance between two worlds that we thought discrete: the depths of the monkey's cortex and the clay, reeds, and parchment surfaces on which the early scribes first etched their scripts.

We can get some insight into the origins of these monkey proto-letters by examining the reason why they appear in the visual field. The most likely hypothesis is that these shapes were selected, either in the course of evolution or throughout the course of a lifetime of visual learning, precisely because they constituted a generic "alphabet" of shapes that are essential to the parsing of the visual scene. The shape T, for instance, is extremely frequent in natural scenes. Whenever one object masks another, their contours almost always form a T-junction. Thus neurons that act as "T-detectors" could help determine which object is in front of which.

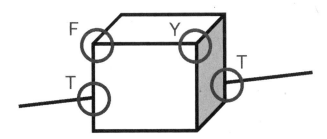

Other characteristic configurations, like the shapes of a Y and an F, are found at the places where several edges of an object meet—they characterize an object's sharp corners and their orientation. The shapes J and 8 result from yet another set of object contours—when the object has curves and holes. All of these fragments of shapes belong to what is known as "non-accidental properties" of visual scenes because they are unlikely to occur accidentally in the absence of any object. If you throw a bunch of matches on the floor, it is unlikely that two of them will meet to form a T-junction, and it is even less probable that three of them will arrive at the configuration of the letter Y. Consequently, when one of these shapes appears on the retina, the brain can safely assume that it corresponds to the contour of an object present in the outside world.

If the cortex finds it useful to encode non-accidental properties it is no doubt because their combinations tend to be extremely invariant to changes in size, angle of vision, and light. When one picks up a coffee cup and rotates it in one's hand, across a wide range of viewing angles the edges of the cup always form two opposing F-junctions. Even with one eye closed, it is virtually impossible to find the only angle at which the edge and sides of the cup are at right angles, so that the F pattern vanishes—usually, another pair of F's immediately appears at the bottom:

In many cases, the list of ways in which the edges meet is an invariant that constrains object identification regardless of angle of presentation. Our primate nervous system seems to have discovered this invariant property and used it to encode shapes. In fact, visual scenes have many other non-accidental properties. Parallelism is one of them: it is unlikely that an image will contain two parallel segments unless they are the edges of a three-dimensional object. Other invariants have to do with spatial organization: if an object contains a hole, its projection on the retina will probably include a closed O-shaped curve. Visual invariants such as these are so distinctive that they have been firmly integrated into our nervous system. According to the California psychologist Irving Biederman, our memory does not store fully detailed visual images of objects. It merely extracts a sketch of their non-accidental properties as well as their organization and spatial relations.[147]

Their extraction allows us, at first, to reconstitute the elementary parts that constitute the object's three-dimensional structure (surfaces, cones, sticks . . .), and later to assemble them into a complete representation of the object's shape. This code has the advantage of remaining consistent in the face of random rotations, occlusions, and other deteriorations of the image.

To support his hypothesis, Biederman gathered evidence to the effect that human object perception relies more on non-accidental properties than on other aspects of the image. For instance, if one starts with a line drawing of an object, and then deletes half of the contours, the impact on perception depends on whether its non-accidental properties are preserved or not:

- If one only deletes the line segments that link two vertices, and leaves the non-accidental junctions intact, the object remains easily recognizable.
- If all the non-accidental properties are deleted, recognition becomes virtually impossible (figure 3.7).[148]

Likewise, when we have to decide if two objects are identical or not, the differences are obvious when they bear on non-accidental properties (for instance a letter "O" versus a figure eight). They are quite difficult to spot if they concern metric properties alone such as size (for instance an uppercase "O" versus a lowercase "o").[149] In collaboration with the neurophysiologist Rufin Vogels, Biederman further demonstrated that many neurons in the inferior temporal cortex of the macaque monkey resist metric distortions of the image, provided that the transformations leave non-accidental properties intact.[150]

In summary, shapes that resemble Western letters, such as T, F, Y, or O, were adopted by inferior temporal neurons because they collectively formed an optimal code, invariant to image transformations, and whose combinations could represent an infinity of objects. It is probable that other shapes were added to this alphabet because of their biological relevance. For instance, Tanaka has observed that some neurons code for a black dot on a white background—an eye detector, clearly an essential device in a social species like ours. Other neurons are sensitive to hand or finger shapes. Primarily, however, the inferior temporal cortex relies on a stock of geometrical shapes and simple mathematical invariants. We did not invent most of our letter shapes: they lay dormant in our brains for millions of years, and were merely rediscovered when our species invented writing and the alphabet.

Figure 3.7 Complex objects are recognized through the configurations of their contours. At the places where they join, these contours form reproducible configurations shaped as T, L, Y, or F. If these junctions are erased, the images become much more difficult to recognize (left column), whereas deletion of an equivalent amount of contour that spares the junctions causes much less difficulty (middle column) (after Biederman, 1987). As soon as an organized set of contours is present in the image, even if it does not form a coherent whole, our visual system cannot help but perceive it as a three-dimensional object (bottom).

The Acquisition of Shape

What remains unclear is whether the cortical alphabet of primitive proto-letters is truly inscribed in our genes, or if it emerges anew in each child as the result of a learning process. Some of the shapes in the alphabet are so useful for vision that they were probably prewired in our visual system in the course of evolution. In the first few months of life, babies are already sensitive to faces and to object occlusion. Thus faces, eyes, and T-junctions may indeed be part of an "innate" shape lexicon. From an evolutionary standpoint, this would give primates a major advantage in their early interactions with other members of their species and the environment.

It is hard to see, however, in what way the human genome, with at most 30,000 genes, could contain the detailed instructions necessary to wire neural detectors for a large number of basic shapes, all the way down to face profiles . . . or fire extinguishers! Furthermore, the well-known plasticity of the inferior temporal cortex makes this improbable. Even if we allow for an initial genetic bias, it seems likely that most of the neurons involved in object recognition become selective because of interaction with a structured visual environment. We are constantly bombarded by millions of incoming images that provide primary data for our brain's statistical learning algorithm. In the course of development, and probably throughout life, the synaptic contacts in our visual system change constantly to encode the fragments of images of greatest relevance to us. It is certainly no coincidence that we start to teach a child to read at a very early age when cortical plasticity is at a maximum. By immersing children in an artificial environment of letters and words, we probably reorient many of their inferior temporal neurons to the optimal coding of writing.

A great many experiments have shown that neurons become progressively tuned to the objects a monkey is taught to discriminate—even when the objects used for training are meaningless wires or fractals, far from any shape that the animal might encounter in its natural environment.[151] Neurons seem to acquire preferences by learning to detect noteworthy conjunctions of salient features. A recent experiment by Chris Baker, Marlene Behrmann, and Carl Olson nicely illustrates this point.[152] These researchers taught monkeys to recognize sticks with characteristic shapes at their two extremities, for instance a square at one end and a trident at the other. After training, several neurons had become sensitive to their precise combinations. When they were shown only one end of the object, they responded feebly. In brief, the neuronal discharge to the whole picture was greater than the sum of the individual responses to each part. This is direct proof that neurons learn to respond to new visual combinations. Within the immense catalog of possible shapes, some neurons learn to fire exclusively for combinations that appear frequently. Other conjunctions exist in our neural inventory, but only potentially. Until the monkey is taught them, they are not explicitly encoded by higher-level neurons.

The Learning Instinct

Cerebral plasticity has been demonstrated so frequently that it is sometimes considered a self-evident property of the cortex. In reality, the capacity to learn is the result of a sophisticated evolutionary process. In many cases, learning is undesirable—a baby that would have to learn to breathe or feed wouldn't survive very long. However, we are born in a complex world that cannot be completely predicted in advance, and thus our brain cannot be fully prewired. In such cases, evolution eventually came up with a useful trick: learning. In some brain circuits and at some points during development, part of the nervous system adapts to external constraints. Naturally, the learning mechanism itself is innate and is ultimately governed by sophisticated genetic mechanisms. There is thus nothing contradictory in talking about rigid learning mechanisms, or even the "instinct to learn," to quote Peter Marler.[153] The old antagonism between nature and nurture is a myth—all learning rests on rigid innate machinery.

The clearest example of the limits of brain plasticity comes from binocular vision—the blending of information arising from our two eyes. A narrow window of plasticity, which lasts a few weeks in cats, a few months in nonhuman primates, and a few years in humans, allows for fine adjustment of connections in the primary visual area. This plasticity is used to align the two visual maps: neurons learn how to relate the inputs from our two eyes. At the end of this critical period, the circuit freezes. Children who squint during the critical period suffer from lifetime impairments of vision—they lose the ability to perceive depth by exploiting the small mismatch between the two images (stereovision). In this instance, nature only leaves a short window of time to nurture.

The plasticity of the inferior temporal cortex that allows us to learn to recognize new objects is not fundamentally different. At all levels in our visual system, our cortex is programmed to look for correlations in the incoming sensory data and store the resulting conjunctions. An object's identity is thus preserved in the form of multiple combinations of correlations that occur in a precise set of nerve cells. The sight of a fire extinguisher, for instance, activates several groups of neurons that code for the body, the handle, the hose, the standard color red, and probably a number of other shape primitives. This reproducible configuration of active neurons is then stored and stabilized through an increase in the strength of the synapses that bind them together into a stable set. Each of these primitives must, in turn, have been previously learned as a conjunction of more elementary features at the lower level. Our ability to recognize objects thus ultimately rests on a pyramid of neurons and a hierarchical learning scheme.

A new problem emerges at the top of the hierarchical pyramid. At this level, some neurons respond to several views of the same object—for instance the profile and the front view of a face, or even a person's name (Jennifer Aniston). It is not clear how such a subtle form of invariance is acquired. Learning by coincidence no longer works. We cannot simply bind together the neurons that are activated at a given moment, because we never see the full face and profile of the same person at the same time (except in a Picasso).

In the natural environment, the many possible views of the same object are often seen in succession. How does a single neuron learn that these views all map onto the same object? Yasushi Miyashita, from Tokyo University, discovered a plausible neuronal mechanism that may well account for this temporal learning. He showed that some neurons in the inferior temporal cortex are sensitive to correlations across time.[154] When a monkey sees unrelated fractal shapes repeatedly, each of these is encoded by a distinct group of neurons. However, if these shapes are temporally linked, either because they always occur together within the same temporal sequence or because they are presented in pairs, the cortex allocates them to the same neurons. This means that one neuron ends up responding to two entirely different images whose only shared property is their occurrence in close temporal succession (figure

3.8). In brief, the inferior temporal cortex would appear to have been aptly named since it detects temporal conjunctions in arbitrary images and assigns them to the same neurons.

This learning mechanism can generate highly abstract and invariant visual representations. Whenever we see Jennifer Aniston, regardless of whether we are near or far, have a front view or a profile, our temporal neurons manage to maintain a stable representation of her even if the views change constantly. The temporal correlation mechanism may well play an essential role in reading. For instance, it should have no trouble uncovering the association of lowercase and uppercase letters such as "a" and "A." At a much higher level, it will also detect the correlation between letters and speech sounds and thus contribute to establishing an efficient grapheme-to-phoneme conversion route.

Neuronal Recycling

To be a well-favored man is the gift of fortune, but to write or read comes by nature.

—WILLIAM SHAKESPEARE, *MUCH ADO ABOUT NOTHING*

From the preceding outline it is apparent that our primate vision is neither prewired at birth nor devoid of any structure and left at the mercy of the outside world. The general architecture of the visual system is tightly constrained and is identical in all of us, but the detail of how each neuron responds depends on the particular visual events to which we are exposed. Sophisticated statistical learning techniques detect the regularities of the outside world. Our brain is built so that nonaccidental properties, like the near-alignment of several bars on the retina, the presence of T- or L-junctions, or a repeated succession of two images, are quickly extracted and stored in our cortical connections.

Our "instinct to learn" plays a crucial role in our capacity to learn to read. Synaptic plasticity, which is extensive in children, but also exists in adults, allows our primate visual cortex to adapt, in part, to the peculiar problems raised by letter and word recognition. Our visual system has inherited just enough plasticity from its evolution to become a reader's brain.

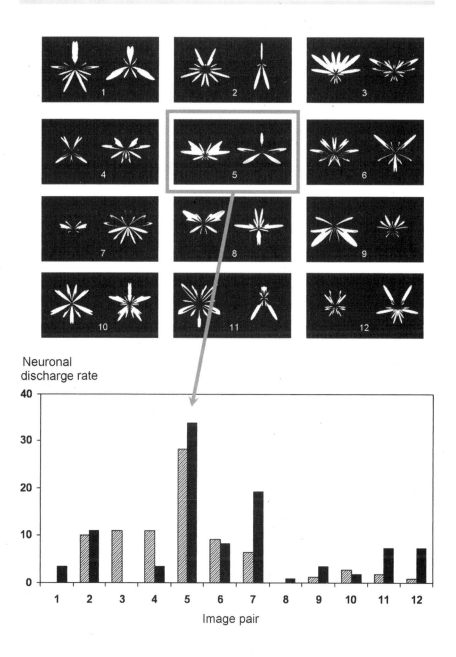

Figure 3.8 Inferior temporal neurons can learn to respond to arbitrary shapes such as these fractal images. Sakai and Miyashita (1991) trained macaque monkeys to associate these images in pairs. After learning, individual neurons started to respond equally well to the two members of each pair—a conventional association reminiscent of the arbitrary cultural link between uppercase and lowercase letters.

When a preschooler first enters school, his brain is already pre-acclimatized to letter and word recognition. Like any other primate, his ventral temporal cortex contains a forerunner of the alphabet. Object recognition already operates along a combinatorial principle, based on a neuronal alphabet of elementary shapes that I have called proto-letters, some of which are very similar to our letters.

My radical proposal is that it is only because this preadaptation of the primate inferior temporal cortex exists that we can learn to read. We would not be able to read if our visual system did not spontaneously implement operations close to those indispensable for word recognition, and if it were not endowed with a small dose of plasticity that allows it to learn new shapes. During schooling, a part of this system rewires itself into a reasonably good device for invariant letter and word recognition.

According to this view, our cortex is not a blank slate or a wax tablet that faithfully records any cultural invention, however arbitrary. Neither is it an inflexible organ that has somehow, over the course of evolution, dedicated a "module" to reading. A better metaphor would be to liken our visual cortex to a Lego construction set, with which a child can build the standard model shown on the box, but also tinker with a variety of other inventions.

My hypothesis disagrees with the "no constraints" approach so common in the social sciences, according to which the human brain is capable of absorbing any form of culture. The truth is that nature and culture entertain much more intricate relations. Our genome, which is the product of millions of years of evolutionary history, specifies a constrained, if partially modifiable, cerebral architecture that imposes severe limits on what we can learn. New cultural inventions can only be acquired insofar as they fit the constraints of our brain architecture. Cultural artifacts can deviate considerably from the natural world in which we have evolved—nothing in the wild looks remotely like a page in a book. However, each of them must find its "ecological niche" in the brain, or a neuronal circuit whose initial function is close enough and whose flexibility is sufficient to be converted to this new role.

A classic Darwinian concept, defined as "exaptation" by Stephen Jay Gould,[155] comes to mind. Exaptation refers to the conversion, in the course of evolution, of an ancient biological mechanism to a role different from the one for which it originally evolved. The minute bones, deep in the ear, that seem so perfectly designed to amplify incoming sounds are an excellent example— Darwinian evolution whittled them out of the jawbones of ancient reptiles. In a much-cited article, François Jacob pictured evolution as a tireless tinkerer who keeps a lot of junk in his backyard and occasionally assembles pieces of it to create a new contraption.[156] In my hypothesis, cultural invention arises similarly from the recombination of ancient neuronal circuits into new cultural objects, selected because they are useful to humans and stable enough to proliferate from brain to brain.

In the case of cultural learning, brain tinkering takes place more rapidly than the slow tempo of biological evolution. The invention of a new cultural tool may only take a few weeks or months (even if several generations are needed for it to spread to a broad population). Furthermore, the creation of cultural objects rests on neuronal learning mechanisms that do not require any change in the genome. Because of these fundamental differences between biological and cultural evolution, I would like to introduce a new term, "neuronal recycling," to qualify the cultural changes that occur in our brains.[157]

By neuronal recycling, I mean the partial or total invasion of a cortical territory initially devoted to a different function, by a cultural invention. The word "recycling" is intended to evoke short-term change taking place over only a few months. The Merriam-Webster dictionary defines it as "to pass again through a series of changes" and "to adapt to a new use." In French, which is my native language, the verb *se recycler* applies to students or employees who take a refresher course or train for a new job better adapted to the job market. Neuronal recycling is also a form of reorientation or retraining: it transforms an ancient function, one that evolved for a specific domain in our evolutionary past, into a novel function that is more useful in the present cultural context.

The word "recycling" also makes it clear that the neuronal tissue that supports cultural learning is not a blank slate, but possesses properties that limit its range of applications. Recycled glass or paper cannot be turned into any object. These materials have intrinsic physical properties that make them more suitable for certain uses than for others. Similarly, each cortical region or network, because of its connectivity, genetic biases, and learning rules, possesses intrinsic properties that are only partially modifiable during the cultural acquisition process. If my neuronal recycling hypothesis is correct, cultural learning never totally undoes these preexisting biases—it merely works around them. As a result, we cannot expect cultural objects to be infinitely compliant and adaptable. The range of human cultural variability is circumscribed by the constraints of our neuronal networks.

Birth of a Culture

I speculate that neuronal recycling plays an essential role in the stabilization of what we call "culture," namely the set of shared mental representations that define a given group of human beings. According to the evolutionary biologist Richard Dawkins, much as reproduction replicates genes, human societies propagate "memes"—elementary units of culture that range from a recipe for lemon meringue pie to more fundamental components such as writing or religion.[158] Susan Blackmore, who has become a major advocate of this idea, speaks of the human brain as a "meme machine," ideally suited to copy cultural representations from one person to the next.[159] However, as has been pointed out by the French anthropologist Dan Sperber, this "memetic" model as a theory of cultural propagation is simplistic and Lamarckist at heart.[160] It reduces children to faithful "copycats" who imitate their peers and family. Accurate imitation is needed to ensure the reliable replication of memes and their cultural stability on a scale of decades or centuries. But this concept of passive imitation does not withstand careful scrutiny. Unlike a blank slate or a Xerox machine, the brain does not simply make copies of its environment. Rather, our brain structures are tightly constrained and only a small element of plasticity allows them to adapt to the cultural representations that surround us.

In essence, the learning brain acts as a filter that selects and constrains the cultural representations that will be propagated. The child's brain is exquisitely attuned to some environmental features, but can be enormously resistant to others. For instance, normal children, unless they are autistic, do not spontaneously become calendrical calculators who can recite the days of the week for every date, even though our culture, unlike many others, is obsessed with timekeeping. Every child, however, in the first few months of life, quickly learns to recognize faces, voices, native language, and a sense of empathy for others—indeed, these competences can be acquired even under conditions of drastic sensory deprivation. The neuronal recycling hypothesis leads us quite logically to postulate the existence of "cultural attractors," universal foci of competence that are shared by all humans, explain the stability of the major features of human cultures and prevent the drifting that would inevitably take place if children were merely attempting to imitate their peers.

In the human species, cultural selection is further amplified by its intentional character. As stressed by the primatologist David Premack, *Homo sapiens* is the only primate with a sense of pedagogy. Only humans attend to the knowledge and mental states of others in order to teach them. Not only do we actively transmit the cultural objects we find most useful, but—as is particularly apparent with writing—we intentionally perfect them. More than five thousand years ago, the first scribes hit upon an extraordinary potential deeply embedded in our brain circuits: the possibility of conveying language through vision. This initial idea was then perfected by generations of scribes. A long chain of teaching tradition links us to these early writers who worked diligently, from one generation to the next, to make their invention easier to assimilate by our primate visual system.

In the final analysis, according to Dan Sperber, the reproduction of a cultural invention bears more similarity to an epidemic than to an imitation process. An immense range of cultural representations constantly solicit our attention, much as viruses ceaselessly challenge our immune system. Only some of these cultural objects become endemic because they find a resonance in our brains, not unlike a virus that finds a breach in our defense system. Whenever this occurs, we actively and intentionally transmit these cultural inventions to others, at a speed made possible by efficient instruction. Ultimately, the stable cultural representations that define the core of a human group are thus those that can be rapidly incorporated into the architecture of the human brain, because they find an echo in preexisting circuits capable of efficient neuronal recycling.

When a new cultural invention finds its neuronal niche, it can multiply rapidly and invade an entire human group. A new period of cultural stability then ensues, until yet another invention arrives on the scene to disrupt the equilibrium. In a nutshell, this is how cultures appear, proliferate, and ultimately die off.

As we shall see further along, the metaphor of a "cultural epidemic" is particularly well suited to writing, with its primary sources of infection found in the Fertile Crescent (ancient Sumeria), China, and South America, and selective periods of proliferation punctuated by long periods of stasis. If reading has thus taken possession of our minds, to such an extent that it now constitutes an essential feature of our literate culture, it is because it has found its natural cerebral niche in the occipito-temporal cortex and its connections.

The ability of this cortical region to recognize words and transmit their identity to other areas results from a two-step evolutionary process:

- The slow emergence of efficient mechanisms of invariant object recognition that have appeared in the course of mammalian evolution.
- The rapid cultural adaptation of writing systems to fit this cortical niche, in the course of cultural evolution over the past five thousand years.

This view holds that the letterbox area of the brain initially evolved to recognize natural images, but not the shapes of letters or words. Nonetheless, evolution endowed it with a capacity to learn, and thus to turn itself into a reading device. Our writing systems have progressively discovered and exploited the elementary shapes that this region is capable of representing.

In brief, our cortex did not specifically evolve for writing—there was neither the time nor sufficient evolutionary pressure for this to occur. On the contrary, writing evolved to fit the cortex. Our writing systems changed under the constraint that even a primate brain had to find them easy to acquire.

Neurons for Reading

If the neuronal networks initially destined to visual object recognition adapt to reading acquisition, we still have to arrive at a convincing description of how neurons in the letterbox area recognize written words in less than one-fifth of a second. What type of neural code is inscribed in the cortex of an expert reader? Does each letter, syllable, and word get assigned its own neurons? How are these neurons laid out on the cortical surface?

A detailed map of the reader's neurons has yet to be charted. Human brain imaging techniques cannot, at the present time, visualize individual neurons. However, our vast knowledge of other primates' brains allows us to speculate about the nature of the neural code for reading. Although theoretical models often greatly underestimate the true complexity of the nervous system, they can provide a framework for new experimental work and lead to the invention of new imaging studies. It is with this idea in mind that my colleagues and I have proposed a tentative model of the neuronal architecture for reading (figure 3.9).[161]

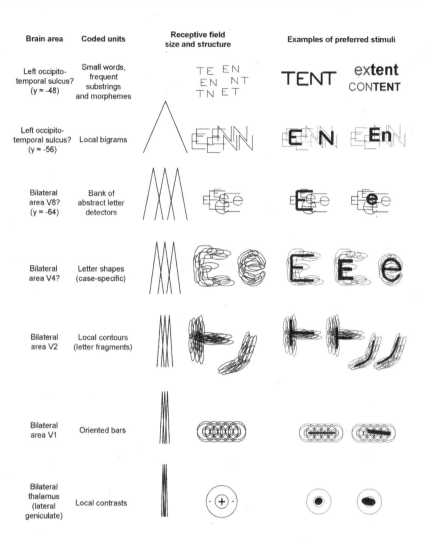

Figure 3.9 A hypothetical model of the neuronal hierarchy that supports visual word recognition. At each stage, neurons learn to react to a conjunction of responses from the immediately lower level. At the bottom of the pyramid, which is shared by word and image recognition, neurons detect local contrasts and oriented bars. As one climbs further up, neurons become increasingly specialized for reading. They detect letters, letter pairs (bigrams), then morphemes and small words. At each stage, the receptive field broadens by a factor of two or three, while the neuronal response becomes increasingly independent of the word's location and of the details of the image (after Dehaene et al., 2005). *Used with permission of* Trends in Cognitive Science.

Our model starts from the well-known fact that the ventral visual system is organized as a hierarchy going from the occipital pole in the back of the brain to the anterior regions of the temporal lobe. As I explained earlier, when one moves from one hierarchical level to the next, the size of the neurons' receptive field increases by a factor of two or three, which means that they respond to increasingly larger portions of the retina. In parallel, the complexity of the features that make the neurons fire also increases, as does their invariance for size, location, and lighting.

We must now imagine how this neuronal architecture might be altered if it were bombarded with written words and had to extract the most salient regularities from the incoming visual flow. At the input level, in the first visual cortical area (V1), neurons are relatively simple: they only recognize lines in a narrow field of the retina. As first discovered by Hubel and Wiesel, each neuron, within the very small retinal zone to which it responds, prefers the sight of a small line to any other visual stimulus. Since letters and words are made up of such lines, it is safe to assume that the acquisition of reading does not fundamentally change the early coding scheme. Most neuronal recycling probably occurs farther downstream, in regions that code for complex properties of the visual image. Nonetheless, it is possible that even the most primary stages of visual processing undergo changes in the brains of expert readers. Shapes such as T and X, because they are very common, may already be processed by area V1. Indeed, laboratory experiments, in both monkeys and human adults, have shown that intensive training can affect even the earliest stages of visual processing in the primary visual cortex.[162] The acquisition of reading, at an early age where the child's brain is even more plastic, is likely to cause similar changes in early visual areas.

The majority of letter shapes, however, are probably encoded by neurons farther downstream, in the next two areas labeled V2 and V4. By combining several elementary lines, neurons in area V2 act as elementary contour detectors. At the next step, in area V4, combinations of these combinations allow neurons to respond selectively to simple shapes that presumably include letters. Even before the acquisition of reading, many neurons already code for shapes such T, L, X, or O. With this alphabet of elementary shapes as a starting point, learning to identify other letters is no doubt a simple matter. Letter acquisition may start in the anterior areas of the ventral temporal cortex, where Tanaka finds cortical columns that respond to learned shapes in the macaque monkey. Later on, as reading becomes more automatic and we learn to read very small print, the identification process may progressively work its way back to the posterior areas of the brain.[163]

If neurons in the V4 area only recognize single conjunctions of curves, they can only code for a single letter shape. How, then, do we learn that the same letter can appear in upper- and lowercase? My colleagues and I speculate that this abstract knowledge does not arise from the V4 area, but is only achieved at the following stage by combining the activation of several V4 shape detectors. Our model assumes that the upper- and lowercase of a letter are recognized as representing the same letter in a different guise—a step that requires cultural learning—in the V8 visual area in both hemispheres. This is the location where our functional MRI experiments revealed the capacity to detect a repetition of the same letters written in upper- or lowercase.

Remember that at each stage, neurons gain spatial tolerance: as they combine responses from many detectors with different receptive fields their responses become progressively less sensitive to changes in location and size (figure 3.9, column farthest to the right). In area V8, however, invariance is incomplete and does not extend to the entire retina. Letter detectors must thus be placed at each location where letters can appear during reading. There are probably dozens of neuronal columns coding for the letter "A," each of which only responds to the letter if it occupies a precise retinal location.

Bigram Neurons

At the next step, when responses from several neurons tuned to letters are combined, we arrive at neurons sensitive to letter conjunctions. Such neurons, for example, might signal the presence of the letter "N" one or two letters to the left of the letter "A"—a very useful feature if one is to separate similar strings such as "AND" and "DNA."

If we follow our simple rule of thumb, the size of the receptive field should be multiplied by two or three at each step. One might therefore expect conjunction neurons to code for groups of one, two, or three letters. Why would pairs of letters be privileged? To answer this question, we have to consider how the nervous system weighs up invariance, selectivity, and the need to maximize information carried by each neuron. A neuron that responds to a triplet of letters should only be able to do so at a single location. It therefore should only convey a limited piece of information, of little use for anything other than a few words (perhaps frequent grammatical words such as "the," "not," or "was"). At the other extreme, a neuron that only codes for a single letter at any of three possible locations should find itself frequently activated, but not very informative, since it would not contribute anything concerning the letter's position in the string.

With this reasoning in mind, my colleagues and I have proposed that the most useful letter combination to which neurons should attend is a "bigram"—an ordered pair of letters such as "E left of N." It is easy to wire a neuron so that it responds selectively to this letter pair but can tolerate some shift in the location of its component letters. As shown in figure 3.9, all that needs to be done is to collect activation from several partially overlapping detectors for the letters "E" and "N," while ensuring that the E's appear largely on the left and the N's on the right. The resulting "EN" detector then has greater location invariance than any of its underlying letter detectors.

If bigram neurons work in this way, they should tolerate the presence of a few intermediate letters inserted in between their two preferred letters. For instance, the bigram detector for "EN" should respond to the words "**enter**," "**rent**," and "**hen**," but also to "**mean**" and "**fern**," in which an intruding letter slips in between the "e" and the "n." This property emerges from the structure of the receptive fields of bigram neurons. In order to accommodate shifts in word location, these neurons must collect inputs from a range of letter detectors at the level immediately below theirs. The receptive fields of these letter detectors are spread over part of the retina, and they have no way of knowing whether one or two other letters have slipped in. Thus bigram neurons should respond to a given letter pair even if it is spatially extended. Their response can be best described as a preference for what might be called an "open bigram," or a specific pair of letters, possibly separated by one or two irrelevant letters.

No one has ever seen bigram neurons. Their existence is the matter of an educated guess, based on what we know about the primate visual system. For the time being, they are a purely theoretical construction that cannot be tested directly with our somewhat rudimentary imaging techniques. Thus bigram neurons are as hypothetical as the neutrino was in 1930 when Pauli postulated the existence of this particle and noted sadly that it was, in principle, impossible to detect (it was subsequently observed twenty-six years later).

Why do serious scientists propose seemingly gratuitous and speculative theories such as these? Even if they are untestable at present, their role is both to lead the way to future research and to bring coherence to past experiments. One of the most rewarding times in scientific endeavor occurs when a scientist realizes, with an unforgettable "aha," that one of his theoretical constructions actually makes a concrete prediction that fits neatly with experimental data. Nothing boosts our confidence more than the unexpected convergence of different lines of research onto the same model. In our case, it turns out that two researchers, Jonathan Grainger and Carol Whitney, working under a totally different premise from ours, have also been led to hypothesize that written words are coded by a list of bigrams.[164] They reached this prediction when they were looking for an abstract code that could explain how we perceive word similarity. However, their experimental observations fit remarkably well with our hypothesis of bigram neurons. Like the neutrino, the hypothesis of bigram neurons would therefore seem to shed light on a number of experimental enigmas and begin to bridge the gap between two previously distinct disciplines, neurophysiology and the psychology of reading.

Grainger and Whitney reported a series of priming experiments where they studied whether the prior presentation of a string of letters speeded up the reading of a second string. For instance, seeing "garden" makes reading "GARDEN" easier. This suggests that the two strings of letters share a common code. In fact, it is not necessary to repeat all a word's letters for priming to occur. The presentation of a subset of letters, like "grdn," is just as effective to expedite "GARDEN" as the word "garden." However, control strings in which letters are inserted or jumbled, such as "gtrdvn" or "dngr," prompt no priming.[165] This means that the letter strings "grdn" and "GARDEN," at some processing stage, share the same code while "dngr" does not. Obviously, this code is insensitive to the deletion of a few letters, but is concerned with their order. Additional experiments show that the code also resists the inversion of two consecutive letters. Thus the string "bagde" accelerates reading of the subsequent word "BADGE" as efficiently as if the word had been repeated.[166]

The preceding observations are particularly damaging to the many models of reading which maintain that strings are coded as arrays of letters.[167] In the context of that scheme, "bagde" and "badge" should not be more similar than "barte" and "badge," since in both cases two letters out of five are different. But this is wrong. The strings "bagde" and "badge" are in fact *very* similar, to the point where they are frequently confused. In a real-life example of this effect, a clothing company called French Connection UK proudly prints the initials "FCUK" on its sweaters and T-shirts. The company owners no doubt found the acronym attractive because they discovered that readers' brains spontaneously linked it to an unprintable English word! These abstract similarity effects are so powerful that we experience little difficulty in raednig etnrie sneetnecs in wihch the ltteers of eervy wrod hvae been miexd up, ecxpet for the frsit and the lsat ltteers.[168]

There must, of course, be some kind of code that is robust enough to resist this alphabet soup. After working hard on the riddle, Grainger and Whitney finally came up with the idea of open bigrams. They noted that words could be encoded not as a list of letters, but as a list of the *pairs* of letters they contained. In the scheme they proposed, the word "badge" is coded by a list of ten bigrams: BA, BD, BG, BE, AD, AG, AE, DG, DE, and GE. If two consecutive letters are switched, as in "bagde," only one of the bigrams is changed (DG becomes GD) and 90 percent of the code remains unchanged. This similarity explains why we can still read the word "bagde" when two of its letters are inverted. When irrelevant letters are inserted or replaced, as in "barte," the code changes more dramatically. The string "barte" only shares 30 percent of its bigrams with "badge," thus explaining the absence of any priming effect. The code is less affected by the inversion or the deletion of a few letters than by the insertion of unacceptable ones. All of the bigrams in the string "grdn" are contained in the word "garden," while less than half of those in "gtrdvn" belong in this word. The resistance of the bigram code to partial deletions and local inversions accounts for our ability to raed snetneecs in which the psoitoin of ltteers is mxeid up.

Another advantage of the bigram code is that it is insensitive to changes in location and size. Even if the word "badge" is moved, enlarged, or reduced, the "b" will remain to the left of the "d," and the "d" to the left of the "e." In other words, bigrams are not affected by the exact location and size of a printed word. This coding scheme guarantees that invariance is respected.

On the basis of all these arguments, Grainger and Whitney proposed that written words are coded by the exhaustive list of their bigrams. Close scrutiny, however, reveals a few problems with this formal proposal. First, it predicts that it should be easy to read words in which the l e t t e r s a r e   s p a c e d, even in an i rr eg ul a r fashion. Indeed, this transformation preserves all the bigrams, and even makes letter order more apparent. Nevertheless, these strings are hard to read. A further hitch is that the list of bigrams is ambiguous since it does not assign a unique code to each and every word. The words "nana" and "anna," for instance, contain the same set of bigrams: AA, NN, AN, and NA. If our visual system only noticed bigrams, we should not be able to distinguish these words. We should even readily accept improbable spelling like "naananan" that has the same bigrams as the word "anna."

It is at this point that the biological properties of bigram neurons come to the rescue of Grainger and Whitney's formal proposal. A bigram neuron cannot respond to its preferred letter pair irrespective of where it appears on the retina. Like any other neuron, its receptive field is limited. Thus the neuronal bigram code has spatial selectivity and must be replicated at several points in the visual field. For instance, some bigram neurons react to the letter pair NA at the beginning of "nana," but not at the end—and therefore not to the stimulus "anna" where the letters "na" fall outside their receptive fields. Collectively, bigram neurons assign a unique code to each specific word.

The local bigram detector hypothesis also explains why the spacing of letters leads to slower reading. The neurons' receptive fields increase only by a factor of two or three at each step up the visual pyramid. As a result, the letters in bigram detectors can only tolerate a small shift of about two or three letter positions (see figure 3.9). Thanks to their limited receptive field, bigram neurons only fire if the first letter of a pair is less than two letters away from the second. For instance, a neuron coding for the pair AM can react to the words "ham," "arm," and "atom," but not to "alarm" or "atrium."

This reasoning leads to a very simple prediction: as we progressively s p a c e the letters in a word, reading performance should be resistant to a modest amount of spacing, but should collapse when spacing reaches two characters. Fabien Vinckier, Laurent Cohen, and I have verified this rule. As soon as letter spacing exceeds two characters, word recognition abruptly ceases to be fast and parallel.[169] You can test this for yourself in the following sentences, where letter spacing slowly increases. Y o u r   b i g r a m   n e u r o n s   e a s i l y t o l e r a t e   a   s p a c i n g   o f   o n e   l e t t e r,   o r   e v e n   a l e t t e r   a n d   a   h a l f .   H o w e v e r ,

```
a   s       s   o   o   n       a   s       t   h   e       s   p   a   c   i   n   g
e   x   c   e   e   d   s       t   w   o       c   h   a   r   a   c   t   e   r   s   ,
y   o   u   r       p   e   r   f   o   r   m   a   n   c   e       c   o   l   -
l   a   p   s   e   s   :       r   a   p   i   d       r   e   a   d   i   n   g
b   e   c   o   m   e   s       i   m   p   o   s   s   i   b   l   e   .       Y   o   u
f   i   n   d       y   o   u   r   s   e   l   f       i   n       t   h   e
s   h   o   e   s       o   f       a       b   e   g   i   n   n   i   n   g
r   e   a   d   e   r   —   t   h   e       b   i   g   r   a   m       n   e   u   -
r   o   n   s       i   n       y   o   u   r       l   e   t   t   e   r       b   o   x
a   r   e   a       h   a   v   e       s   t   o   p   p   e   d
r   e   s   p   o   n   d   i   n   g   !
```

A Neuronal Word Tree

In the event that bigram neurons do exist, what kind of neuronal code should we expect at the next stage up the visual pyramid? By putting together several bigrams, the neurons higher up in the hierarchy should respond to complex combinations of letters. Their receptive fields should allow them to detect strings of up to five letters in the reader's language. Which strings should be preferentially encoded remains largely a matter of speculation. In chapter 1, I explained how reading is organized along two parallel pathways, the spelling-to-sound and spelling-to-meaning routes. Higher-order visual neurons should therefore extract letter strings that are meaningful to one route or the other. Some neurons should be concerned primarily with frequent graphemes that map onto specific sound patterns, such as "ough," "ain," or "ing." Others should react to short grammatical words ("but," "then," "have"), word roots ("think-," "giv-"), prefixes ("anti-," "pre-") or suffixes ("-ing," "-tion," "-ese"). At this level, it would be reasonable to expect a cerebral coding of morphemes, the smallest linguistic units to have semantic meaning

We must bear in mind, however, that at this stage the visual system cares only about spelling. It merely detects frequent letter strings that convey information needed for comprehension and pronunciation. The neuronal model that I am sketching thus predicts that "pseudo-morphemes"—strings of characters that look like word roots but in fact bear no relation to word meaning—should be extracted just like real morphemes. Our visual system should blindly break down the word "department" into the prefix "de-," the root "part," and the suffix "-ment," and it should file it with words like "part," "depart," and "departure," even if these words do not have the same meaning. A great many experiments, both in French and in English, confirm that this kind of blind decomposition occurs: the extraction of an approximate word tree is a key automatic and unconscious step in visual word recognition, to such an extent that the presentation of "hard" can prime the reading of a semantically unrelated but superficially similar word like "hardly."[170]

In chapter 1, I stressed that words behave like trees, where the leaves are letters and the branches the various combinations of letters extracted by our visual system. This tree diagram now takes on a very literal meaning as it coincides with the anatomical tree made up of the millions of neurons in the letterbox area and their billions of interconnections. Each of them makes its small contribution to word recognition, from initial line detectors up to the neurons sensitive to bigrams and morphemes. At no point in this cortical hierarchy is a word ever represented by a single neuron. At each level, a sizable set of active cells represents the vast collection of features that collectively paints a pointillist portrait of the printed word. Like a hundred-year-old oak tree, this neuronal tree is well structured but flexible and built to resist changes in its natural environment. A word shift on the retina is like a gust of wind: it shuffles the leaves and the small branches, but leaves the trunk and main branches entirely unaffected. In this way, the neuronal tree that represents a word manages to remain invariant in the face of changes in size, position and shape, and even the local ordering of its component letters.

The neuronal tree that I sketched in figure 3.9, with its neat hierarchical structure, is probably too simple for the burgeoning redundancy of our nervous system. For the sake of simplicity, I only included excitatory connections, which would imply that neurons only excite each other. In fact, some neurons also inhibit their neighbors and suppress their firing. Inhibitory connections could play an important role in word recognition—for instance, they may signal the absence of a given letter in the input string. Furthermore, I assumed that each neuron only connects to the level immediately above, in an ever-ascending, feed-forward manner. In the real-life nervous system, however, other connections, called "lateral" or "horizontal," are known to link neurons *within* a given visual area. In the primary area V1, these connections help active neurons support each other when they all vote for the same visual interpretation (for instance, when neurons code for different portions of the same contour). At the letter and bigram levels, such lateral connections may encode the probabilistic relations between consecutive letters, such as the fact that the letter "q" is almost always followed by the letter "u."

Finally, we should not forget that the visual cortex contains a huge number of "descending" projections that go from a given area to the *preceding* one in the visual hierarchy. Those feedback connections may preselect certain combinations of letters depending on context. They may help to remove certain ambiguities, particularly in deciphering handwriting. Handwritten words often cannot be decoded serially, by first recognizing the letters and then assembling them into words. Remember the *honey bees' sweet nectar,* in which the letters "e" and "c" have exactly the same shape in "bee" and "nectar." In such cases, letter identification comes after word identification and not before. With the support of sentence context, the word "nectar" probably sends a top-down vote in favor of the letters "n," "e," and "c."[171] The constant interplay of bottom-up and top-down connections provides a neuronal implementation of Selfridge's "pandemonium," which we examined in chapter 1. Populations of neurons literally act as an assembly of daemons that constantly send messages in all directions, thus passing on the fragmentary data at their disposal to each other until the whole group converges toward an agreement.

How Many Neurons for Reading?

The hierarchical model I just sketched might appear to lead to an explosion in the number of neurons required for reading. If each stage in our visual system represents all the possible neuronal combinations occurring at the preceding stage, the number of neurons required might logically appear to increase beyond all reasonable bounds. Luckily this does not in fact need to be the case.

Simple solutions to the combinatorial problem posed by my model do exist. On the one hand, at each processing stage, neurons only need to extract letter combinations that occur frequently and are relevant to reading—a small subset of all possible combinations of letters. At the bigram level, for example, there should be detectors for the letter pair EN, which is usual in English, but it seems unlikely that the visual system would dedicate neurons to pairs of letters that never occur, such as QR or JB. In support of this hypothesis, a frequency bias has recently been observed with functional MRI—the more frequent the bigram, the stronger the activation in the letterbox area.[172] It could also explain why this area responds more to words and plausible pseudowords such as "enchible" than to random strings of consonants such as "zhgsieq."[173]

On the other hand, as neurons respond to increasingly complex combinations of visual features, their invariance is greater and they therefore drop other distinctions that are irrelevant for reading. The discrepancy between uppercase and lowercase letters, for instance, ceases to matter at an early stage in the visual stream (presumably at area V8 and beyond). For bigram neurons, likewise, the larger number of possible letter pairs would be offset by the decrease in the number of positions encoded, because the receptive fields of the neurons increase at each step.

We should also bear in mind that eye movements, in the course of reading, always draw the relevant words into a narrow area of the visual field, close to the fovea and mostly to the right of it. When we learn to read, only the neurons that code for these locations are given the opportunity to convert to letter and bigram detectors. Indeed, only words presented at the center of gaze, or slightly to the right of it, and at an angle close to horizontal, are efficiently processed by the ventral occipito-temporal pathway.[174] Thus only a limited number of neurons are concerned.

In the end, can we come up with a rough estimate of the amount of cortex needed for visual word recognition? If we consider that there are twenty-six letters in the Western alphabet and approximately twenty retinal locations where each of these letters can appear during reading, roughly five hundred columns of neurons are needed for the initial bank of letter detectors. Assuming a conservative value of ten columns per square millimeter of cortex, this processing stage should fit into half a square centimeter of cortex. At the next level, assuming that only three approximate locations are coded by bigram neurons (word beginning, middle, or end), and that the two hundred most common bigrams are enough to represent the majority of words, we reach a similar estimate of about six hundred columns of bigram neurons. Again, these would occupy a cortical patch of roughly half a square centimeter. Finally, at the next level up, where bigrams are combined into longer strings, it seems plausible that a collection of about five hundred prefixes, suffixes, roots, and complex graphemes suffices to represent the few tens of thousands of words whose spelling is stored. Here again, this would mean half a square centimeter of cortex.

These orders of magnitude, although very rough, make a simple prediction: whether words are encoded by letters, by bigrams, or by morphemes, each of these representations should occupy about the same cortical surface— a small patch of half a square centimeter, which corresponds roughly to a disc of about eight millimeters diameter. This, in turn, predicts that the cortical activation evoked by a word should move forward in the visual brain by approximately eight millimeters as the information progresses from one level to the next (figure 3.9). Strikingly, brain imaging confirms the existence of a spatial progression. Reading activates a narrow band of cortex, several centimeters long, extending from the back of the brain to the front of the left occipito-temporal sulcus. Functional subdivisions have now been detected in this strip.[175] The neuronal code clearly becomes more abstract as it progresses toward the front of the brain.

My estimate of the number of neurons needed for reading is obviously very rough. It fails to take into account several factors. According to Keiji Tanaka, neighboring neurons *within* a cortical column can code for variants of the same basic shape. For instance, nearby neurons in the monkey might code for a disc, an ellipse, and a potato-shaped object. If this idea is extended to reading, it might predict the existence of "combinatorial columns," where the various neurons would code, for instance, for all the possible combinations of a given consonant and different vowels (ta, te, ti, to, tu . . .) or vice versa (sa, na, ta, ra, la, ca . . .). If this were the case, the cortical code would be so compact that only a few dozen cortical columns would be enough to represent any written word. Since it would occupy less cortical space, this solution would allow for an intermingling of cortical columns for words, objects, and faces. Brain imaging and intracranial recordings indeed suggest that the preference of a cortical region is rarely categorical.[176] The letterbox area thus probably includes a mixture of cortical columns, some of which respond to letters while others also react to faces, tools, and a variety of other visual shapes.[177]

In brief, the scheme that I have proposed is compatible with the amount of cortex available for visual word recognition. An explosion in the number of neurons dedicated to reading can be avoided. All that is needed is an efficient learning mechanism that allocates neurons sparingly to the most frequent, informative, and invariant combinations of letters.

Simulating the Reader's Cortex

A precise picture of the neuronal architecture for reading will probably only be achieved by simulating reading acquisition on a computer whose architecture resembles that of the cortex. At present, however, such a simulation remains an inaccessible dream. A supercomputer far more powerful than any machine available today would be needed to duplicate the impact of thousands of hours of reading on the millions of neurons in the letterbox area. For the time being, we can only hope that future progress in computing power, together with the emergence of new computers dedicated to neural network simulations, will ultimately lead to a proper simulation of the reader's brain.

In the meantime, it is worth reflecting on how a simulation could incorporate all the world's many different writing systems. In principle, the brain architecture for Western writing should apply just as well to Chinese writing. When exposed to Chinese, it should develop neurons attuned to frequent characters as well as to their internal components. Indeed, most Chinese characters include a small number of semantic and phonetic markers that are in turn made up of a few standardized strokes—a nested set of visual regularities that could easily be captured by a hierarchical learning algorithm. Behavioral experiments already show that Chinese reading rests on such a hierarchical and combinatorial scheme.[178] Chinese equivalents for bigram neurons, however, remain to be discovered.

Even among readers of the Western alphabet, neurons probably become sensitive to units of different sizes. Languages where spelling is transparent, such as Italian, only require a modest bank of letters and bigram detectors. Languages where spelling is opaque, however, such as English or French, need a more extensive set of neurons in order to encode spelling units as large and as complex as "ough," "tion," or "ould," which map onto speech sounds.[179] Such visual regularities, which go beyond the level of letters, can only be extracted by neurons positioned high in the visual hierarchy. Thus our model predicts that the reading of English should call on more extended and more anterior brain regions than reading Italian. This is exactly what Eraldo Paulesu and his colleagues observed, as we saw in figure 2.20.[180] Similarly, in Japanese, differences in brain activity for the Kanji (which requires memorizing thousands of characters) and Kana (which only contains forty-six syllables) writing systems could be attributed to the learning of visual inventories whose size and complexity are strikingly different.[181]

Cortical Biases That Shape Reading

Although we now know that reading acquisition systematically converges onto the same region of the cortex, the reason for this precise localization is still shrouded in mystery. Why is the letterbox area systematically located in the left hemisphere and why is this true for all readers and all writing systems the world over? What allows this area to be segregated from the other areas involved in object and face recognition?

Although it is endowed with considerable plasticity, the visual cortex should not be thought of as a "blank slate" on which learning is imprinted.[182] The very fact that it is organized into reproducible subterritories, whose layout is identical in all humans, implies that our cortex is not equipotent. Early predispositions must shape the reader's visual brain and explain the consistent localization of reading. Even before they recycle into letter and bigram detectors, these neurons probably share intrinsic properties of visual sensitivity and of projection toward language areas that make them particularly appropriate for reading.

A first bias was recently identified by Uri Hasson, Rafi Malach, and their colleagues at the Weizmann Institute.[183] Using brain imaging, they showed that preference for central versus peripheral images differs from one visual region to another. A beautiful gradient of preference, five or six centimeters long, cuts across the whole ventral visual cortex (figure 3.10). Each point on the visual cortex, even quite far from the primary visual area, has a characteristic preference for a given location on the retina. In both hemispheres, regions located on the sides of the brain prefer fine visual stimuli presented close to the fovea, while regions closer to the brain's midline prefer input from the periphery of the visual field.

The origins of this gradient remain unknown, but its sheer size and smoothness are reminiscent of the gradients that form during embryonic

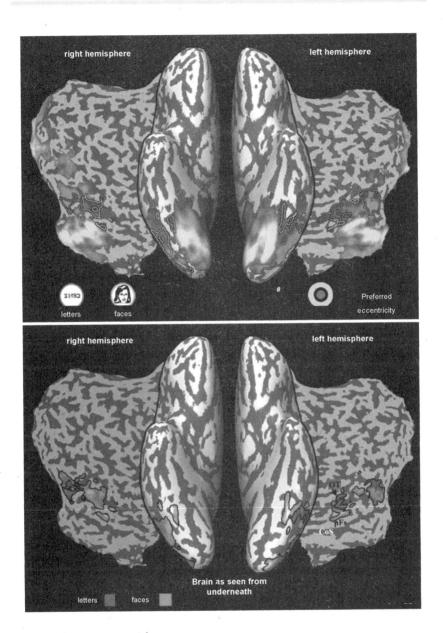

Figure 3.10 Gradients of visual specialization may explain why the occipito-temporal "letterbox" area is always found at the same cortical location. A major gradient spans the entire length of the ventral visual cortex (top): lateral regions prefer images presented at the center of the retina, the fovea (in dark gray), while regions closer to the brain's midline prefer images presented on the periphery (in light gray). Reading and face recognition, both of which require minute analysis, land in the cortex biased toward the fovea (after Hasson et al., 2002). *Used with permission of* Neuron.

development. Think, for instance, of how the size of the vertebrae smoothly decreases as one moves down an animal's spine toward the tail. This type of smooth variation is thought to result from a classic biological mechanism—a change in the concentration of chemical messages called morphogens, which diffuse through a developing organism and specify its basic spatial plan.[184] The concept dates back to the mathematician Alan Turing, who was intrigued by the beautiful shapes that can be found in nature. Turing demonstrated mathematically that if substances react chemically as they spread through biological tissue, they can induce spatially organized patterns. When a morphogen diffuses, starting at a fixed spatial location, it creates a gradient of decreasing concentration that serves as a surrogate spatial marker. For instance, if morphogen levels were higher at the center of the brain than on its sides, and if this concentration modulated the attraction of nerve fibers coming from the fovea, the result would be a genetically biased network. This system would preferentially process high-resolution stimuli using the lateral regions of cortex—precisely the spatial layout that was observed by Uri Hasson and his colleagues.

How does this retinal preference gradient relate to reading? Hasson observed that in all individuals, the letterbox area always falls at a precise location on this gradient—a place characterized by its massive preference for fine-grained foveal images. The nearby region that prefers faces is also biased toward the fovea, whereas the region that is concerned with houses and outdoor scenes prefers the periphery of the visual field.

From these observations, we can arrive at a simple scenario that explains why reading develops where it does. From birth on, visual neurons are already biased—some are best suited to learning fine-grained visual details, thanks to their preferred connection to the fovea, while others prefer coarse-grained stimuli. The finer-grained neurons have a decided advantage over other neurons for the recognition of letter strings, where a high degree of visual precision is required. As a result, the lateral regions of visual cortex, where these neurons are more numerous, preferentially engage in reading.

The gradient of retinal preference, however, provides only one of the three axes of a coordinate system that specifies the location of the letterbox area. Other biases are needed to single out this area. As we have already discussed, there is a second innate gradient, going from back to front, which roughly defines the hierarchical level of each visual area (figure 3.5). Back in the occipital pole, neurons respond to simple fragments of the input image, while more anterior sectors of cortex prefer increasingly complex and structured objects.[185] In this respect, it is not surprising to find that the visual word form area occupies a relatively extended strip of cortex, whose back end responds to simple letters while the front responds to complex word fragments.[186]

Finally, another source of bias may come from the distinct properties of the two hemispheres. Why does reading rely on the left visual brain, while faces always preferentially land in the right hemisphere? This lack of symmetry, or "symmetry-breaking" as they say in physics, could be of visual origin. It is well known that the left hemisphere has the upper hand in the discrimination of small local shapes, while the right preferentially deals with global shapes.[187] Another factor might be the lateralization of language. Even at birth, speech processing preferentially engages temporal and frontal regions of the left hemisphere. Reading acquisition probably selects visual regions whose projections to the language areas are abundant and direct. The left ventral temporal region would thus have an edge over its right-hemisphere twin, whose axons have to make a lengthy detour through the corpus callosum before reaching the left hemisphere.

The hierarchical model in figure 3.9 tentatively proposes that the lateralization of visual word recognition to the left hemisphere is a progressive process. Low-level regions have to be bilateral, because early visual neurons in each hemisphere only respond to letters presented in the opposite half of the visual field. As progress is made into the brain, the receptive fields broaden and one begins to find neurons capable of responding to either side of the visual field in both hemispheres. I believe that whenever there is this potential redundancy between the hemispheres during reading acquisition, visual and linguistic biases cause the left hemisphere to gain ascendancy over the right. As a result, in the progress up the visual hierarchy, an increasing number of neurons specialized for reading should be found in the left hemisphere.

If my scenario about innate cortical biases is correct, there is no prewired area for reading, but several genetic biases create a gamut of neuronal preferences for different types of visual stimuli. During reading acquisition, visual word recognition simply lands in the cortical location where neurons are most efficient at this task. In all humans, the intersection of genetic gradients creates a single "sweet spot" for letter strings—the letterbox area.

What happens if this area, for one reason or another, ceases to be available?

Lesion caused by a surgical intervention at the age of four

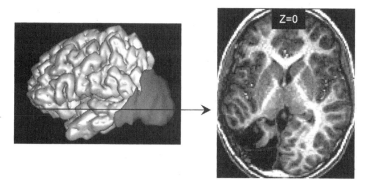

Reading network observed at the age of eleven

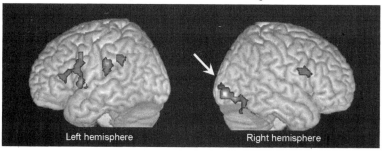

Displacement of the visual word form area

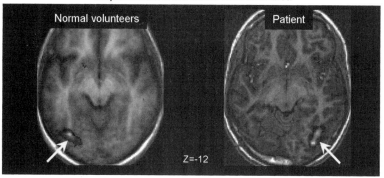

Figure 3.11 During development, children's reading networks are much more flexible than those of adults. In this young patient, the area normally associated with visual word recognition, in the left occipito-temporal region, was removed surgically at the age of four (the lesion is clearly visible in the top images). In any adult, such a lesion would cause a severe reading impairment. Yet this young girl learned to read without much difficulty. Seven years later, the reading network activated normally, but the visual word form area had swapped hemispheres: it now lay within the right occipito-temporal region, at a location exactly symmetrical to its usual place (bottom) (after Cohen et al., 2004). *Used with permission of* Annals of Neurology.

Learning should ensure that other less efficient neurons take over. My colleague Laurent Cohen and I recently tested this idea on a young patient. At the age of four, this girl had suffered from epileptic fits so severe that they necessitated the removal of the entire left visual brain. The surgical lesion clearly encompassed the normal location of the letterbox area (figure 3.11). In spite of this major surgery, the girl subsequently learned to read normally, although she was slightly slower than normal. When we scanned her brain at the age of eleven, we finally understood how she had managed to learn to read so well. Although all of her spoken language areas remained intact in the left hemisphere, she compensated for the loss of the letterbox area by recognizing written words using the right hemisphere, which had not been damaged by surgery. Furthermore, reading selectively activated a region exactly symmetrical to that of normal readers: in the lateral occipito-temporal sulcus, at the same location as the normal letterbox area on both the lateral and anterior-posterior axes—but in the right hemisphere (figure 3.11).[188] Her letterbox area had simply swapped hemispheres!

Other researchers have observed a similar activation of the right hemisphere in a "hyperlexic" child who was remarkably precocious in learning to read.[189] A right-sided visual word form area is also seen in adult patients with pure alexia who recover from a stroke affecting the left visual system.[190] Thus the contribution of this region to reading is not exceptional. The visual system behaves as if its biases implemented a kind of "waiting list" of areas best suited for reading. If the optimal region is no longer available, the second on the list—which appears to be the symmetrical area in the right hemisphere—takes over.

No other experimental finding could better illustrate the fundamental conclusion that there is no predefined area that evolved for reading. Learning to read involves tinkering with our primate brain. Using all the means available, our brain recycles the best-suited regions of our visual cortex to the novel task of recognizing words.

CHAPTER 4

Inventing Reading

The neuronal recycling hypothesis implies that our brain architecture constrains the way we read. Indeed, vestiges of these biological constraints can be found in the history of writing systems. In spite of their apparent diversity, all share a great many common features that reflect how visual information is encoded in our cortex. The neuroscience of reading sheds new light on the twisted historical path that finally led to the alphabet as we know it. We can consider it as a massive selection process: over time, scribes developed increasingly efficient notations that fitted the organization of our brains. In brief, our cortex did not specifically evolve for writing. Rather, writing evolved to fit the cortex.

> At the Egyptian city of Naucratis, there was a famous old god whose
> name was Theuth. . . . He was the inventor of many arts, such as
> arithmetic, calculation, geometry, astronomy, draughts and dice, but his
> great discovery was the use of letters.
>
> —PLATO, *PHAEDRUS*

 he idea that writing is a gift of God recurs in cultures all around the world. The Babylonians thought that all forms of magic, including writing, came from Ea, the god of all wisdom. Assyrians revered Nabu, son of Marduk, who taught mankind arts and crafts, ranging from architecture to writing. In the Hindu religion, Ganesh, the elephant-headed god of wisdom, invented writing—he even broke one of his tusks to use it as a pencil! In the Bible it was Yahweh himself who handed Moses the Ten Commandments written in his holy hand.

Writing has a very magical quality—not because of anything divine about its origins, but because it greatly increased our brain's capacities. It is close to miraculous that *Homo sapiens*, a mere primate, was able to dramatically increase its memory by making a few marks on paper. This transformation was not predestined. Sheer luck gave us a cerebral network that links visual and language areas and is plastic enough to recycle itself and recognize the shapes of letters. Even this recycling process is tightly constrained—only a localized circuit seems to possess the optimal properties needed for reading. Even after this circuit has been converted to reading, it still possesses most of the properties inherited from its evolution. Biological inertia keeps the upper hand over cultural innovation.

My proposed neuronal recycling hypothesis thus leads to a radical conclusion: brain constraints have marked the history of writing, and continue to have an impact on its acquisition. Our knowledge about the cortical circuits for reading should allow us to shed some light on the invention and learning of reading. This hypothesis leads us to ask two new questions: How did humans discover that their visual cortex could be turned into a text comprehension device? How does this recycling process recur in the brain every time a child learns to read?

In the next three chapters of this book, I attempt to address these issues and examine three simple but profound consequences of the neuronal recycling hypothesis:

1. **The evolution of writing.** If our brain organization places a drastic limit on cultural variations, some striking cross-cultural regularities should be apparent in all past and present writing systems. These regularities should ultimately be traced back to cerebral constraints.

2. **The evolution of human abilities.** Learning to read and write dramatically increased the capacity of human memory and the volume of communication between humans, but this development may also have a price. If competition exists between new cultural acquisitions and evolutionarily older ones, do we lose some of our capacities when we learn to read?

3. **The acquisition of reading.** The neuronal recycling model also predicts that the ease with which children learn to read may depend on the amount of cortical recycling required, and on the way teaching methods fit with the structure of our cerebral networks.

The Universal Features of Writing Systems

"Let's make the ssh-noise of a snake. Will this do?" And she drew this:

"There," she said. "That's another s'prise-secret. When you draw a hissy-snake by the door of your little back-cave where you mend the spears, I'll know you're thinking hard; and I'll come in most mousy-quiet. And if you draw it on a tree by the river when you are fishing, I'll know you want me to walk most most mousy-quiet, so as not to shake the banks."

"Perfectly true," said Tegumai. "And there's more in this game than you think. Taffy, dear, I've a notion that your Daddy's daughter has hit upon the finest thing that there ever was since the Tribe of Tegumai took to using shark's teeth instead of flints for their spearheads. I believe we've found out the big secret of the world."

—RUDYARD KIPLING, "HOW THE ALPHABET WAS MADE"

We do not generally ask ourselves how our letters got their shapes. Is there a Kipling-like "just-so story" behind every writing system? Are the shapes of individual characters simply a matter of historical contingency, or do their universal features reflect cerebral organization? At first glance, the enormous diversity of the world's writing systems seems to provide an easy answer (figure 4.1). What properties are common to the roundness of Indian scripts, the geometrical rigor of Greek uppercase letters, the curves and dots of Arabic writing, and the diversity of Chinese characters whose simple strokes are confined to a virtual square?

The neuronal recycling hypothesis predicts that human creativity is bounded by brain architecture—an idea that stands in sharp contrast to cultural relativism, which considers that cultural variations are essentially unlimited. If there is any truth to neuronal recycling, our genetic makeup severely curbs the set of writing systems that can be learned. This prediction leads us to question the apparent boundlessness of cultural diversity. If we scratch the surface of the more obvious superficial cultural variations, we should uncover evidence for universally shared deep structures.[191]

In this new light, the diversity of the existing writing systems bears examination. Even a cursory look reveals that they share a great many features:

- All provide the retina's fovea with a high-density concentration of optimally contrasted black-on-white marks. This format probably optimizes the amount of visual information that our retina and visual areas can transmit in a single eye fixation.
- All rely on a small inventory of basic shapes whose hierarchical combinations generate sounds, syllables, or entire words. Chinese and Kanji char-

Greek

α β γ δ ε ζ η θ ι κ Α Β Γ Δ Ε Ζ Η Θ Ι Κ

Cyrillic

а б в г д е ж з и ф А Б В Г Д Е Ж З И К

Armenian

Ա Բ Գ Դ Ե Զ Է Ը Թ Ժ Ի Լ Խ Ծ Կ Հ Ձ Ղ Ճ Մ

Hebrew

פ ק ץ ע ס נ ן מ ם ל כ ך י ט ח ז ו ה ד ג ב א

Arabic

ء ؤ ئ ة ث ب ج ز س ص ط ع ف ل م ن ه و ى

Hieroglyphs

Devanagari (India)

अ आ इ ई उ ऊ ऋ ॡ ऍ ऑ क ख ग घ ङ द ध ल व ह

Bengali (India)

অ আ ই ঈ খ ৯ এ ও ক খ গ ঘ ঙ ছ ঞ ট ঠ ন ত থ

Oriya (India)

ଃ ଅ ଆ ଇ ଉ ର ଊ ଏ ଓ କ ଖ ଘ ଙ ଛ ଟ ଡ ଠ ଥ ଦ ଲ

Tamil (India)

அ இ ஈ உ எ ஜ க ங ண த ம ந ழ வ ள ரு அ

Thai

ก ข ค ฅ ฆ ง จ ฉ ฑ ด ถ ท ธ ป ผ ภ ร ล ศ ใ ฦ แ

Hangul (Korea)

人 ㅒ ㅎ ㅔ ㅎ ㅺ ㅅ ㅆ △ ㅇㅁ ㅉ ㅊ ㅍ ㅎㅎ ㅚ ㆍ

Kana (Japan)

ア イ オ ケ コ サ ス シ タ ッ テ ナ ハ フ カ モ ヱ ヶ

Chinese

专 乇 亇 亣 六 古 土 坷 喚 壇 涼 焱 屟 嵩 麃 惏

Figure 4.1 In spite of their obvious diversity, all writing systems share numerous visual features—highly contrasted contours, an average number of about three strokes per character, and a reduced lexicon of shapes that constantly recur, even in unrelated cultures.

acters are no exception—even if there are many thousands of them, each character is made up of only two, three, or four basic shapes, in turn composed of a few strokes. This hierarchical organization fits nicely into our visual system's pyramid of cortical areas. Vision neurons use a combinatorial principle to encode units of increasing size and invariance.

- All writing systems assume that the location and size of characters is irrelevant. No culture ever teaches this principle to children: all are born with cortical mechanisms for translation and size invariance. That is not the case with rotation, however. All writing systems impose a specific orientation to reading. This universal feature is probably due to the ancient scribes' awareness that invariance for rotation is limited. Because our visual neurons only tolerate about 40 degrees of rotation, we could never learn to read efficiently in all orientations without first assigning a prohibitively large number of additional neurons to each viewing angle beyond 40 degrees.

- All writing systems, finally, tend to jointly represent sound and meaning. It is as though the ancient scribes were aware that our letterbox area's connections make it a hub that projects shape information both toward the superior temporal regions coding for speech sounds and to the middle and anterior temporal regions coding for meaning. Whether sound is privileged rather than meaning is one of the main sources of differences between writing systems (see figure 2.20). In all of them, there is always some statistical correlation between written marks and speech sounds—but the size of the speech unit transcribed ranges from whole words (in Chinese or Japanese Kanji) to syllables (in Japanese Kana), phonemes (in alphabetic writing systems), or even isolated phonetic features (in the case of Korean Hangul writing). Brain physiology does not regulate this domain, but the choice of the speech unit to be rendered ultimately determines the number of written symbols and hence the complexities of reading acquisition.

A Golden Section for Writing Systems

Marc Changizi and his Caltech colleague Shinsuke Shimojo have undertaken a careful analysis of the cross-cultural regularities in the world's writing systems.

They studied the detailed visual organization of each character in a total of 115 writing systems of all styles and from all ages. Their work even looked back in time at the "Linear B" characters of ancient Crete and the old Scandinavian runes, as well as the Etruscan alphabet and the International Phonetic Alphabet.[192] The analysis revealed many unsuspected regularities. First, as is clear in figure 4.1, most characters are composed of roughly three strokes (curves that can be traced without ever lifting or stopping the pen). Variability around this mean is rather low—our capital letters, for instance, have either one stroke (C, I, J, O, S, U), two strokes (D, G, L, P, Q, T, V, X), three strokes (A, B, F, H, K, N, R, Y, Z), or four strokes (E, M, W), but never more. When a writing system requires more signs—Etruscan had twenty-three letters, the International Phonetic Alphabet counts 170—new signs are created by inventing more basic strokes so that the mean stays close to three strokes per character. The International Phonetic Alphabet, for instance, uses characters like ɕ, ɜ˞, ɲ, ɷ, ʃ, and ʉ. Although they may seem exotic to us, they are no more complicated than our letters. They simply rely on a different set of primitive curves.

I would like to propose that the magic formula of three strokes per character was chosen by our forefathers because it corresponds to the way in which the neurons' receptive field increases across the hierarchy of visual areas. As we saw in chapter 3, the receptive field grows by a factor of two or three at each step in the cortical pyramid. Furthermore, the size, complexity, and invariance of the encoded visual units also increase along with it, in the same proportion. In all writing systems, the world over, characters appear to have evolved to an almost optimal combination that can easily be grasped by a single neuron, through the convergence of inputs from two, three, or four types of curve-detecting neurons at a level immediately preceding it in the pyramid.

Because this scheme is replicated at all levels in the visual system, it would be tempting to extend Changizi and Shimojo's analysis to other levels of the cortical hierarchy. At one step lower in the visual system, it would seem plausible that each stroke in itself consists in a conjunction of two, three, or four line segments. One step higher, in our alphabetic languages, multiletter units such as word roots, prefixes, suffixes, and grammatical endings are almost invariably two, three, or four letters long. Similarly in Chinese, most characters consist in a combination of two, three, or four semantic and phonetic subunits. Visually speaking, all writing systems seem to rely on a pyramid of shapes whose golden section is the number 3 plus or minus 1.

Artificial Signs and Natural Shapes

Marc Changizi also discovered a second major trait common to all the world's writing systems. In all of them, the arrangements or configurations of individual strokes tend to be the same. Their frequency follows a universal distribution that closely parallels the features of natural scenes.[193]

To understand Changizi's law, we must take a look at how the strokes of a pen meet to shape distinct characters. If we disregard orientation, two strokes always make a **T**, an **L** or an **X**. Three strokes can make a great many configurations like **F**, **K**, **Y**, **Δ** . . . Changizi proposed that these shapes be considered from a purely topological standpoint, barring rotation or distortion. He then simply counted how many times each of them occurred on a page—are there more **Δ**'s or more **Y**'s? An elegant regularity appeared: in all writing systems, the frequency with which the different configurations were observed was a constant (see figure 4.2). For instance, **L** and **T** shapes always occurred most often, followed by **X** and **F**, which in turn were much more common than **Y** or **Δ**.

This universal distribution of signs is not a matter of luck. When sticks are randomly thrown on the ground their intersections do not obey Changizi's frequency law—**X**'s, for instance, are much more common than **L**'s and **T**'s, unlike what is seen in writing.

Surprisingly, however, the Changizi distribution is replicated when one counts how often these shapes are found in the natural world. When objects touch each other, are superimposed, or hide one another, their outlines frequently arrive at configurations like **T** and **L**. An **X** is not as frequent, unless a twig crosses a branch or a line. It is even more unusual to find three lines forming a **Δ**. When the number of configurations are averaged over hundreds of images, they neatly correlate with the universal distribution of written symbols (figure 4.2).

In summary, whether by design or thanks to some extraordinary intuition, the first scribes appear to have been aware, from the beginning, that the shapes they chose should be the easiest to read. Everywhere on the planet, they appear to have settled on characters whose shapes resemble those found in the environment—and are thus easily represented by our brains. This extraordinary regularity fits nicely with the prediction of the neuronal recy-

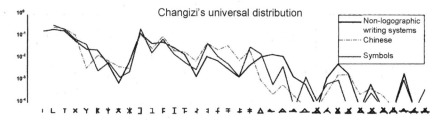

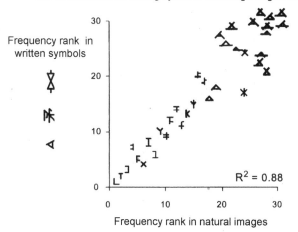

Figure 4.2 All writing systems, whether alphabetical, syllabic, or logographic, rest upon a small set of configurations whose frequencies trace out a universal curve (top). The most frequent configurations are also those that recur most often in natural images (bottom), many of which tend to be coded by inferior temporal neurons (after Changizi et al., 2006).

cling hypothesis. All cultures select signs whose learning requires minimal cortical change. During evolution as well as in the first years of life, our neurons became finely attuned to the characteristic configurations of the environment. Later, writing took the same route. Every time a new system is invented, it converges by trial and error onto the characteristic shapes of "proto-letters" that are already coded deep in the primate's visual cortex.[194]

Overall, the analysis of writing systems underlines the fact that letter shape is not an arbitrary cultural choice. The brain constrains the design of an efficient writing system so severely that there is little room for cultural relativism. Our primate brain only accepts a limited set of written shapes.

Prehistoric Precursors of Writing

How did humanity find out that its visual system could be recycled to code speech in writing? Although we cannot, in this book, dig deeply into the paleoanthropology and history of writing, it is nonetheless interesting to out-line how this extraordinary invention developed. Indeed, the neuronal recy-cling hypothesis sheds some new light on the series of successive mental upheavals that the discovery of writing must have caused.

The oldest cave paintings, at the Chauvet cave in southern France, for instance (33,000 years before the present), already bear evidence of sophisti-cated graphic forms. Very early on, the first *Homo sapiens* discovered that they could evoke the recognizable image of an object or an animal by a few strokes on a bone, in clay, on a cliff or a cave wall—and that drawing the major contours was enough.

This discovery may seem insignificant, yet it must have played an essential role in the invention of writing. It does away with the need for a laborious reproduction of the three-dimensional shape, or for the careful reproduction of its surface in two dimensions (although our ancestors were also quick to discover those art forms that we now call sculpture and painting). A few strokes sufficed to sketch a contour that even the untrained eye instantly rec-ognized as a bison or a horse. I see in this invention what might perhaps be man's first intentional manipulation of his nervous system. Most ganglion cells in the retina are insensitive to large surfaces with a homogeneous color. They prefer to discharge in response to the contours of objects—and they respond in a rather similar way whether this contour is a single line (as in a drawing) or the junction of two surfaces (as in a real-life scene). Drawing only works because we have cells that are endowed with this property. With engraving and drawing, humanity invented the first form of neuronal self-stimulation.

An age-old tradition speculates that writing began with the pictorial rep-resentation of natural shapes. The view was already popular in the eighteenth century, as attested by this entry on "Writing" in Diderot and d'Alembert's encyclopedia:

This way of communicating our ideas by marks and figures initially consisted in drawing quite naturally the images of things; thus, in order to express the idea of a man or of a horse, one would represent their shapes. The first attempt at writing was, as one can see, a simple painting: one knew how to paint before knowing how to write.

Archaeological evidence, however, contradicts this simple idea. In most painted caves, drawings of animals cohabit with a rich set of nonfigurative shapes: series of dots, parallel lines, checkerboards, abstract curves . . . Contrary to Diderot and d'Alembert's intuitions, symbolism thus seems to be as ancient as art itself. The outlines of these first symbols sometimes resemble our letters: it would be fascinating to see if they follow Changizi's universal distribution.

Dating from the same period are Paleolithic bones engraved with series of lines. These tally marks most probably served as elementary counting or calendar devices. Such symbols, even if we cannot decode them, go to show that the men and women who etched them were not different from us. They were *Homo sapiens*, and with language-enabled brains like ours, there was no reason why they could not try to express their abstract ideas in writing.

Among the many precursors of writing also figure the painted hands that abound in a number of prehistoric caves. In some cases, strangely enough, some fingers are missing. Interpretations have ranged from self-mutilation to disease. In my opinion, however, the most likely interpretation is that fingers were folded to indicate some sort of symbolic code. Certain finger configurations recur with greater frequency and match the distribution of the animal species painted on the same cave walls.[195] These paintings might therefore represent a rudimentary sign language used by hunters to silently indicate the number, nature, and movements of their prey. Present-day hunter-gatherers still rely on similar signs to avoid alerting game. Amazing though it may seem, the hands on cave walls may thus bear testimony to the presence, in some prehistoric societies, of an advanced two-tiered symbol system: first the association of an arbitrary hand position to objects or actions, then its permanent fixation by painting.

From Counting to Writing

In *Gesture and Speech*, André Leroi-Gourhan stressed how important language and oral tradition must have been in the "reading" of cave art:

> Originally, figurative art was directly linked to language and was much closer to writing (in the broadest sense) than to a work of art. It was a symbolic transposition, not a carbon copy of reality. . . . In both signs and words, abstraction reflects a gradual adaptation of the motor system of expression to more and more subtly differentiated promptings of the brain. The earliest known paintings do not represent a hunt, a dying animal, or a touching family scene, they are graphic building blocks without any descriptive binder, the support medium of an irretrievably lost oral context. . . . They are really "mythograms," closer to ideograms than to pictograms and closer to pictograms than to descriptive art. . . . Ideography in this form precedes pictography, and all Paleolithic art is ideographic.

He added this essential note:

> A system in which three lines are followed by a drawing of an ox or seven lines by a drawing of a bag of corn is also readily conceivable. In this case phonetization is spontaneous, and reading becomes practically inevitable. This form of pictography is probably the only one that existed at the time of the birth of writing, and writing was bound to merge immediately with this preexisting ideographic system.[196]

Corroborating Leroi-Gourhan's statement, in Mesopotamia (present-day Iraq), the birthplace of writing, number symbols played an essential role in the emergence of the written code.[197] At many sites dating back to 8000 BC, archaeological excavation has unearthed small clay objects bearing abstract shapes of cones, cylinders, spheres, half-spheres, and tetrahedrons. According to Denise Schmandt-Besserat, they are "calculi," small markers used for counting and calculation. Some represent units, others multiples of the arithmetic bases 10 and 60 (10, 60, 600, 3600 . . .). A sophisticated accounting system was in use in the Middle East over an extended period of at least five thousand years going from 8000 to 3000 BC. At the city of Susa, in around 3300 BC, an intriguing transitional form appeared with the calculi placed inside hollow clay envelopes bearing notches of various shapes. The shape and number of these notches often matched the calculi found inside, thus constituting a full-blown symbolic notation of number. Progressively, the contents of the envelope, which only served as proof of the accuracy of the accounts, disappeared while the notation remained. The carving of a numerical token, followed by a symbolic portrayal of the object, was enough to guarantee the transmission of a message ("twenty goats"). And thus the idea of writing was born.

Counting also played a role in the birth of an independent writing system in pre-Columbian South America. Here, the calculation of time cycles in a calendar system served as the major incentive. As early as 2000 BC, the Olmec used carved symbols called glyphs to represent units of time such as days, years, and other important ritual cycles. The Zapotec culture, around 600 BC, extended the system by using signs for dates, places, and celebrities involved in historical events.

In brief, as suspected by André Leroi-Gourhan, the coding of abstract ideas such as number or time played an essential role in the emergence of writing. They perhaps also contributed to the very idea that concepts could be put in writing. The first symbols were often abstract geometric shapes devoid of any pictorial content. They only provided access to a small visual lexicon. In ancient Sumeria and Egypt, pictography appeared, at least transitorily, as a simple way of enriching the lexicon of written forms. Its main asset was that it was easy to read. No scribe was needed to explain that the drawing of an ear of wheat was the symbol for harvest.

It is amusing to note that the signs in the first written scripts used a whole gamut of shapes that physiologists have since found to be coded by single neurons in the primate's visual cortex. Before converging onto the letterbox area, the cortical location best suited to link visual shapes to language, cultural evolution seems to have systematically explored the space of possible shapes encoded in the visual cortex. Egyptian hieroglyphs can be seen as a catalog of stimuli capable of activating the patchwork of ventral cortical regions: animals (⬠ 🕊 ⬠ 🦆 🦅 🐛 🦉 🦢 🦤 🐟 🐝), objects and tools (📏 🕯 🏺 🕯 📦), body parts (👁 ⬠ 👄 ∧ and even ⬠ or ⬠), body postures (🧍 🧎 🙆 🧍 🧍 🚶), and simple geometric shapes (△ □ ◁ ⌐).[198] It is worth noting, however, that the first scribes spontaneously avoided two categories of visual representations that are well represented in cortex. The first is the category of places, houses and landscapes. Only a few hieroglyphs refer to places, and then only in a very stylized fashion that fails to evoke any sense of space or depth (⊏⊐ ⧄).

The other underexploited category is faces: only Mayan writing relied on a variety of "mug faces" to denote syllables:

On the cortical surface, places and faces occupy extended and well-separated areas, but both are very far from the letterbox area in the left hemisphere. The place area, present in both hemispheres, lies close to the brain's midline, while the face area is principally found in the right hemisphere—thus maximally distant from the seat of language in the left hemisphere. Did the first scribes spontaneously discover that these two categories failed to provide an efficient connection to the language areas? The near absence of faces among written symbols could be taken as another indirect proof that brain architecture constrained the evolution of writing.

The Limits of Pictography

In his *Fifth Ennead*, as he commented on Plato's *Phaedrus* (which I quoted in this chapter's epigraph), the philosopher Plotinus (AD 205–270) stated his boundless admiration for ancient Egyptian writing:

> The wise men of Egypt, when they wished to signify something, did not use the forms of letters which follow the order of words and propositions. They did not imitate sounds or the enunciations of philosophical statements. Rather, they

drew images and inscribed in their temples one particular image for each particular thing. . . . Every image bears a kind of knowledge and wisdom.

Unfortunately, Plotinus was wrong. If hieroglyphs, like a cartoon, had indeed provided direct access to meaning, we wouldn't have had to wait for the French Egyptologist Jean-François Champollion to decipher them. The concept of a universal pictographic system that men and women of all cultures could understand without training is utopian. In the evolution of writing, the pictographic stage was so brief that one can even wonder if it ever existed at all. From the start, writing was used to express abstract ideas. To serve this purpose, it adopted a great many arbitrary conventions that required a lot of learning. As a result, writing and reading were rapidly reserved to an elite.

The early scribes faced a number of concrete problems. The first was to write as quickly as possible. In the tombs of some wealthy Egyptians, hieroglyphs were amazingly detailed sculptures, including superb portrayals of bird and flower species. Creating these minute works of art was, however, too slow for everyday use. A faster and simplified writing system now known as "demotic" (literally, the scripture of the people) was soon introduced. In all countries where writing was widespread, stylization led to a quick move away from pictography to a simpler set of conventional symbolic characters.

The forms of writing that were adopted depended in part on the materials available. The Sumerians wrote on soft clay tablets with a finely sharpened reed pen. The only shape they could easily produce was a wedge (called *cuneus* in Latin). Their script, made of various arrangements of this basic wedge form, evolved into "cuneiform" writing. By arranging the wedges in different ways, they created dozens of characters so stylized that the initial pictogram became unrecognizable (figure 4.3). The symbol for an expanse of water, for instance, was initially two parallel waves. In cuneiform script, it became a large wedge and two smaller ones, a pure convention that every scribe had to commit to memory.

Chinese followed on a similar track. In their earliest known inscriptions, dating back to the Shang dynasty (1000 BC), characters were incised on bone or tortoiseshell for oracular divination. The character for "horse" resembled

From pictography to the first cuneiform characters

From Proto-Sinaitic to Phoenician: hieroglyphs give way to a limited set of letters

From Phoenician to Greek: letter rotation and emergence of vowels

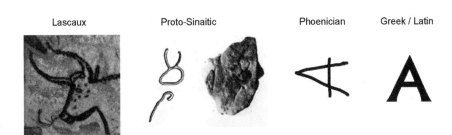

Lascaux	Proto-Sinaitic	Phoenician	Greek / Latin

Figure 4.3 Convention and simplification are two essential factors in the evolution of writing. In Sumerian (top), the first conventional characters, whose pictographic origins are obvious, quickly evolved into abstract symbols, largely because they had to be traced in soft clay with wedges. Similarly, Proto-Sinaitic writing (middle) adopted a small set of conventional pictures to represent the consonants of Semitic language. During their adoption by the Phoenicians and the Greeks, these shapes were further simplified and rotated by 90 or 180 degrees, under the influence of changes in the direction of writing. They ultimately became the letters of our alphabet. Each of them, such as the letter A (bottom), can be seen as the end point of a cultural evolution that tended toward greater simplicity while maintaining a core shape that could be recognized easily by our inferior temporal neurons. *Used with permission of Robert Fradkin.*

the noble quadruped: 馬 Quickly, however, stylization took over. Who could tell that there is a horse behind the present-day character, 馬 which converged onto its final shape as early as the third century and was further simplified recently into 马? It is estimated today that only 2 percent of Chinese characters contain a recognizable pictographic content.[199]

The stylization underlying all existing writing systems is at the root of orthography, which literally means "drawing right." As long as writing was based on drawing a recognizable picture, its exact shape could vary. Once written symbols became a matter of convention, there was only a single way to spell them properly, or a single "orthography."

A second factor that drew writing away from pictography was the problem of drawing pictures of abstract ideas. No picture could possibly depict freedom, master and slave, victory, or god. Frequently, an association of ideas did the trick. In cuneiform writing, a divinity was a star; an egg next to a bird stood for giving birth; the profile of a face with the mouth touching a bowl meant a ration of food. Unfortunately, clever as they were, these conventions only meant anything to the trained eye—the direct connection from picture to meaning was lost.

Another trick consisted of exploiting the similarity between certain sounds to draw what were essentially visual puns. This is known by historians as the rebus principle. It involves the use of a pictogram to represent a syllabic sound. This procedure converts pictograms into phonograms. For the Sumerian scribes, the word for "life," pronounced *til*, was illustrated by an arrow, which was pronounced *ti*.[200] This kind of transcription of meaning progressively gave way to writing sounds. In Sumerian, the drawing of a plant, pronounced *mu*, was first adopted to denote *mu*, a year, then *mu*, a noun, then grammatical words like the possessive *mu* = "mine." Finally it became the conventional sign for any *mu* syllable, even when it appeared within another word.

With the rebus principle, the Sumerians and the Egyptians gradually created an array of symbols that could transcribe any speech sound in their languages. Indeed, a subset of hieroglyphic characters forms what is traditionally called, somewhat inappropriately, the hieroglyphic alphabet: a few dozen characters that represent all the consonants in ancient Egyptian. This system was used to transcribe proper nouns such as "Alexander" or "Cleopatra." Its very simplicity allowed Champollion to crack the hieroglyphic code. Similarly, the Sumerians used a larger set of characters to represent all their syllables.

They invented a cuneiform sign for every possible combination of a single vowel [V] or of a vowel surrounded by consonants [CV, VC, or CVC].

The Egyptians and the Sumerians thus came very close to the alphabetic principle, but neither managed to extract this gem from their overblown writing systems. The rebus strategy would have allowed them to write a word or sentence with a compact set of phonetic signs, but they continued to supplement them with a vast array of pictograms. This unfortunate mixture of two systems, one primarily based on sound, the other on meaning, created considerable ambiguity. It was impossible to tell, for instance, from a star, if the scribe meant "God," "star," or merely the corresponding sound. With the wisdom of hindsight, it is clear that the scribes could have simplified their system vastly by choosing to stick to speech sounds alone. Unfortunately, cultural evolution suffers from inertia and does not make rational decisions. Consequently, both the Egyptians and the Sumerians simply followed the natural slope of increasing complexity. For hundreds of years, rather than simplify their writing systems, the two civilizations tried to lift ambiguities by adding an increasing number of characters. Cuneiform notation added "determinative" ideograms to clarify the concept of the accompanying signs. Each marked the semantic categories of words: city, man, stone, wood, God, and so on. For instance, the character for "plow," accompanied by the determinative "wood," meant the agricultural tool. The same character, with the determinative "man," stood for the plowman. Determinatives also helped specify the meanings of words written in syllabic notation—a useful trick since any given syllable often corresponded to several homophone words (much like "one" and "won").

It is striking that Egyptian writing adopted an almost identical solution. Hieroglyphs included a sophisticated system of determinatives for categories such as man, water, fire, plant, action, and even "abstract concept." In this elaborate scheme, a word ended up being represented by a mixture of three clues: sound, category, and explicit meaning. For instance, the verb "go out," which was pronounced *pr*, was written with no fewer than three signs:

- A house, which was also pronounced *pr*, served as a phonetic element.
- A mouth, pronounced *r*, served as a phonetic supplement. It confirmed that the symbol for house should be read for its pronunciation and not for its meaning.

• Finally, a pair of legs served as a determinative expressing the idea of motion.[201]

Very similar logic governed the evolution of Chinese characters. They too are composed of subunits that include both meaning and phonetic markers. For instance, the character for "sunshine," pronounced *qing*, is made up of the characters for "sun" and "green." The first is an obvious semantic marker, while the second, pronounced *qing*, indicates proper pronunciation.

Although all these pictorial combinations are very elaborate, they are not much more complicated than the logic that leads us to distinguish "cellar" and "seller," "I" and "eye," or "but" and "butt." We too rely on spelling conventions that involve a mixture of morphemes and phonemes. Our spelling may seem complicated, but it actually clarifies the ambiguities of phonetic notation. It uses special written forms for different meanings of words that sound the same, such as "lessen" and "lesson," "pea" and "pee," or "horse and "hoarse." In "board" versus "bored," the word ending "-ed" acts as a label for a past participle, not unlike the markers used in ancient Egyptian or cuneiform.

In brief, a mixed writing system combining meaning and sound was independently adopted by various cultures. But why do mixed writing systems appear to constitute such a stable attractor for societies throughout the world? The reason for this probably lies at the crossroads of multiple constraints: the way our memory is structured, how language is organized, and the availability of certain brain connections. Our memory is poorly equipped for purely pictographic or logographic script, where each word has its own symbol. It would be impossible for us to memorize a distinct sign for each of the 50,000 words in our lexicon. The mere notation of sounds would be equally unsatisfactory. Most languages contain such a vast number of homophones like "wretch" and "retch" that a purely phonological writing system would suffer from serious ambiguities. Reading would be comparable to decoding a rebus—it wood bee two in knee fish hunt. A mixed system using fragments of both sound and meaning appears to be the best solution.

The mixed writing system also has the vast advantage of being particularly well suited to the connectivity of the letterbox area. By projecting to both the middle temporal and frontal regions coding for word meaning, and to the superior temporal and inferior frontal areas involved in auditory analysis and articulation, this region probably plays a pivotal role in the distribution of the phonological and semantic fragments of information present in writing. If this connectivity did not preexist in all human beings, it is probable that our Sumerian and Egyptian ancestors would have had to invent a very different visual code.

The Alphabet: A Great Leap Forward

From its inception in Mesopotamia, the "virus" of writing spread quickly to the surrounding cultures. The favorable breeding ground that it found in the human brain no doubt helped it greatly. The epidemic, however, remained confined, in all societies, to a small group of specialists. The complexity of this invention curbed its capacity to spread. Some estimates hold that Sumerian once used nine hundred signs, a number that later fell to five hundred as pictography progressively gave way to increasingly syllabic writing. In Egypt too, the education of a scribe involved learning a total of about seven hundred signs. Even in present-day China, scholars must learn several thousand signs. As recently as the 1950s, the rate of illiteracy in the adult Chinese population was close to 80 percent—before radical simplification and massive investment in education brought this figure down to about 10 percent.

Before the writing virus became pandemic, it had to mutate. As with biological evolution, this mutation—the alphabetic principle—probably emerged in small groups of people on the fringes of mainstream society. The first traces of an alphabetic system, called Proto-Sinaitic, date from 1700 BC and were uncovered in the Sinai peninsula, close to the turquoise mines first worked by the pharaohs of the Middle and New Kingdoms. This writing system borrowed the shapes of several Egyptian characters, but used them to represent a Semitic language. Signs no longer referred to meaning, but to speech sounds alone, and in fact solely to consonants. In this way, the inventory of written symbols was dramatically reduced: two dozen signs were enough to represent all the existing speech sounds with perfect regularity.

The seeds of alphabetic writing germinated because the scribes were writing for a new language. This allowed them to jettison old rules and converge onto rationalized transcription. They were no doubt inspired by the surrounding Egyptian and cuneiform systems, but they had a unique opportunity to remove any historical mishaps, extract essential principles, and distill them into their simplest form. Finally, they gave up on ideographic writing and concentrated exclusively on the abstract notation of speech sounds and word roots.[202]

The scribe's new language belonged to the Semitic family, which today includes Arabic, Amharic, and Hebrew. The morphology of these languages is peculiar in that it emphasizes consonants. There is little doubt that this feature helped the invention of the alphabet. In Semitic languages, the roots of many words consist in a fixed grid of consonants within which vowels can vary. In Hebrew, for instance, the root *gdl*, which expresses the general meaning of "big," can be declined as *gadol*, big (masculine); *gdola*, big (feminine); *giddel*, to raise; *gadal*, to grow; *higdil*, to enlarge; and so on. The inventors of Proto-Sinaitic writing exploited this peculiarity. They understood that all they had to do was represent the abstract consonant grid. Their language did not require a large inventory of signs for every possible syllable—the reader could get the gist of a message by simply looking up the two dozen shapes in the consonant lexicon.

The choice of a shape for each consonant was also guided by a very simple and mnemonic idea: each shape stood for a word that began with the corresponding consonant. This clever idea is known as the acrophonic principle (literally, "using the sound at one end"). Thus the consonant *b* was represented by the outline of a house, called *beth* in most Semitic languages. It gave its name to the letter beta in Greek. Likewise, the glottal stop, a consonant unique to Semitic languages, that leads the word *'aleph* (ox), was represented in Proto-Sinaitic writing by an ox head. This shape, stylized and rotated, became the letter alpha (α) in the Greek alphabet, and then our own letter A. If we turn over a capital A, the head and two horns of the original ox can easily be recognized. The acrophonic principle is the sole reason why Greek letters have curious names (alpha, beta, gamma, delta . . .). In fact, they are the distorted Semitic names of the two dozen images that gave our letters their shape, name, and the pronunciation of their first consonant.

Each of the letters that we routinely use in our Roman alphabet thus contains a small, hidden drawing dating back four thousand years. An "m" symbolizes waves (*mem* or *mayyūma*), an "n" is a snake (*nahašu*), an "l" a goad (*lamd*), a "k" a hand with outstretched fingers (*kaf*), an "R" a head (*res*) . . . I like to think that cultural evolution took the same route as the neurophysiologist Keiji Tanaka—both whittled down drawings until only their essential shape remained (compare figures 3.6 and 4.3). Writing went from the first realistic drawings of ox heads in the Lascaux cave, to Semitic writing that transformed these heads into a few strokes, and then on to the Phoenicians and Greeks, who turned them into the modern letter A. In much the same way as Tanaka simplified shapes until he reached the simplest one that still triggered the monkey's visual neurons, writing gradually evolved toward a simplified system that could still be immediately recognized by the neurons in our inferior temporal cortex. Cultural evolution, through trial and error and a progressive selection across many generations, arrived at a small inventory of minimal and universal letter shapes. Although they can vary from one country to the next, they all originate in the same set of basic features that are present in natural scenes and that our visual system grasps most easily.

With the invention of the alphabet, writing became more democratic. Years of learning were no longer needed to become a scribe. Anyone with a reasonable degree of motivation could learn to read and write an alphabet of only about twenty letters. Writing and the alphabet quickly spread worldwide. In Ugarit, on the Syrian coast, far from its birthplace, the alphabetic principle was adapted to the technology of clay tablet writing. Inscriptions dating back to the thirteenth century BC using a cuneiform alphabet have been found on the site of the ancient Canaanite city. This alphabet was Semitic. It indirectly gave birth to all the world's alphabets: first Phoenician, which gave rise to Greek, Cyrillic, Latin, and probably all of the Indian writing schemes; then Hebrew script, which remained essentially unchanged to this day; and finally, Aramaic, which is the source of modern Arabic with its two hundred million contemporary readers.

Vowels: The Mothers of Reading

The Phoenicians introduced another major contribution: the explicit notation of vowels. In Semitic, only consonants were transcribed. This feature made reading complicated and led to ambiguities, since letters only specified the roots of words and not their full pronunciation. Phoenician script, following on the tracks of the Hebrew and Ugarit alphabets, introduced the representation of vowels by adding symbols called *matres lectionis* ("mothers of reading"). Initially, the *matres lectionis* were simply consonants converted into vowels. This conversion probably came about because the pronunciation of some words had drifted.[203] For instance, the Semitic word *panamuwa*, initially written with the consonants PNMW, was progressively distorted in spoken language into *panamua*, then *panamu*. Because the word was still written PNMW, the final consonant W came to be the transcription of the sound *u*. Similarly, the consonant "j" (*jodh*) became the official transcription of the vowel *i*—the same kind of speech distortion that makes us pronounce the capital of Slovenia, Ljubljana, as *Liubliana*.

The Phoenician system, however, was not perfect. It failed to represent all vowels, and was ambiguous because the same symbols stood for both consonants and vowels. It was the Greeks who finally created the alphabet as we know it. In doing so, they adopted the Phoenician letter names ('*aleph, beth, gimmel* . . .), although by now these words were meaningless. Unfortunately, several Phoenician consonants did not exist in Greek. For instance, the Greeks could not pronounce the glottal stop indicated by an apostrophe and which is the first letter of the word '*aleph* (depicted by the letter A). Unwittingly, Greek speakers dropped this first sound and incorrectly pronounced this letter *aleph*. Thus the letter "A" came to denote the vowel *a*, even though this letter initially depicted a consonant in Semitic languages. A similar story accounts for the letters iota, omicron, and upsilon, all of which are Semitic consonants converted into Greek vowels. The system continued to evolve for several centuries. New sounds were still appearing in Greek. As a result, new letters emerged to denote the long *o* (literally *omega*) and the long *e* (the letter eta, borrowed from the Phoenician consonant *heth* that no longer existed in Greek). Complex graphemes were also created: the conjunction of "o" and "u" was selected to denote the sound *u*, while the letter upsilon ended up representing the sharp sound *ü* as in the French *sur* or the German *über*.

Although these adjustments were slow, a new principle was under way. For the first time in the history of mankind, the alphabet allowed the Greeks to have a complete graphic inventory of their language sounds. Writing had been stripped of its pictographic and syllabic origins. The Greeks had discovered the smallest units of spoken language, phonemes, and had invented a notation that could transcribe them all. By trial and error, cultural evolution had converged onto a minimal set of symbols. These were compatible with our brain, both because they could be easily learned by the letterbox area, and because they established a direct link to speech sounds coded in the superior temporal cortex.

CHAPTER 5

Learning to Read

Learning to read involves connecting two sets of brain regions that are already present in infancy: the object recognition system and the language circuit. Reading acquisition has three major phases: the pictorial stage, a brief period where children "photograph" a few words; the phonological stage, where they learn to decode graphemes into phonemes; and the orthographic stage, where word recognition becomes fast and automatic. Brain imaging shows that several brain circuits are altered during this process, notably those of the left occipitotemporal letterbox area. Over several years, the neural activity evoked by written words increases, becomes selective, and converges onto the adult reading network.

These results, although still preliminary, are rich in implications for education. Above all, we now understand why the whole-language method deluded so many psychologists and teachers, even though it does not fit with the architecture of our visual brain.

I wish you to gasp not only at what you read
but at the miracle of its being readable.

—VLADIMIR NABOKOV, *PALE FIRE*

*A*lthough it took many generations of scribes to arrive at our present-day writing systems, our children must master them in only a few years. Whether they speak French, English, Chinese, or Hebrew, two or three years of teaching will allow them to begin to decipher words. During those crucial years, writing, which was only a jumble of marks on paper, suddenly acquires a meaning. Research into the inner mechanisms of this learning process, from psychology down to the brain level, has vastly progressed in the last ten years. This renewed understanding of reading mechanisms has important implications for the improvement of teaching, which we will review in this chapter.

The neuronal recycling hypothesis claims that writing is progressively anchored in the child's brain because it finds an appropriate niche for itself in circuits that are already functional and only need to be minimally reoriented. A cerebral process involving trial and error, similar to the cultural experimentation that occurred during the evolution of writing, must take place within the visual and linguistic circuits of the child's brain. A central prediction is that reading gradually converges onto the left occipito-temporal letterbox area. As the child becomes an expert reader, this brain region should become progressively specialized in writing. Its communication should also increase with the outlying temporal, parietal, and frontal language areas. Developmental psychology and, more recently, brain imaging have begun to shed light on the stages that punctuate this process.[204]

Curiously, the neuronal recycling hypothesis leads us to focus on the first year of life, before reading is even an issue. If the recycling model is correct, children learn to read only because their brains already contain the required architecture—thanks either to evolution or to earlier learning. Before children are exposed to their first reading lesson, their prior linguistic and visual development should play an essential role in preparing their brains for this new cultural exercise.

The Birth of a Future Reader

The infant's view of the world was once thought to be a giant "blooming buzzing confusion." Its brain was considered highly immature and barely organized—massive inputs and years of statistical learning would be needed to structure it. Recent research, however, refutes this simplistic "constructivist" theory. Sure enough, the brain remains plastic well into adolescence, thus opening a window of opportunity for learning and education. Even in the first year of life, however, the two main faculties that will later be recycled for reading are already being put into place: speech comprehension and invariant visual recognition.

In the first few months of life, infants demonstrate surprising language competencies. A few days after birth, they easily perceive linguistic contrasts like the difference between *ba* and *ga*.[205] Furthermore, they pay special attention to the rhythm of their native language,[206] heard in utero during the last months of pregnancy. Their language skills already rely on the network of left-hemisphere areas that will be active in our adult brains when we process speech. The left superior temporal region analyzes speech sounds, and the temporal lobe is organized hierarchically into a series of areas presumably capable of extracting phonemes, words, and sentences. Even the left inferior frontal region, called Broca's area, which was traditionally thought to host advanced speech production and grammatical skills, already activates when three-month-olds listen to sentences.[207]

There is little doubt that these brain regions are genetically predisposed to form a network that allows for language acquisition. Of course learning also plays a significant role. During the first year of life, the infant's speech areas become specialized for the language in its environment.[208] At the age of six months, the representation of vowels, which can be considered a "vocalic space" unique to each language, is progressively distorted as it adapts to the native language's vowels. At roughly eleven to twelve months, consonants converge toward the appropriate targets. This is the time when Japanese babies cease to perceive the distinction between *r* and *l*—and we cease to perceive foreign contrasts like the various *t* sounds that are crucial to Indian languages, but not to ours.

The infant brain, like a young naturalist or a statistician, systematically extracts, sorts, and classifies segments of speech. It detects regularities in speech inputs, decides which sound transitions are acceptable, and eliminates all others. Very rapidly, the phonotactic rules of language are in place: a Polish baby accepts the string of consonants *p*, *r*, and *ch*, although this sequence grates on English ears. Babies also compute which bits of speech occur most often: these will become the first words in their input lexicon.[209]

At the end of the second year, a child's vocabulary grows at the amazing rate of ten to twenty new words per day. At the same time, he establishes the basic grammatical rules of his language. At the age of five or six, when children are exposed to their first reading lessons, they already have an expert knowledge of phonology. They also possess a vocabulary of several thousand words, and have mastered the basic grammatical structures of their languages. These "rules and representations"[210] are implicit. The child is not aware of his expertise and cannot account for it. Nonetheless, this knowledge no doubt exists in an organized set of speech circuits that are on standby for the written word.

Organization of the child's visual system takes place concurrently. Even in its first few months, the infant's visual system is already sophisticated enough to parse the visual scene into objects and track them as they move, even if they are briefly concealed at some point.[211] In the first year of life, infants discriminate among objects using their contours, their texture, and their internal organization. They even notice the difference, moreover, between concave and convex objects.[212] When they see an object from several different viewpoints, they make sophisticated inferences as to its three-dimensional shape.[213] This amazing competence seems to rest on an interpretation of how the object's edges meet and form T-, Y-, or L-junctions.[214] An all-essential system, which will later be recycled into a letter-recognition device, thus seems to be laid down in the first year of life. In their second year, babies recognize stripped-down versions of objects, thus indicating that they are capable of abstracting essential elements of shape from an image.[215]

The development of face perception has also been studied in great depth. At birth, infants pay special attention to faces. At two months, faces already seem to activate the region of the occipito-temporal cortex used by adults.[216] This early specialization is progressively enhanced in the course of the first year as babies learn to recognize the faces around them. At about nine months, their face system becomes specialized to human beings and loses the ability to distinguish the faces of other primates.[217] In their second year, they also learn to recognize a face out of context.[218] This specialization process continues during the next ten years or more, albeit more slowly.[219]

At present, we do not know at what age the patchwork of cortical specialization for faces, places, and objects finally reaches its adult arrangement— some research suggests that this cortical specialization may start as early as the first year of life.[220] What is certain is that at around the age of five or six, when a child begins to read, the key invariant visual recognition process is in place, although it is still maximally plastic. This period is thus particularly conducive to the acquisition of novel visual shapes like letters and words.

Three Steps for Reading

Sir, he hath never fed of the dainties that are bred in a book; he hath not eat paper, as it were; he hath not drunk ink: his intellect is not replenished; he is only an animal, only sensible in the duller parts.

—SHAKESPEARE, *LOVE'S LABOUR'S LOST*

What did we do to become "paper eaters" and "ink drinkers," or in other words expert readers? In 1985, the British psychologist Uta Frith introduced a model of reading acquisition that has become a classic and distinguishes three main learning stages.[221] This is of course a theoretical simplification, since in fact the three stages are not rigidly partitioned. The child moves through them constantly over several months or years, and future theories of reading may eventually manage to capture this continuity within a unified network of neurons where representations gradually emerge thanks to a fixed learning rule.[222] Meanwhile, however, Frith's three simple steps provide a rough outline of the massive changes that occur in the child's mind. If nothing else, from the standpoint of pedagogy, they provide a very useful description of the child's learning curve.

According to Frith, the first reading stage, which occurs around the age of five or six, is "logographic" or "pictorial." The child has not yet grasped the logic of writing. The visual system attempts to recognize words as though they were objects and faces. It relies on all the available visual features: shape, color, letter orientation, curvature . . . At this stage, which often predates formal teaching, the child typically recognizes his name and perhaps a few other striking words such as brand names (*CocaCola*).

The size of this sight vocabulary varies significantly across children. Some manage to memorize a few dozen words. Others, particularly in transparent languages like Italian, exhibit a very brief, often undetectable pictorial stage. In any case, the recognition of whole words as pictures is only an artificial form of reading. Frequent errors reveal that the child does not decode the internal structure of words, but only exploits a few superficial cues. He is easily taken in by visual resemblance (*Chinchilla* is read as *CocaCola*), fails to recognize known words in a new guise ("COCA-COLA" in uppercase letters), and does not generalize to related strings like "coco."[223] These features suggest that the child's brain, at this stage, is attempting to map the general shape of words directly onto meanings, without paying attention to individual letters and their pronunciation—a sham form of reading.

Becoming Aware of Phonemes

In order to move beyond the pictorial stage, the child must learn to decode words into their component letters and link them to speech sounds. The development of a grapheme-to-phoneme conversion procedure is characteristic of the second stage in reading acquisition, the phonological stage. At this point, whole words cease to be processed. The child learns to attend to smaller constituents such as isolated letters and relevant letter groups ("ch," "ou," "ay" . . .). He links graphemes to the corresponding speech sounds and practices assembling them into words. He can now even sound out unfamiliar words.

Sometimes the child knows the names of letters (*ay, bee, see, dee* . . .). Unfortunately, this knowledge, far from being helpful, may even delay the acquisition of reading. To know that "s" is pronounced *ess*, "k" *kay*, and "i" *eye* is useless when we try to read the word "ski." Letter names cannot be assembled during reading—the hookup only concerns phonemes. But phonemes are rather abstract and covert speech units. A true mental revolution will have to take place before the child finds out that speech can be broken down into phonemes, and that the sound *ba* is made up of two such units, the phonemes *b* and *a*.

The first years of reading instruction lead to the emergence of an explicit representation of speech sounds. The key stage is the discovery that speech is made up of atoms or phonemes, which can be recombined at will to create new words. This competence is called "phonemic awareness." Studies by the psychologist José Morais have shown that the discovery of phonemes is not automatic. It requires explicit teaching of an alphabetic code.[224] Even adults, if illiterate, can fail to detect phonemes in words.

To support this amazing conclusion, Morais and his colleagues invited dozens of Portuguese adult volunteers to their laboratory. Half of them had just learned to read, while the others, who came from the same socioeconomic background, were illiterate. These volunteers were invited to play several kinds of speech games. They were asked questions like: If you remove the first sound in the word *Porto*, what do you get? (*orto*). Do the words *table* and *tower* start with the same sound? (yes). How many sounds are there in the syllable *tap*? (three).

The results from this experiment were clear. Illiterates systematically failed whenever a game required attention to be directed to the phoneme level. Additional experiments narrowed down the problem. The illiterates did not have any trouble with the discrimination of speech sounds—for instance, they could easily hear the difference between *da* and *ba*. They could also play with syllables and rhymes and recognize that *table* and *label* ended with the same sound. They were only unable to pick up the smallest speech constituents, or phonemes. They failed to see that the same phonemes recurred at different locations within words or that there is a shared *t* in *tuna*, *stop*, and *foot*. They even failed at a simple substitution game in which one player says a word and the other must repeat it after swapping the first sound with a *p*: when the first says *Dallas*, the second must respond *palace*, and so forth.

The Reverend Spooner, dean of New College, Oxford, became famous in England for his slips of the tongue (he once reprimanded a student for "**h**issing his **m**ystery lecture" and "**t**asting two **w**orms"). Thanks to José Morais's work, we now know that this minor art form requires alphabetic literacy.

Decoding spoonerisms such as "Our Lord is a shoving leopard" puts considerable emphasis on phonemic awareness—it is amusing to note that the Sumerians, with their syllabic writing, were probably unaware of such phonemic puns.

The deep effects of phonemic awareness prove how profoundly the acquisition of the alphabetic code changes our brains. Learning the alphabet gives us access to a verbal fluency beyond the reach of illiterates. It also leads to profound cultural differences. Chinese adults who can only read traditional Chinese fail on phonemic awareness tests, while those who have also learned the alphabetic Pinyin writing readily succeed.[225]

These discoveries also have implications for the child's brain. When a young student learns to decipher an alphabetic script, the brain's visual areas learn to break down the word into letters and graphemes. This means that some of the speech areas must adapt to the explicit representation of phonemes. These modifications must be highly coordinated if an efficient letter-to-sound conversion route is to emerge.

Graphemes and Phonemes: A Chicken and Egg Problem

We do not fully understand the causal chain that links visual and linguistic acquisition. Must a child first analyze speech inputs into phonemes in order to figure out the meaning of letters? Alternatively, does the child understand the nature of the letter code before he discovers that speech is made up of phonemes? This is probably just another "chicken and egg" problem. The two types of learning are so tightly linked that it is impossible to tell which comes first, the grapheme or the phoneme—both arise together and enhance each other.

A wealth of data indicates that the preschoolers who are most fluent in phonological games such as rhyming also learn to read more quickly.[226] Furthermore, practice with speech sound manipulations at an early age improves both phonemic awareness and reading scores. These findings lead many researchers to conclude that phonemic awareness is a prerequisite to reading acquisition. In other words, the discovery of phonemes precedes that of graphemes.

Recently, however, Anne Castles and Max Coltheart[227] have stressed that the causal link from phonemic awareness to reading acquisition is not as solidly established as one might think. Most experiments in this area were performed with children who already knew some letters and might have benefited from this knowledge to segment speech. Castles and Coltheart also draw attention to the massive influence that spelling can have on phonemic processing. If you ask children how many sounds they hear in a word like "rich," they say three (*r-i-ch*), but in the word "pitch" they hear four (*p-i-t-ch*)—in fact, both words have four phonemes, but the influence of the written code dramatically affects the ability to perceive them.[228] Likewise, if you ask a nine-year-old to remove the *n* sound in a spoken word like *bind*, chances are that he will say *bid* rather than the correct *bide*, again an error influenced by spelling.[229] In brief, knowledge of the written code clearly permeates phonemic awareness tasks. The case of illiterates and Chinese readers demonstrates that without explicit teaching of the alphabetic code, conscious manipulation of phonemes does not emerge.

In the final analysis, the relation between grapheme and phoneme development is probably one of constant reciprocal interaction or "spiral causality." The acquisition of letters draws attention to speech sounds, the analysis of speech sounds refines the understanding of letters, and so on in a never-ending spiral that leads to the simultaneous emergence of the grapheme and phoneme codes.

The appearance of the two codes can be directly gauged by monitoring a child's reading mistakes. In all languages, the phonological stage is marked by peculiar regularization errors. The beginning reader can turn a few letters into sounds, but typically fails when a word is even mildly irregular. He will, for instance, say *kay* instead of *key*.[230] Another symptom is the syllable complexity effect: a first-grader may be able to read simple syllables with a consonant followed by a vowel (CV), but typically experiences increasing difficulty as the number of consonants grows (CVC, CCVC, and so on). Complex words such as "strict" (CCCVCC) cannot be deciphered by a novice.[231]

Those findings indicate that reading acquisition progresses from simple to complex rules. A child first learns to spell out letters whose pronunciation is regular, such as "b" or "k." Then he progressively learns to decode increasingly complex and infrequent graphemes. He discovers relevant consonant clusters such as "bl" or "str." He also memorizes special letter groups, for instance the suffix "-tion," which is read *shun*, or the irregular morpheme for the verb "walk" (which is read *wok*). The expert reader is, above all, a well-read man or woman who implicitly knows a large number of prefixes, roots, and suffixes and effortlessly associates them with both pronunciation and meaning.

The Orthographic Stage

When a child attains a certain level of expertise, he reaches Uta Frith's third, orthographic stage. Progressively, a vast lexicon of visual units of various sizes settles in. It includes a rich body of information on the frequency of these units and their neighbors. At this stage, reading time is no longer determined primarily by word length or by grapheme complexity. Rather, it becomes increasingly influenced by how often a word has been encountered: rare words are read more slowly than frequent ones. The number of neighbors also emerges as a major factor. A word like "vain," which is surrounded by high-frequency neighbors such as "rain" or "pain," is read more slowly. All of these effects reflect a gradual establishment of the second reading pathway, the lexical route, which progressively supplements letter-to-sound decoding.

Probably the clearest feature of the orthographic stage is that word length gradually ceases to play a role. At the phonological stage, children slowly decipher words sequentially, one letter at a time. As a result, reading time increases with the number of letters in a word.[232] At the orthographic stage, as reading becomes increasingly fluent, this length effect slowly vanishes. It is essentially absent in expert adults—we all read words using a parallel procedure that takes in all letters at once, at least in short words (eight letters or fewer).

In summary, growing parallelism and efficiency are characteristic of the orthographic stage. An increasingly compact word code appears that represents the entire letter string in a single snapshot. This neuronal analysis, organized like a hierarchical tree, can now be effortlessly transmitted in parallel to brain regions that compute meaning and pronunciation.

The Brain of a Young Reader

The psychology of reading infers the existence of three principal stages in reading acquisition, but their cerebral counterparts have yet to be defined.

Does each stage have its own distinct brain signature? Is a set of active regions unique to each particular phase of reading development?

The neuronal recycling hypothesis postulates that the visual system of the young reader specializes gradually. This hypothesis therefore leads to clear predictions at the brain level. During the first pictorial stage, when children treat words like pictures, no clear specialization should be present, and both hemispheres should contribute to reading. With growing expertise, activation should be progressively more focused, and it should slowly converge to the left occipito-temporal letterbox area where visual word recognition is always housed in expert adults.

If one could zoom down to the scale of single neurons or cortical columns, one would see a major upheaval in the neuronal microcode. According to the recycling view, each reading lesson leads to a neuronal reconversion: some visual neurons, previously concerned with object or face recognition, are committed to letters; others to frequent bigrams; yet others to prefixes, suffixes, or recurring words. In parallel, the neural code for spoken language is also in flux. Somehow, as phonemic awareness emerges, the code explodes into a more refined structure where phonemes are explicit. Finally, if we could track nerve fibers during development and sort them depending on function, we would see a regular, comblike projection appear that links each visual unit to its corresponding pronunciation.

Before I go any further, I would like to make it clear that imaging tools today do not allow us to track reading progress directly in the child's brain. We will probably have to wait many years before this becomes possible. The first images of the child's brain were only obtained with functional MRI in the late 1990s and developmental imaging still remains a major challenge.

Ethics is not the main issue—MRI has been used for more than twenty years in pediatric hospitals, even on newborns, and no side effects have ever been detected. Below a well-known threshold, exposure to magnetic fields is safe, provided that no metallic objects are brought too near to the magnet. During safety tests, some animal species have spent their entire lives in very strong magnetic fields, without any adverse effects on their biological development.[233]

The real problem with developmental imaging is methodological. It is extremely difficult to obtain noiseless data, on a millimeter scale, from a young child's brain in the space of a half-hour experiment. Only a few laboratories have acquired this expertise. They generally arrange for children to come in on a first visit to explore a mock scanner, get accustomed to its noise, and try on the peculiar headphones and goggles that they will have to wear. In addition, they must be trained to remain perfectly still. Colleagues at the Sackler Institute in New York rely on a clever device that consists of a video screen that plays the children's favorite cartoons, but comes to an abrupt halt whenever their heads move. Once fully invested with their mission, described to them as a space shuttle trip, it is amusing to see these youngsters enthusiastically participate in an MRI experiment.

The few results obtained to date show that, as early as the age of seven, the normal reading network begins to activate at the sight of text.[234] The letterbox area is already visible in its adult location in the left occipito-temporal cortex. The lateral temporal regions are also active. Unfortunately, this research only provides a single snapshot, and does not track the full development of reading acquisition. In order to follow the evolution of these activations in time, longitudinal studies will have to be performed with the same children scanned several times as they learn to read. Such studies have not yet been conducted, presumably because families are hesitant to participate in projects extending over several years that require monthly visits to a laboratory.

At present, our only solid developmental data come from "cross-sectional" studies where groups of children were tested at various ages and their age and performance correlated with brain activity. In the last ten years, Bennett and Sally Shaywitz, at Yale Unversity, with Ken Pugh, now at Haskins Labs, have tested hundreds of good and mediocre readers of all ages. Their results show a very clear evolution. As reading improves, activation in the left occipito-temporal region increases at the precise location of the adult letterbox area.[235] Activation in this region correlates more neatly with a child's reading score than with his age. It constitutes a genuine cortical signature of reading acquisition, not just a deterministic effect of brain maturation.

In a smaller group of children, slightly less homogeneous in age, Guinevere Eden, a researcher from Georgetown University, did not find a left occipito-temporal activation increase during reading acquisition. However she did observe a clear *decrease* in *right* occipito-temporal activation, at a location in the right hemisphere that was exactly symmetrical with the letterbox region.[236] This finding fits with the idea that learning involves a selective pruning process. Initially, written words, like any other visual image, lead to a bilateral activation pattern. Activation then progressively tapers down to a narrower focus that is presumably optimal. Strikingly, this observation fulfills a prediction made by Samuel Orton, the founding father of the psychology of reading and dyslexia. As early as 1925, he asserted that "the process of learning to read entails the elision from the focus of attention of the confusing memory images of the non-dominant [right] hemisphere."[237]

Progressive focalization can be seen most clearly in recordings of electrical and magnetic activity in children's brains.[238] Prior to reading, when a child sees a written word, there is no trace of the quick convergence towards the left hemisphere that occurs in adults at around 170–200 milliseconds and reflects invariant recognition of letter strings. If anything, during the earliest reading stages, the *right* occipito-temporal region appears to differentiate words from consonant strings. This right-sided activation may constitute the brain's base for Uta Frith's "pictorial" stage that occurs at the onset of reading acquisition. During this period, the child memorizes the snapshots of a few familiar words and recognizes them because of their overall shape, much as one recognizes familiar faces. A few years later, in second grade, a burst of activity appears for printed words when compared with meaningless strings of geometric shapes.[239] Activation here is so broad and bilateral that a major portion of the visual system, in both hemispheres, seems to be recruited for reading. When a clearly lateralized response from the left letterbox area first appears, at around the age of eight,[240] specialization is still far from complete.[241] Even in ten-year-olds, the negative waveform, which arises around 170 milliseconds after a word, seems to be seen only for words that are frequent and well known by the child. This response is not evoked, as in adults, by pseudo-words that respect spelling rules, such as "moulton" or "fost." The letterbox area only reaches full maturity at the beginning of adolescence—provided, of course, that the child reads regularly enough to become an expert.

Learning to read does not simply alter the response of the visual cortex to written words. Brain imaging also reveals considerable changes in the language areas of the left hemisphere. Striking modifications have been observed in the superior temporal sulcus and the left inferior prefrontal cortex (Broca's area) associated with phoneme perception and articulation. Activation increases in direct proportion to the child's acquisition of "phonemic awareness,"[242] the capacity to consciously manipulate the elementary sounds of language mentally. In the spoken language networks, these two areas are probably the main anchor points for reading.

The Illiterate Brain

Literacy drastically changes the brain—literally! A fascinating brain imaging experiment proves this fact unequivocally. It shows that measurements of brain activity are profoundly different in the brains of illiterates. Although the experiment does not bear specifically on the developing brain, it sheds indirect light on the acquisition of reading by revealing the long-term impact of literacy on brain circuitry. In this case, the adult brain helps us understand the child's brain.

To visualize the major upheaval produced in the brain by reading acquisition, Alexandre Castro-Caldas and his colleagues compared the brains of literate and illiterate adults.[243] A major difficulty with such studies, however, lies in the selection of the participants. Illiteracy can be a sign of social exclusion or of genetic or neurological disease, conditions which in and of themselves can cause profound changes in brain activity. The ideal situation, from the researcher's viewpoint, is the comparison of two normal and otherwise identical subjects—a reader and a nonreader. In order to recruit volunteers who had not been exposed to reading but came from the same social and cultural environment, Castro-Caldas focused on a situation that used to be relatively common in some Portuguese families. In the 1930s, parents were often too poor to send all of their children to school. As a result, an elder sister stayed at home to take care of the younger children. The comparison of two sisters in a family where this occurred is ideal for the evaluation of the impact of schooling and literacy on brain organization.

In the context of an international collaboration, twelve Portuguese women, half of whom were illiterate, flew to the Karolinska Institute in Stockholm for anatomical MRI and a PET scan. During the scan, they were asked to repeat Portuguese words and pseudo-words. This procedure revealed that although both groups had comparable vocabulary sizes, the illiterates had trouble repeating the pseudo-words. Instead of faithfully repeating the meaningless stimuli, they frequently confused them with words that they already knew— for instance reading *capeta* as *cabeza* (head), or *travata* as *gravata* (tie). This interesting behavioral result fits with the idea that reading increases phonemic awareness. When we learn the alphabet, we acquire the new ability to carve speech into its elementary components. We become aware of the presence of phonemes in what initially sounded like a continuous speech stream. The well-read acquire a universal phonemic code that facilitates the storage of speech sounds in memory, even if they are meaningless. In the absence of this analytic code, illiterates can only rely on coarse analogies of words they already know—a strategy that severely impairs their memory for pseudo-words.

In illiterates, brain imaging showed that outside of the right prefrontal region involved in memory retrieval, the difference between hearing a word and hearing a pseudo-word was minimal—thus confirming that pseudo-words were assimilated to words. In the literate women, however, schooling had profoundly modified the responses to pseudo-words. The most important change was seen in the anterior insula, a region very close to Broca's area, where activation is also observed in children during reading acquisition. Other differences were seen throughout the left hemisphere. The literate brain obviously engages many more left-hemispheric resources than the illiterate brain—even when we only *listen* to speech.

Most strikingly, literacy did not only alter brain activity during language listening tasks, but also affected the anatomy of the brain. The rear part of the corpus callosum, which links the parietal regions of both hemispheres, had thickened in the literate subjects.[244] This macroscopic finding implies a massive increase in the exchange of information across the two hemispheres—perhaps explaining the remarkable increase in verbal memory span in literates.

In *Phaedrus*, Plato imagines an exchange between the Egyptian king Thamus and Theuth, the Ibis-headed god who gave mankind the gift of writing: "This invention," said Theuth, "will make the Egyptians wiser and give them better memories; it is a specific both for the memory and for the wit." King Thamus, barely convinced, retorts:

This discovery of yours will create forgetfulness in the learners' souls, because they will not use their memories; they will trust to the external written characters and not remember of themselves. The specific which you have discovered is an aid not to memory, but to reminiscence, and you give your disciples not truth, but only the semblance of truth; they will be hearers of many things and will have learned nothing; they will appear to be omniscient and will generally

know nothing; they will be tiresome company, having the
show of wisdom without the reality.[245]

Four thousand years later, psychological studies of illiteracy have decided
in Theuth's favor, and refute both Thamus's and Plato's arguments—learning
to read clearly improves verbal memory. Illiterates can remember the gist of
stories and poems, but their verbal working memory—the temporary buffer
that stores instructions, recipes, names or phone numbers over short periods
of time—is vastly inferior to ours. Even if Castro-Caldas's results on the cerebral correlates of illiteracy will have to be extended to a larger group of participants with more precise imaging methods, his findings are the first to
demonstrate how schooling and literacy alter the native capabilities of our
brains. Education inoculates us with the reading virus. It spreads quickly to
our language system and enhances our verbal memory. When children learn
to read, they return from school "literally changed." Their brains will never be
the same again.

What Does Reading Make Us Lose?

There is no doubt that the main effect of literacy is positive: learning to read
induces massive cognitive gains. If the neuronal recycling hypothesis is right,
however, the brain pays a price for literacy. Reading invades the neuronal circuits destined for another use and probably brings about the loss of some of
the cognitive abilities that were handed down to us by evolution.

This argument about the cost of reading rests on the observation that cortical reorganization is probably, at some level, a "zero-sum game." With some
rare exceptions, the number of cortical neurons is fixed in infancy. The day
some become dedicated to word recognition, they are probably no longer
available for other purposes. Thus, reading acquisition possibly reduces the
cortical space available for our other mental activities. The neuronal recycling
hypothesis makes us wonder if our illiterate ancestors had visual skills that we
have now lost.

Before I go any further with this hypothesis, I would like to stress that it is
speculative and is not experimentally grounded. Any evidence that exists on
the loss of brain competencies brought on by reading is ambiguous and very

scarce. It is not even clear that our cortical envelope is firmly fixed and cannot evolve. Although the number of our neurons is finite, the synapses can definitely change. Even in the adult brain, learning can still drastically alter neuronal connections. The tree of dendrites and axons that define how many cells can communicate with a given neuron varies with the complexity of the environment. These fluctuations open up a universe of plasticity whose boundaries remain unclear. Learning is unlikely to be a genuine zero-sum game because outside stimulation can lead to significant increase in the density of neuronal connections.

It is not clear either that a single neuron cannot simultaneously participate in more than one function. In fact, there are a few clear examples where the same "associative" neurons in the prefrontal, temporal, and parietal cortex participate in different neuronal assemblies. A parietal neuron, for instance, can exhibit a preference for a certain number of objects and at the same time also be concerned with the direction of visual motion—two seemingly unrelated functions.[246]

Learning also refines the neural code. In an untrained organism, there is considerable redundancy, with many neurons performing the same rough discrimination. Learning often leads to more exact representations, where each neuron responds precisely to a narrow range of stimuli. When a monkey, for instance, learns a fine manual discrimination task, its neurons respond to a narrower band of skin and the cortical map that represents tactile inputs from the relevant fingers becomes enlarged and more precise. Musical training affects humans similarly. Pianists and violinists have larger and more precise cortical maps of their left hands, particularly when they started to play in early childhood.

During tactile training, the finger map encroaches on nearby territories normally dedicated to the arm or to the face. This invasion could have a negative impact on the perception of other body parts, but the opposite could also be true: learning could lead to a positive transfer. What neurons have learned for function A could turn out to be useful for function B. The manual dexterity of violinists could perhaps increase their ability to learn to type or to sew.

In brief, it is unclear whether the net effect of learning is invariably positive, or whether it can also be negative. In the case of reading, hours spent on recognizing letters probably increase our visual acuity. Indeed, readers have a better perception of geometric shapes than illiterates.[247] For the time being, such comparisons appear only to reveal the positive effects of literacy. However, no serious studies have yet been undertaken to see which cognitive functions might be negatively affected by reading. Many fundamental questions surrounding the cortical precursors of reading remain unsolved. What did our letterbox area do before we learned to read? Did this region play a significant role in evolution? Does it have an original and well-defined function that disappears in expert readers?

Years of experience with hunter-gatherers in the Amazon, New Guinea, or the African bush lead anthropologists to marvel at the aborigines' ability to *read* the natural world. They decipher animal tracks with amazing ease. Meticulous inspection of broken branches or faint tracks in the dirt allows them to quickly figure out what animal has been around, its size, the direction in which it went, and a number of other details that will be invaluable for hunting. We are essentially "illiterate" about all these natural signs. It is possible that reading animal tracks is the cortical precursor for reading. If evolution has yielded bodily specializations as refined as the eagle's eye or the leopard's leap, it no doubt can modify the predator's visual brain. The intense selective pressure imposed by millions of years of interaction between predator and prey may have led to a cortical specialization for reading animal tracks.

Chinese historical tradition contains a reply to my speculations. In a legendary account, animal tracks inspired the invention of Chinese writing. In the reign of Emperor Huang Di, around 2600 BC, one of his ministers, Cang Jie, decided that the footprints left in the dirt by various bird species constituted a small set of easily recognizable shapes—and he used them to create the first Chinese characters.

This legend is reminiscent of the mystic metaphor of the "book of Nature," which captured the imagination of many theologians and scientists, including Galileo himself. Perhaps we slowly learn to decipher the natural world as though it were an open book written by God. In his *Introduction to the Symbol of the Faith*, Luis of Granada, a sixteenth-century theologian, writer, and preacher, wrote:

> What is this entire visible world if not a big and wondrous book that you, Lord, have written and offered to the eyes of all nations of the world? . . . What then are all the creatures

of this world, so beautiful and perfect, if not as though they were richly illuminated letters that proclaim the elegance and wisdom of its author?

Sir Thomas Browne, an eighteenth-century English author, reiterated these beliefs:

The finger of God hath left an inscription upon all his works, not graphical, or composed of letters, but of their several forms, constitutions, parts, and operations, which, aptly joined together, do make one word that doth express their natures.

If we come down from these mystic heights and return to a real life and the brain imaging laboratory, we have to give serious thought to how we might test the hypothesis of a cortical competition between word reading and animal track recognition. Research into this topic first requires the selection of a number of natural images like faces, human bodies, animals, footprints, trees, plants, minerals, rivers, and clouds. We would then have to observe the activation patterns they induce on the surface of the visual cortex. The real challenge lies in conducting this experiment with two distinct groups of similar ages, one literate and the other illiterate—while controlling for all other aspects of their visual environment.

Although it would be hard to carry out, this experiment would be well worth attempting. It would show directly whether the acquisition of reading disturbs the preexisting patchwork of visual activation. Some cortical territories may shift, shrink, or even disappear as the response to written words emerges. Perhaps the response to footprints would be attenuated, shifted toward the right hemisphere or simply wiped out by the acquisition of reading.

At the time of writing, none of these ideas has yet been tested. My research team hopes to conduct experiments along these lines in the next few years—but a major hurdle will involve the identification of a population of illiterate subjects who can be scanned easily and under stringent ethical conditions. Meanwhile, the notion of a cortical competition induced by neuronal recycling can be studied for domains other than reading. All of us know people who are passionate about some limited field of visual expertise like cars or birds and who are ready to invest as much time on them as some of us spend on reading. Does a car buff who can instantly tell the difference between a Studebaker Gran Turismo and an Alfa Romeo Giulietta have a reorganized brain? Does a new brain area develop when bird-watchers amass knowledge about the dozens of species of avocets or woodcocks? Above all, does expert training of this kind have a cognitive price?

Some of these problems are now being examined by a network of "experts of expertise" managed by the Canadian researcher Isabel Gauthier and her colleagues. In addition to car and bird experts, they have also convinced a few paid volunteers to spend hours studying funny little virtual characters called "greebles," a family of shapes designed to challenge human visual recognition:

Brain imaging has revealed that all of these forms of visual expertise increase occipito-temporal responses to the corresponding objects. The visual brain of a car buff is better activated by the sight of a Studebaker than by that of an avocet. In the brains of bird-watchers, perhaps unsurprisingly, the reverse occurs.[248]

More controversially, Gauthier and Michael Tarr have claimed that this kind of expertise systematically invades and interferes with cortical regions previously dedicated to face processing. The claim is that cars and birds compete with faces because all are handled in the same cortical "expertise" area.[249] Brain activity evoked by faces is in fact slightly reduced when these compete with the presentation of objects from other domains of expertise, either cars or greebles.[250] In a test of face perception, car or bird experts even lose a few points when their efficiency for face recognition is measured by evaluating the brain's capacity to integrate face parts into a homogeneous whole ("holistic" perception).

Although these results are contested, they do suggest that the notion of cortical competition, a direct consequence of the neuronal recycling hypothesis, could have a grain of truth to it. When we spend time on reading or ornithology, we also trade in cortical space. This obviously reduces the brain resources available for other skills—and our face perception abilities may suffer.

When Letters Have Colors

A black, E white, I red, U green, O blue: vowels,
I shall tell, one day, of your mysterious origins

—ARTHUR RIMBAUD, "VOWELS"

Data from an unexpected source confirms that reading acquisition competes with prior cortical representations. Some people, like the French poet Arthur Rimbaud, are convinced that the letters "A," "E," or "I" have well-defined colors that can be seen exactly in the mind's eye. The experience is called synesthesia—a strange intermingling of sensory modalities that constitutes further evidence for neuronal recycling.[251]

In synesthetes, the five senses are no longer separated, but appear to cross-activate each other. Synesthesia can take various forms. One person sees color and motion each time he hears a voice or listens to music. Another sees a halo of color around certain letters or digits. A third thinks that numbers occupy fixed spatial locations on an internal continuum that curls around like a twisted line in two or three dimensions.[252]

These strange reports are more than figments of a poet's imagination. Synesthetes are generally not cranks, but levelheaded people. Several well-known artists (Kandinsky, Messiaen, Nabokov), scientists (Richard Feynman, Nikola Tesla), and a great many anonymous human beings figure among them. All insist that they merely report what they see. Rimbaud himself is said to have confided to his friend Ernest Delahaye about his famous poem "Vowels," "I believe that I sometimes saw or felt in this manner, and I say it, I talk about it, because I find it just as interesting as anything else."

Recent experiments show that synesthesia behaves like a perceptual illusion. At intervals as long as several years, synesthetes always select the same shades of color to describe their feelings about colored letters—while non-synesthetes, even after some training, often choose different tints from those they originally selected. Synesthetes also have talents that cannot be faked.[253] The average person has trouble finding a hidden digit 2 in a sea of 5s because their shapes are very similar, but digit-color synesthetes see it immediately because the two digits appear in different colors (figure 5.1).

Learning seems to play an important role in synesthesia. To a synesthete, not all objects and shapes appear colored—in most cases, only letters and digits are affected. These are precisely the symbols that are learned late in childhood. Cases are also on record of synesthesia for many other domains of human culture such as spoken words, language sounds, symphonies, musical instruments . . . and even the taste of dishes.

That synesthesia appears exclusively for learned cultural objects suggests a profound link with the necessary nature of neuronal recycling. My recycling hypothesis predicts that every new cultural object must compete with representations established earlier in life, as it searches for a "neuronal niche" on the surface of the cortex. This neuronal competition may create temporary confusion at the cortical level.[254] In the case of reading, as we have seen, the activation evoked by written words is initially diffuse. It gropes around within

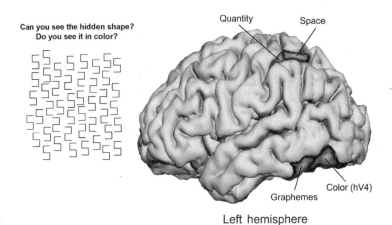

Figure 5.1 Synesthesia—a crosstalk of sensory modalities—can be linked to a partial failure in neuronal recycling. Some synesthetes claim to see digits and letters in colors of variable vividness—for instance "A" is red, "E" is blue, and so on. Tests show that this is not solely a vague subjective report, but a genuine visual illusion: for synesthetes, the difference between the digits 2 and 5 pops out, because they see them in distinct colors. When these people see digits or letters, the left occipito-temporal "letterbox" area is activated, but stimulation also overflows into the neighboring V4 area associated with color vision, thus creating a neural code that confounds letters and colors. Other synesthetes see numbers in space, a type of association that might relate to the overlap of parietal areas for quantity and space. *Used with permission of Ed Hubbard.*

the visual system before it finally concentrates on the left occipito-temporal letterbox area. Once this has occurred, the rest of the brain can learn that the neural signals arising from this area are now coding for words, not colors. In synesthetes, however, I speculate that this choice is never fully decided. It is as though the neurons could not quite decide if they should dedicate themselves to letters or to colors.

Thanks to brain imaging, Edward Hubbard was able to confirm part of this scenario. He scanned several adult synesthetes who associated letters with colors and observed an unusual overlap of cortical activations.[255] The sight of letters generally activates a well-defined area whose rear portion, interestingly enough, lies very near another visual region called area V4. This area plays a central role in color perception. In fact, a classic way to isolate area V4 consists of getting volunteers to compare paintings made of many colored rectangles, like those of the Dutch painter Piet Mondrian, with the presentation of similar black-and-white pictures. In synesthetes, Hubbard found that the activations induced by letters and by color patches overlap quite noticeably. It was as if their neuronal recycling was incomplete and the division of labor within the visual cortex had been prematurely aborted. The synesthete's cortical patchwork appeared to have remained stuck in an intermediate state of specialization, where letters failed to concentrate on a narrow cortical patch. Why? Perhaps the synesthete's cortex relies on distinct cortical plasticity rules that prove inappropriate when the brain needs to revise its cortical code. Hubbard and Vilayanur Ramachandran speculate that synesthesia might be due to a mutation of the genes involved in the pruning of synapses during brain development.[256] Indeed, this phenomenon runs in families along a pattern that might suggest genetic transmission, possibly linked to the X chromosome.

If my idea is correct, however, synesthesia is not just a genetic anomaly but a problem that occurs systematically during cortical recycling. A transient form may therefore be present in all children for a brief period, before the cortical maps for letters and colors converge onto their final locations. The hypothesis that all children are synesthetes is not unfounded. Even young infants perceive, for instance, a clear relation between sharp sounds and small pointed objects.[257] One of my children, when he was seven, clearly saw the digits 1, 2, 3, 4, 5, and 9 in color. A few months later, the colors had slightly changed, and at the age of eight they suddenly vanished. "I remember seeing the digits in colors," he said, "but now I can't." To the best of my knowledge, no systematic study has ever examined whether all children temporarily experience a similar phenomenon. It merely seems plausible that the transient association of symbols and colors, in children, can serve as a marker of the ongoing reorganization of the cortex during the acquisition of reading.

From Neuroscience to Education

What we have seen so far is that the acquisition of reading entails massive functional changes in children's brains. They must first discover phonemes, then map letters onto sounds, and then establish a second lexical reading route. Learning to read implies a literal search for a proper "neuronal niche" for written words in the patchwork of cortical areas for face, object, or color perception.

From a practical standpoint, it is essential to examine whether we can take advantage of these scientific advances to improve teaching. Does our growing understanding of reading lead to clear indications concerning optimal teaching methods? Do some educational techniques ease the transition toward the adult state better than others?

A great deal of caution is needed here. My own impression is that neuroscience is still far from being prescriptive. A wide gap separates the theoretical knowledge accumulated in the laboratory from practice in the classroom. Applications raise problems that are often better addressed by teachers than by the theory-based expectations of scientists. Nevertheless, brain imaging and psychological data cannot be detached from the great pedagogical debates. Relativism notwithstanding, it simply is not true that there are hundreds of ways to learn to read. Every child is unique . . . but when it comes to reading, all have roughly the same brain that imposes the same constraints and the same learning sequence. Thus we cannot avoid a careful examination of the conclusions—not prescriptions—that cognitive neuroscience can bring to the field of education.[258]

To define what reading is *not* is a good starting point. As overtrained readers, we no longer have much perspective on how difficult reading really is. We tend to believe that one glance at a word will allow its immediate and global identification in a single step. Nothing could be further from the truth. The brain does not go straight from the images of words to their meaning. An entire series of mental and cerebral operations must occur before a word can be decoded. Our brain takes each string apart, then recomposes it into a hierarchy of letters, bigrams, syllables, and morphemes. Effortless reading simply serves to show that these decomposition and recomposition stages have become entirely automatic and unconscious.

With this definition in mind, the goal of reading instruction becomes very clear. It must aim to lay down an efficient neuronal hierarchy, so that a child can recognize letters and graphemes and easily turn them into speech sounds. All other essential aspects of the literate mind—the mastery of spelling, the richness of vocabulary, the nuances of meaning, and the pleasures of literature—depend on this crucial step. There is no point in describing the delights of reading to children if they are not provided with the means to get there. Without phonological decoding of written words their chances are significantly reduced. Considerable research, both with children and with illiterates, converges on the fact that grapheme-phoneme conversion radically transforms the child's brain and the way in which it processes speech sounds. This process whereby written words are converted into strings of phonemes must be taught explicitly. It does not develop spontaneously, and must be acquired. Reading via the direct route, which leads straight from letter strings to their meaning, only works after many years of practice using the phonological decoding route.

Reading Wars

Cognitive psychology directly refutes any notion of teaching via a "global" or "whole language" method. I have to stress this point forcefully because pedagogical strategies of this kind were once very popular and have not lost their appeal for some teachers. These methods teach children to recognize direct associations between written words or even whole sentences and their corresponding meanings. The technique involves the child's immersion in reading, and the hope is that he will acquire reading spontaneously like a natural language. Extreme advocates of the whole-language or whole-word approach explicitly deny the need to teach the systematic correspondence between graphemes and phonemes. They claim that this knowledge will appear by itself as the result of exposure to the correspondences between words and meanings.

Although its postulates may seem strange, the whole-language approach was grounded in a generous principle. It refused drill, which was thought to turn children into automata who could only drone out silly sentences like "Pat the cat sat on the mat." The whole-language movement was vigorously opposed to phonics because it considered that this training detracted from understanding text, which was the primary goal of reading instruction. Whole-language advocates placed emphasis on text comprehension by quickly giving children access to meaningful stories. The claim was that children found it more fun to discover phrases than words, spelling rules, or boring letter-to-sound decoding. They would be empowered if they could "build their own learning environment" and spontaneously discover what reading was all about; never mind if they initially played at riddles and read "the kitty is thirsty" instead of "the cat drank the milk." For the supporters of the whole-language approach, the child's autonomy and the pleasure of understanding were what counted most, over and above the accuracy with which individual words could be decoded.

The dispute between advocates of whole-language learning and the proponents of phonics instruction plagued schools and education policymakers around the world for at least the last fifty years. In the United States, the "reading wars" culminated in 1987 when the state of California, as part of a new language-arts curriculum, passed bills favoring the whole-language approach over basic decoding skills. In a matter of a few years, reading scores in California plummeted. In 1993 and 1994, scores collected by the National Assessment of Educational Progress found that the reading skills of three out of four children in the state were below the average for their grade. Whole-language instruction was largely blamed for the disastrous scores, and a major backlash ensued. Radical reform led most schools to return to the systematic teaching of letter-sound correspondences, thus implementing the conclusions of an influential review by the National Reading Panel.[259]

The whole-language approach has today been officially abandoned. Nonetheless, I suspect that the issue is still alive in many a teacher's mind because whole-language advocates are still firmly entrenched in their position. They are convinced that their approach is best suited to children's needs. In France as well as in the United States, efforts to reconcile the two camps have led to the adoption of an unhealthy compromise called "mixed" or "balanced reading" instruction.

A great many teachers are so confused by the constant swings back and forth from one educational approach to the other that they borrow at random from all the existing methods. Whole-language has been officially scorned, but either out of inertia or habit it still is surreptitiously present in reading manuals and teacher instruction programs. Even if grapheme-phoneme correspondences are now the main focus, activities dating back to the whole-language approach are still present in the classroom. These include pairing a word with an image, recognition of the overall contour of words, and sight recognition of the child's first and last names (figure 5.2).

It would be all too easy to lay the blame for this dispute on educators. In fact, both teachers and psychologists must bear the burden of guilt because

Figure 5.2 In spite of its inefficacy, the whole-language method continues to inspire educators throughout the world. The tests shown here are still used in some classrooms during the first few weeks of reading instruction. Although the child cannot yet sound out letters, he is asked to pair a whole word with the corresponding image (top). He of course makes gross errors such as calling a cat "pig," suggesting that he is unaware of the alphabetic principle. The child is also taught to attend to the overall contours of ascending and descending letters. Note that even the teacher errs on the word "boat." These exercises are in no way related to how our brain recognizes words.

they once were convinced advocates of whole-language reading.[260] As early as the eighteenth century, Nicolas Adam's *Pedagogical Dictionary* (1787) spread the progressive ideas of the abbot of Radonvilliers, who complained that syllabification "torments" children by forcing them to "memorize a great number of letters, syllables, and sounds that are meaningless to them." Instead of this torture, he argued in favor of "amusing children with whole words" written on cards. At around the turn of the nineteenth century, the psychologist James McKeen Cattell, at the University of Pennsylvania, announced that he had discovered that written words were recognized faster than their individual letters. This finding seemed to uphold the whole-language approach. His work was extended by the Swiss psychologist Edouard Claparède, for whom the acquisition of reading relied on the "syncretic" perception characteristic of young children. In the 1930s, the Belgian physician and psychologist Ovide Decroly incorporated these ideas into his "ideo-visual" reading method, which then spread to several European countries and received the seal of approval from psychologists as eminent as Jean Piaget and Henri Wallon.

The emphasis on the global shape of words also invaded the world of typography, where the term "bouma" (named after the Dutch psychologist Herman Bouma), was coined to refer to the contours of words. In the hope of improving readability, typographers intentionally designed fonts that created the most distinctive visual "boumas."[261]

The Myth of Whole-Word Reading

It is amazing to think that scientists and educators could have joined forces to support a conclusion that we now know was wrong. The scientific arguments that led to the fallacious idea of whole-word reading and the way they have been refuted by recent experiments merit careful examination. A recent article lists at least four points that constitute the clay pillars of the whole-word construction.[262]

1. **Reading time does not depend on word length.**
   It takes us almost the same time to read short and long words, regardless of number of letters (within an interval of about three to eight letters). This is a proven fact that, on the surface, appears to suggest that word

recognition does not rest on the systematic breakdown of words into component letters. Subjectively, indeed, word reading does seem to be an elementary and immediate operation that processes a letter string like an undivided whole. Modern scientific studies of reading, however, demonstrate that this thinking is wrong. Even though word length has no impact on adult readers, it does not mean that our brain does not pay attention to letters. The explanation is much simpler: our visual system simply processes all letters simultaneously and in parallel rather than one after another. In young children, however, the process is different. During the first few years of reading acquisition, reading time is strictly related to the number of letters in a word. This word length effect takes years to vanish. The massive impact of the number of letters on young children's reading time provides clear evidence that reading is not a global, holistic process—especially during the early years.

2. **Recognizing a whole word can be faster and more efficient than recognizing a single letter.**
This effect, discovered by Cattell, was replicated by Gerald Reicher and popularized in psychology as the "word superiority effect." We examined it in chapter 1 (page 48). The finding is undisputed, but it is no longer considered as proof that whole-word recognition precedes letter extraction. On the contrary, recent analyses indicate that performance in reading a word hidden in visual noise is directly related to the rate at which its component letters are recognized.[263] If we respond faster to a whole word, it may be that our conscious attention is more easily oriented toward the higher levels of the cortical hierarchy. Accessing the identity of individual letters, even if they are extracted earlier during visual processing, demands more effort.[264] A related factor concerns the number of relevant neurons. Many neurons, at several lexical, semantic, and phonological levels, discriminate "head" from "heat," whereas only a limited set of visual detectors discharges in a distinctive fashion for the isolated letters "d" and "t." Of course, this does not mean that word perception precedes letter identification, but that the differences between letters get amplified at higher levels of word processing.

3. **We are slightly faster at reading words in lowercase than in uppercase.**
For supporters of the whole-language approach, the speed at which we

read lowercase rather than uppercase letters reflects the unique visual pattern created by ascending and descending letters such as "f," "l," "g," and "p." These letter shapes create a contour-specific signature for each word— the typographers' "bouma." This contour disappears when the word is written in uppercase letters, which are all of the same size, and our reading speed thus decreases. However, this theory does not work. If we really use contours to recognize words, we should not just be slower, we should quite simply be unable to recognize uppercase words. It should also be impossible to read WoRdS PrInTeD iN MiXeD cAsE, which destroys familiar contours. But we know that these manipulations leave words amazingly readable. As we saw in chapter 2 (pages 88–91), the brain's letterbox area generalizes invariantly across uppercase and lowercase presentations, even under subliminal conditions. Thus the slight slowing observed for uppercase letters is not worth noting. It may just be due to less familiarity with these shapes.

4. **Typographical errors that respect the overall contour of a word are more difficult to detect than those that violate it.**
This is one of the key arguments that convinced typographers to pay attention to word contour. Here is a simple example: if we look for the target word "test," we find it more difficult to detect an error in "tesf" than in "tesg," where the ascending letter "t" has been replaced by a descending letter "g." Again, while this is an undisputed fact, its interpretation has nothing to do with global word shape. It can be entirely attributed to the confusion between individual letters. The "f" in "tesf" is quite similar to the "t" in the target word "test," while the "g" in "tesg" is strikingly different. Experiments that disentangle the two factors prove that it is letter similarity that leads to the confusion between the two letter strings, not whole-word resemblance.[265]

In summary, there is no longer any reason to doubt that the global contours of words play virtually no role in reading. We do not recognize a printed word through a holistic grasping of its contour, but because our brain breaks it down into letters and graphemes. The letterbox area in our left occipito-temporal cortex processes all of a word's letters in parallel. This fast and parallel processing probably explains why well-known and respected psychologists once propounded theories of global or "syncretic" reading. Today, we know that the immediacy of reading is just an illusion engendered by the extreme automaticity of its component stages, which operate outside our conscious awareness.

The Inefficiency of the Whole-Language Approach

Direct experimentation confirms that the whole-language system does not perform effectively. Several researchers have demonstrated this by comparing performance on learning a new writing system using either a whole-language method or a phonics-based approach. One of the most striking experiments was designed by my friend Bruce McCandliss, a professor at the Sackler Institute in New York.[266] Bruce invented a new artificial alphabet that is peculiar both because it is written from bottom to top and because the letters touch each other. As a result, each word is represented by what looks, at first sight, like a continuous pattern of lines. Here are four words written in this alphabet:

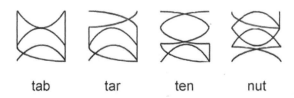

tab   tar   ten   nut

When we first look at these patterns, it is not obvious that they are made up of letters. Only close examination reveals that there is a common pattern for "t" that is repeated four times in this example. McCandliss taught his script to two groups of students divided into a whole-language group and an analytic group. He asked the whole-language group to memorize each word as a global shape, and did not warn them of the presence of letters. The analytic group was told, however, that the words were made up of a sequence of letters written from bottom to top. Thereafter, the two groups were trained identically through repeated exposure to a given shape and the corresponding English name.

The very slight difference in instruction given to the two groups had an impressive cascading impact. After one day of training on a list of thirty words, the whole-language group was actually *better* at recognizing them than the analytic group, who were still struggling to discover the letters. This interesting result fits with reports by many advocates of the whole-language method. They repeatedly assert that their approach gives children a head start. However, this is only true at the beginning, for the first thirty words or so. The acquisition of letter-to-sound correspondences requires greater initial effort, but the results in the long run are more rewarding. Indeed, on the second day, when students learned a new list of thirty words, the whole-language group began to lose ground. They learned most of the new words—but at the expense of the initial list, which they quickly forgot. The same pattern recurred each time a new list was introduced—they had to start again from scratch and lost the little they remembered from the previous day. There is nothing surprising about this. The group was trying to accomplish the impossible task of learning each word independently, as if it were a Chinese character.

Meanwhile, the students who were attending to letters were making slow but steady progress. Every day, their success rate improved. Strikingly, even on the first encounter with a new list of words, they performed better than chance. Their performance, moreover, also improved for the words they had seen earlier, even when they had had no occasion to practice them again! This result is not as puzzling as it might seem. The students' familiarity with the letter-to-sound correspondences allowed them to decipher any word, even when they did not remember having seen it before.

The McCandliss experiment reveals the two major problems inherent in the whole-language approach. First, mere exposure to written words, without any explicit training in letters and sounds, is not enough to allow for the discovery of spelling regularities—at any rate not systematically and fast. Even after thousands of words, it is entirely possible for an uninformed adult to fail to notice that they are made up of a regular alphabet of letters.[267]

Second, although generalization plays an essential role in children's acquisition of reading, the whole-language method provides no ground for generalizing to new words. Even the most dedicated educator cannot teach his pupils all the words in the English language! The mastery of reading lies, above all, in our ability to decode new words. "Self-teaching"[268] is an essential ingredient on the road to independent reading. Once they master the spelling-to-sound correspondences, children can, on their own, decipher the pronunciation of a novel string and associate it with a familiar meaning. With self-teaching, the neuronal links from letter strings to sound and meaning can be progressively automatized without any further formal instruction.

I would like to insist on this last point, because it refutes one of the central claims of the whole-language system—the idea that this method alone is respectful of the child's independence and happiness. In reality, only the teaching of letter-to-sound conversion allows children to blossom, because only this method gives them the freedom to read novel words in any domain they choose. It is therefore misguided to pit the intellectual freedom of a child against rigorous drill. If a child is to learn to read quickly and well, he must be given well-structured grapheme-phoneme instruction. The effort is real, but the payoff in independence is immediate when children discover, often with awe, that they can decipher words they never learned in class.

I would not want to give the impression that my rejection of the entire whole-language method is based solely on laboratory experiments or on theoretical principles. Its efficiency has also been disproved in the classroom. The psychology of reading has taken advantage of the variety of existing teaching methods to measure each of their impacts on children. This line of research, which borders on epidemiology, has scored performance on various standardized reading tests not only as a function of the teaching method used, but also of the children's age, socioeconomic background, and several other variables.

Results coming from these studies support a conclusion that has been validated for American children as well as for those exposed to other alphabetical writing systems.[269] Teaching methods based on a whole-language approach are systematically less efficient than phonics. Performance is best when children are, from the beginning, directly taught the mapping of letters onto speech sounds. Regardless of their social background, children who do not learn letters and graphemes suffer from reading delays. These are often far from negligible and persist for many years, even if adults do finally get over them.

Crucially, contrary to what is frequently claimed by whole-language advocates, tests reveal that children taught with a whole-language approach not only score below their grade level when reading new words, but are also slower and less efficient in sentence and text comprehension. This observation refutes the idea that phonics transforms children into robots unable to attend to meaning. In fact, decoding and comprehension go hand in hand. The children with the best scores in decoding single words and pseudo-words also perform best on sentence and text comprehension.

Whole-language advocates appear to confuse the means and the end. They are correct in saying that decoding is a means, not an end in itself. In that respect, one can only praise the many recent reading manuals that introduce meaningful short stories early on instead of gibberish syllables like "pa ta ka." Nevertheless, fluent decoding is indispensable for comprehension. The faster the speech-to-sound route is automatized, the more a child will be in a position to concentrate on the meaning of what he reads.

A Few Suggestions for Educators

In the final analysis, what can psychology and neuroscience recommend to teachers and parents who wish to optimize reading instruction? The growing science of reading has no ready-made formulas, but it does offer a few suggestions. The punch line is quite simple: we know that conversion of letters into sounds is the key stage in reading acquisition. All teaching efforts should be initially focused on a single goal, the grasp of the alphabetic principle whereby each letter or grapheme represents a phoneme.

In kindergarten, very simple games can prepare children for reading acquisition. At the phonological level, preschoolers benefit from playing with words and their component sounds (syllables, rhymes, and finally phonemes). At the visual level, they can learn to recognize and trace letter shapes. The Montessori method, which requires tracing sandpaper letters with a fingertip, is often of considerable help at this early age. It helps children figure out each letter's orientation, and makes it clear that "b," "p," "d," and "q" are different letters.

After this preparatory stage, children must be taught, without fear of repetition, how each letter or group of letters corresponds to a phoneme. The child's brain does not automatically extract these correspondences by dint of seeing a great many words. It must be explicitly told that each speech sound can be represented in different "clothes" (letters or groups of letters) and that each letter can be pronounced in one of several ways. Because English spelling is complex, introduction of graphemes must occur in logical order. Their presentation must start with the simplest and most frequent ones that are almost always pronounced in the same way, such as "t," "k," and "a." Less frequent graphemes ("b," "m," "f"), irregular ones ("i," "o"), or complex ones ("un," "ch," "ough") can be introduced gradually. Children's attention must be drawn to the presence of these individual elements within familiar words. This can be done by assigning each grapheme a distinctive color, or by moving them around to create new words. It should also be explained that letters unfold in a fixed order, from left to right, with no gaps. The ability to attend to the various subcomponents of words is so essential that this must be taught explicitly by, for instance, covering words with a sliding window that reveals only a few letters at a time.

Of course, learning the mechanics of reading is not an end in itself—in the long run, it only makes sense if it leads to meaning. Children must know that reading is not simply mumbling a few syllables—it requires understanding what is written. Each reading period should end with reading words or sentences that can be easily understood and that the child can repeat, summarize, or paraphrase.

A great many teachers will consider my recommendations redundant and obvious—but it does no harm to specify them. I once tried out reading software that was supposedly "award-winning," where the very first word introduced to the beginning reader was the French word *oignon*, pronounced *onion* almost as in English—probably the most irregular spelling in the French language! Errors as ridiculous as this one clearly show that even the most basic principles of teaching have not yet been absorbed by everyone.

Stressing what parents and teachers should *not* do is equally important. To trace the global contours of words is useless. Likewise, to draw children's attention to ascending and descending letter patterns is not particularly helpful. Exercises like these may even be detrimental to reading, inasmuch as they mislead children into paying attention to the global contour of words. This makes them conclude that they can guess at words without examining their component letters one by one. The contours of the words "eight" and "sight" are almost identical. Children need to understand that only the analysis of letters one by one will allow them to discover a word's identity.

Because of the essential need to avoid distracting the child's attention from the letter level, I am wary of the many richly decorated reading manuals that contain more illustrations than text. Word posters displayed in classrooms all through the school year, with the same words appearing at the same place, can also create problems. Some children, often the most gifted, merely memorize the fixed position of each word and the general layout of the page and no longer attend to the actual letters in the individual words. This strategy can give teachers and parents—and worst of all, the child himself—the illusion that he knows how to read. Illustrations also divert attention from text. Children now live in a world of constant overstimulation and distraction, so that some no longer learn to sustain attention for long periods of time. A return to sober texts, written on a blackboard during class (so that gesture is also memorized) might be beneficial. It might also be worthwhile to remind the child that although reading is hard work, it has its own inherent reward in the decoding and understanding of text.

Going too fast can also be a handicap. At each step, the words and sentences introduced in class must only include graphemes and phonemes that have already been explicitly taught. Reading lessons provide little room for improvisation. A teacher cannot simply decide, at the last minute, to work on a few unprepared words or sentences. A haphazard choice of this kind will be confusing, because it is very likely to require advanced knowledge that the child has yet to learn.

As expert reading adults, we systematically underestimate how difficult it is to read. The words given to beginning readers must be analyzed letter by letter in order to ensure that they do not contain spelling problems that are beyond the child's current knowledge—for instance, unusual pronunciations, silent letters, double consonants, or peculiar endings such as the suffix "-tion." All of these peculiarities, if they are introduced too early in the curriculum, can make children think that reading is arbitrary and not worth studying. As a scientist and a professor myself, I expect the teachers and educators to whom I entrust my children to invest as much obsessive care in the design of lessons as my colleagues and I do when we prepare a psychological experiment.

Finally, guardians of children with reading problems should not give in to despondency. Reading difficulty varies across countries and cultures, and English has probably the most difficult of all alphabetic writing systems. Its spelling system is by far the most opaque—each individual letter can be pronounced in umpteen different ways, and exceptions abound. Comparisons carried out internationally prove that such irregularities have a major impact on learning.[270] Italian children, after a few months of schooling, can read practically any word, because Italian spelling is almost perfectly regular. No dictation or spelling exercises for these fortunate children: once they know how to pronounce each grapheme, they can read and write any speech sound. Conversely, French, Danish, and especially British and American children need years of schooling before they converge onto an efficient reading procedure. Even at the age of nine, a French child does not read as well as a seven-year-old German. British children only attain the reading proficiency of their French counterparts after close to two full years of additional teaching (figure 5.3).

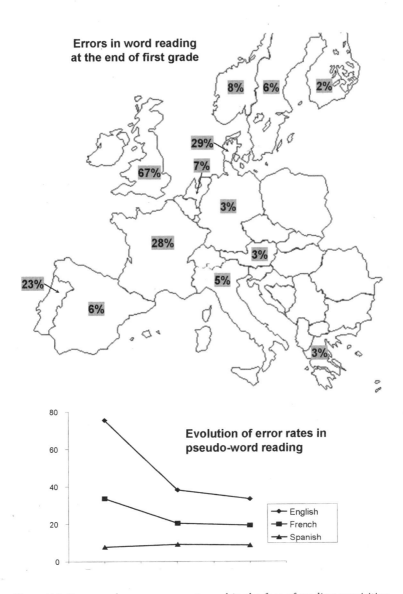

Figure 5.3 European languages are not equal in the face of reading acquisition. Error rates in reading familiar words were measured in fifteen European countries after one year of schooling (top, data from Seymour, Aro, & Erskine, 2003, table 5). Finnish, German, Greek, Austrian, and Italian, whose spelling is transparent, were already read accurately. At the other end of the scale, English is very opaque and children could only read one out of three words. These inequalities were perpetuated in subsequent years, particularly for pseudo-words such as "balist" or "chifling" that can only be deciphered with grapheme-phoneme decoding (bottom, data from Goswami et al., 1998, table 8). It takes one or two additional years of schooling before an English child reaches the reading level of a French child.

Barring major spelling reform, there is not much we can do to simplify the acquisition of reading in English. All we can do is encourage our children to practice reading daily . . . and to remind ourselves that our situation could be worse. In China, reading lessons extend well into the teens, in order to acquire the several thousand characters needed to read a newspaper. Chinese children's plight is all the more surprising in that it could be avoided, since most of them nowadays start by learning the simple alphabetical Pinyin notation, which is acquired in a matter of months.[271]

Teachers can also derive some consolation from bearing in mind that the time spent on learning to read has an extraordinarily profound and useful impact on the child's brain. Try to picture the ceaseless activity of new connections building up after each reading lesson. Every young reader's letterbox area is called on to integrate a hierarchy of neurons coding for letters, bigrams, graphemes, and morphemes. This effort creates tremendous neuronal effervescence throughout the reading circuitry. Hundreds of millions of neuronal wires must find their proper targets within other regions coding for speech sounds and meaning. Whether we like it or not, this neuronal hierarchy is far more complex for English or for French than for transparent languages like Italian. The amount of neuronal recycling required for English is so impressive that we must relentlessly teach children to cope with each of its countless spelling pitfalls—even long after the end of elementary school.

My firm conviction is that every teacher should have some notion of how reading operates in the child's brain. Those of us who have spent many hours debugging computer programs or repairing broken washing machines (as I have done) know that the main difficulty in accomplishing these tasks consists in figuring out what the machine actually does to accomplish a task. To have any hope of success, one must try to picture the state in which it is stuck, in order to understand how it interprets the incoming signals and to identify which interventions will bring it back to the desired state.

Children's brains can also be considered formidable machines whose function is to learn. Each day spent at school modifies a mind-boggling number of synapses. Neuronal preferences switch, strategies emerge, novel routines are laid down, and new networks begin to communicate with each other. If teachers, like the repairman, can gain an understanding of all these internal transformations, I am convinced that they will be better equipped to discover new and more efficient education strategies. Although pedagogy will never be an exact science, some ways of feeding the brain with written words are more effective than others. Every teacher bears the burden of experimenting carefully and rigorously to identify the appropriate stimulation strategies that will provide students' brains with an optimal daily enrichment.

CHAPTER 6

The Dyslexic Brain

Some intelligent and well-rounded children experience disproportionate trouble in learning to read—dyslexia. In most cases,
this problem is linked to an impaired ability to process phonemes, and an entire causal chain from gene to dyslexic behavior is now being uncovered. The brains of dyslexic children
present a number of characteristic anomalies: the anatomy of
the temporal lobe is disorganized, its connectivity is altered,
and several regions are insufficiently activated during reading.
The disorder is suspected to have a strong genetic component,
and four susceptibility genes have been identified, most of
which control neuronal migration, a major event in the construction of the brain during pregnancy. Any disturbance that
affects this episode can lead to a disorganization of the layers
of the cortex.

Do these biological anomalies imply that dyslexia cannot be
cured? Not at all—new remedial intervention strategies are
bringing fresh hope. Based on intensive computerized training, these techniques improve reading scores and lead to partial normalization of brain activity in dyslexic children.

> Mrs Everett produced a book called *Reading Without Tears*. It certainly did not justify its title in my case. . . . I was, on the whole, considerably discouraged by my school days. . . . It was not pleasant to feel oneself so completely outclassed and left behind at the beginning of the race.
>
> —WINSTON CHURCHILL, *MY EARLY LIFE: A ROVING COMMISSION*

*A*ll schoolteachers, regardless of talent and devotion, occasionally meet a child who is resistant to reading. His intelligence is normal, or even above average for mathematics or handicrafts. When it comes to reading, however, he suddenly becomes hopeless, stumbles on every syllable, mixes speech sounds, guesses at words without thinking, quickly gets discouraged . . . and also discourages everyone around him. A visit to the speech therapist often confirms the dreaded diagnosis: this child suffers from dyslexia.

What brain mechanisms lurk behind this familiar term? Can our scientific understanding of reading shed light on dyslexia? Does it result from a "mental block" (assuming that this term has any meaning) or from genuine brain impairment? Which areas, which neurons, which genes are involved? What sort of therapy can be recommended?

This chapter summarizes thirty years of research on dyslexia, from its initial psychological classification, to brain imaging, anatomical, and genetic studies, and finally the recent design of intervention software. Renewed hope of partially curing this common handicap is now emerging, and a neuroscientific revolution is under way as hardly a month goes by without some new finding. Discoveries include the identification of risk factors for dyslexia, the origins of their genetic bases, and an inventory of the relevant chromosomes, genes, and the biological mechanisms they control. Research into all these domains contributes daily to our knowledge of the genetic and neuronal mechanisms of dyslexia.

What Is Dyslexia?

A number of traits define dyslexia: a disproportionate difficulty in learning to read that cannot be attributed to mental retardation, sensory deficit, or an underprivileged family background. This definition makes it clear that not all poor readers are dyslexics. Misdiagnosed auditory deficits, low IQ, a poor educational environment, or simply the complexity of spelling rules can explain some children's reading problems. It is only when all of these possible causes have been eliminated that one would suggest dyslexia.

Current estimates indicate that from 5 to 17 percent of children in the United States suffer from dyslexia.[272] While these numbers seem very high, they depend on the threshold used to define the impairment and are therefore somewhat arbitrary. Within an age group, scores on standardized reading tests trace a perfect bell-shaped curve.[273] If that curve had two peaks, one for normal scores and the other for poor performance, it would be easy to define the population of dyslexic children with fairly objective criteria. However, because there is only one peak in the center, there is no clear boundary separating normal readers from dyslexics. The prevalence of dyslexia depends on the setting of an arbitrary criterion for "normality." In our literate environment, it is estimated that children who score below the 5 to 10 percent level on reading tests are at a severe disadvantage, even if their problem would probably have gone unnoticed two centuries ago, when only a fraction of humanity learned to read.

The degree of arbitrariness in the cutoff point for dyslexia might lead to the erroneous conclusion that dyslexia is a purely social construction, linked to prevalent overmedicalization that tends to confuse behavior and health problems. This is not the case here. A number of indicators point to the cerebral origins of dyslexia. Since the 1950s, it has been found to run in families. On these grounds, vast behavioral genetic studies have been conducted in hundreds of families, particularly at the University of Colorado by John DeFries and his collaborators. The results confirm the genetic heritability of reading abilities. Identical twins, who share essentially the same genetic material, have much more tightly correlated reading scores than fraternal twins of the same sex, who share only half of their genetic background. Within a family, if a child suffers from dyslexia, his siblings have a 50 percent chance of being dyslexic too.

While scientists now believe that dyslexia is often genetically based, it is clearly not a "monogenic" disease linked to the mutation of a single gene. A range of risk factors and a bunch of genes collectively conspire to disrupt the acquisition of reading. It is not surprising to find that a great many genes contribute to a cultural ability such as reading. Expert reading depends on a fortuitous combination of cerebral connections that luckily preexist in our primate brains and take years of training to convert to a new use. One mishap in the circuitry is enough to bring the fragile process of neuronal recycling to a grinding halt.

Phonological Trouble

In spite of considerable research on the topic, consensus on the exact nature of dyslexia is only beginning to emerge. A number of results point to an anomaly in the phonological processing of speech sounds.[274] Additionally and perhaps more controversially, a smaller subcategory of dyslexic children with spatial attention deficits is also being defined.[275]

Modern dyslexia research started in the 1970s, with the first careful psychological studies of the children's disorder. Although the goal of reading is to understand sentences and whole texts, it quickly became clear that what affected dyslexics did not stem from this higher level. Even when these children were asked to read a single word, errors appeared and their reading slowed drastically. Unsurprisingly, those who suffered from a word recognition deficit also experienced severe difficulties in sentence and text comprehension. As a result, impairment at the word processing level seemed to explain the whole range of deficits seen in dyslexics. This simple observation eventually led to the demise of some of the more complex accounts of dyslexia, which attributed the problem to an anomaly in eye movements or to the inappropriate use of sentence context. Progressively, dyslexia research came to focus on the deeper mechanisms of single-word processing.

At the beginning of the twentieth century, the early dyslexia pioneers Morgan, Hinshelwood, and Orton considered it an essentially visual pathology, a form of "congenital word blindness." They believed that the dyslexic's visual system confused letters. This was particularly true for those that are mirror symmetrical, like "q" and "p" or "b" and "d." The average person continues to view dyslexia as a form of visual clumsiness—in the next chapter we will see that for some children this description has a certain ring of truth. However, present-day research no longer grants it much credit. Attention has shifted to the key role of phonological decoding. Most dyslexic children, indeed, seem to suffer from a peculiar deficit that affects the conversion of written symbols into speech sounds. For this reason one of the revealing tests used to tease dyslexics apart from normal readers requires reading meaningless pseudo-words like "cochar" or "litmagon." Even after years of practice, some children still stumble over reading such strings. Even if they finally do succeed, slow reading betrays them—some children require over 300 milliseconds per letter. This speed is comparable to that of adults with pure alexia due to lesions of the occipito-temporal letterbox area.[276]

Dyslexia thus appears to be a reading deficit that can be reduced to a problem with single-word decoding, which is itself due to an impairment in grapheme-phoneme conversion. This cascading analysis can be taken one step further. Nowadays, the scientific community generally agrees that the dyslexics' impairment in letter-to-sound conversion stems from a more fundamental source. The majority of dyslexic children appear to suffer from a deficit in the processing of phonemes—the elementary constituents of spoken words. To some extent, this is a revolutionary idea: a problem that seems restricted to reading would in fact be due to subtle deficits in speech processing.

Although at face value this conclusion seems barely credible, it is supported by a whole array of converging data. When compared with normal children, dyslexics often have trouble with phonemic awareness tests. Impairments are commonly observed in rhyme judgments (for instance, do *won* and *one* rhyme?), in the segmentation of words into component phonemes (such as breaking *cat* into *k, a, t*), and in the mental recombination of speech sounds to form new spoonerisms (for instance turning "crushing blow" into "blushing crow," or "smart fellow" into . . .).[277]

Admittedly, several of these studies run in circles, because phonemic awareness is known to improve as one learns to read. Acquisition of the alphabetic principle, or the understanding that each letter or grapheme must map onto a phoneme, has a major impact on our ability to mentally manipulate phonemes. This aspect, as we saw earlier in chapter 5, is underdeveloped in illiterate adults. Phonemic awareness is an area where it is very difficult to separate causes from consequences.[278] Do children partly become poor readers because they have trouble processing phonemes? Or is their phonemic awareness impaired because they have not yet learned to read? A vicious circle seems to close in on dyslexic children.

This logical circularity can however be prevented if children's performance is monitored over several years. Longitudinal studies indicate that even when phonological competence is measured before school age it can predict future reading scores. A particularly interesting study on this topic was recently conducted in Finland.[279] Heikki Lyytinen and his colleagues measured dyslexic children's competence from birth on—an amazing feat, given that at this early age no one can tell whether a child will later suffer from a reading deficit. The technique involved interviewing several hundred pregnant women to assess their familial predisposition to dyslexia. Even prior to birth, children could thus be classified into two categories, the at-risk group and the control group, each made up of about one hundred children.

Every six months after birth, all of the children were regularly tested and interviewed. Once they reached the age of seven, the official age at which reading acquisition begins in Finland, it became obvious that some of the at-risk children (but not all) did experience major difficulties. It was possible at this point to return to the first tests and search for the clues that, even in the first year of life, predicted potential dyslexics. An auditory discrimination test turned out to be particularly telling. In Finnish, the difference between *ata* (with a short consonant) and *atta* (with a long consonant) marks a categorical boundary. It separates words as distinct as *bog* and *dog* in English. At six months of age, the detection of this contrast was already less marked in the at-risk group. The babies born of dyslexic parents needed greater consonant duration before they could tell the difference between one consonant and another. Recordings of brain potentials also showed a deviation in the left temporal language region that was predictive of subsequent reading difficulties.

These studies reveal a solid link between early phonological abilities and the ease with which literacy will later be acquired. Most dyslexic children seem to suffer, above all, from a faulty representation of speech sounds. Poor functioning at this level prevents precise processing of spoken words and the consequent pairing with visual symbols. In some children, the speech impairment is so drastic that the diagnosis may be different. Pediatricians no longer speak of dyslexia, but of dysphasia or of specific language impairment—yet as soon as these children have phoneme processing deficits, they also tend to suffer from severe reading impairment.

It is important to ask whether dyslexic children only suffer from deficits affecting the processing of speech sounds or whether they have even more fundamental impairments in auditory perception. Several studies suggest that many dyslexics (but not all) are afflicted with both problems, but this issue remains highly controversial. The categorical perception of speech, for instance, is frequently impaired. Many children find it difficult to detect the minute distinction that separates the syllables *ba* and *pa*. In several studies, the deficit also extends beyond language.[280] Discrimination of pure pitch, detection of a brief noise pause, and in particular attention to the order of a fast auditory sequence can be highly impaired. In some studies, these signs were found to correlate tightly with reading scores. Even the visual perception of temporal order can be altered. Finally, basic visual deficits have occasionally been reported, for instance for motion or for contrast perception.[281]

In brief, the diagnosis of dyslexia is often buried in a morass of associated visual and auditory impairments. The trouble with these related difficulties is that they may not play any causal role. Sensory deficits may simply co-occur with dyslexia because the underlying brain pathology spreads to multiple domains. Which associated deficits are coincidental, and which play a genuinely causal role in the disruption of reading acquisition? This issue continues to divide the scientific community. It remains extremely difficult to sort out the causes, consequences, and coincidental associations of dyslexia.

Basic impairments in sound processing may appear to be fundamental inasmuch as their rehabilitation tends to improve reading. A Finnish team has designed a nonverbal remedial technique where children play a simple computerized audiovisual game.[282] On each trial, they hear a series of sounds that vary in pitch, duration, and intensity. They have to decide which visual shape represents each sound best. For instance, a series of ascending sounds may be paired with a row of rectangles of increasing height. After training for only ten minutes a day, two days a week, over a period of seven weeks, reading-impaired children saw their word recognition scores improve spectacularly compared to those of a control group that had not benefited from any special coaching.[283] This innovative remediation technique is reminiscent of musical notation training, where children learn to map sequences of notes onto series of sounds. Interestingly, although well-designed studies on this particular topic remain rare, early musical training does seem to have a positive impact on reading scores.[284]

In spite of these advances, it remains hotly debated whether *all* dyslexic children suffer from phonological deficits, and whether those in turn reduce to lower-level sensory deficits.[285] Part of the confusion probably stems from the fact that an activity as complex and integrated as reading necessarily lies at the intersection of multiple causal chains. Because of this, a great variety of other hypothetical bases for dyslexia have been proposed. Alternative suggestions range from impaired automatization of acquired skills (a competence that involves the cerebellum)[286] to a disorganization of the "magnocellular" neural pathway (which conveys fast temporal, visual, and auditory information).[287] These theories posit that the core deficit in dyslexia far exceeds the sphere of phonological processing. Reading impairment is simply the tree that hides a forest of impairments in motor automation or fast perception.

Recent findings by the French dyslexia researcher Franck Ramus and his collaborators do not, however, agree with this broad view of dyslexia. They reaffirm that, whatever its ultimate biological cause, a core deficit in phonological processing lies at the origin of most dyslexia. Dyslexic children may well suffer from concomitant deficits in motor, visual, or auditory functions, but this is not true for all children and consequently fails to provide a unified explanation for the symptoms of dyslexia. According to Ramus, only the phonological deficit provides a causal link to reading deficits. Even if the additional impairments contribute to dyslexia as a medical syndrome, they do not belong to the causal core that lies at the origins of reading difficulties.[288] Ramus and his colleagues do not deny, however, that a minority of dyslexic children (about one in four) presents a pronounced visual deficit and no phonological impairment.[289] In the next chapter, we will cover cases where dyslexia can be clearly and convincingly related to a primary visual origin.

In conclusion, we should perhaps question the very idea of a single cause for dyslexia. The problem that faces us is complex and cannot easily be reduced to a single well-defined cause. At the interface between nature and culture, our ability to read stems from a fortunate array of circumstances. Reading instruction capitalizes on the prior presence of efficient connections between visual and phonological processors. I therefore think it very likely that dyslexia arises from a joint deficit of vision and language. The weakness iself probably rests somewhere at the crossroads between invariant visual recognition and phonemic processing. As I will now go on to show, brain imaging supports the claim that the crux of the problem often lies at the ·interface between vision and speech, inside the web of connections found in the left temporal lobe.

The Biological Unity of Dyslexia

Is dyslexia a genuinely neurological disease? Only ten years ago, there was still considerable doubt surrounding this question. Pediatricians faced with cases of learning disabilities knew that it was generally useless to ask for a standard MRI exam because at first sight, on clinical examination, brain anatomy in these cases always looked essentially normal. As a result, skeptics often blamed dyslexia on poor teaching, inappropriate family support, or a failure in the education system.

These positions were forced to evolve when several converging cues suggested that dyslexia runs in families and is associated with genetic markers. This observation motivated a more serious search for cerebral mechanisms that could be associated with the disability. In the last decade, thanks to several methodological refinements in brain imaging, the biological foundations of dyslexia have finally been exposed. My friend Eraldo Paulesu, from the University of Milan, in collaboration with French and English colleagues coordinated one of the most conclusive brain imaging studies to date, which contributed notably to the way we now look at dyslexia.[290]

The Paulesu group, being based in Milan, started from what seemed to be a paradox: dyslexia is hardly ever diagnosed in Italy. Their French collaborators also noted that the pathology seems to be less frequent in France than in Britain or the United States. At first sight, these observations seemed to cast doubt on the universal biological origins of the syndrome. The Italians asked themselves if dyslexia was simply a cultural label invented by Anglo-Saxon medicine. They wondered if it applied solely to a fringe of children who were overwhelmed by the opacity of English or French spelling combined with poor teaching.

Paulesu and his colleagues realized that these questions could be viewed differently. An alternative theory could be proposed: in all countries, the same proportions of children suffer from a genetic predisposition to dyslexia, but its symptoms only appear in some cultures. Was it conceivable that dyslexia only turned into a full-blown pediatric problem in countries whose writing systems were so opaque that they put a major stress on the brain circuits linking vision to language?

With this question in mind, Paulesu went on an active search for adult Italians who had suffered from severe difficulties in learning to read. After submitting roughly twelve hundred Italian students to standardized tests, he retained eighteen of them, on the lower end of the scale, who could be considered true dyslexics. Superficially, it was not clear that they were impaired. Their reading skills considerably exceeded those of English and French dyslexics. They could read words and pseudo-words aloud, albeit somewhat slowly. However, when compared to normal Italian readers, their scores were as deviant as those of groups of French and English dyslexics as compared to control subjects in their respective countries.

Once the three groups of Italian, French, and English dyslexics had been formed, the only question that remained was whether their brains shared features that systematically distinguished them from normal readers. With this aim in mind, Paulesu and his colleagues visualized their brain activity using positron emission tomography. Subjects were asked to read target words that were easy to read, even for dyslexics, so that performance was identical in all groups. A control task consisted of watching letter-like geometrical shapes.

The comparison of the dyslexics with their respective control groups revealed a clear anomaly. A whole chunk of their left temporal lobe was insufficiently active. Furthermore, this reduced brain activity was observed at the same location and to the same degree for all three nationalities (figure 6.1). Paulesu's results thus point to a universal cerebral origin for dyslexia, at least for alphabetic writing systems—the left temporal lobe seems to be systematically disorganized.

A recent study challenges these conclusions. It claims that Chinese dyslexics suffer from a very different impairment.[291] MRI suggests that children who experience disproportionate difficulties in reading Chinese characters mostly exhibit reduced brain activity in a region quite different from Paulesu's: the left middle frontal region. Based on these results, the authors of this study conclude that brain networks for reading and their pathologies vary radically

Physiology of dyslexia

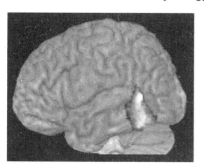

Brain activation during reading

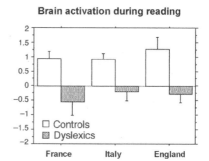

Anatomy of dyslexia

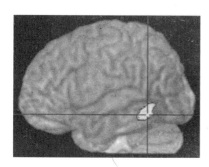

Gray matter density in the middle temporal gyrus

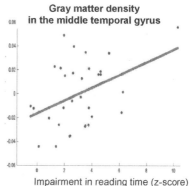

Impairment in reading time (z-score)

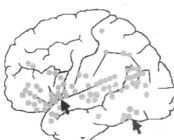

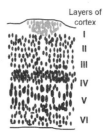

Layers of cortex

Figure 6.1 Anatomical and functional anomalies can be observed in the brains of dyslexic children. Brain activation is reduced during reading in the middle and inferior regions of the left temporal lobe (top, after Paulesu et al., 2001). In the same patients, an anomalous increase in gray matter density, correlated to reading scores, is seen in the left middle temporal region (middle, after Silani et al, 2005). In the rare cases that were autopsied, Albert Galaburda and his colleagues (1985) observed numerous "ectopias"—disorganized groups of neurons that had migrated beyond their habitual layers of cortex (bottom). These anomalies were particularly dense in the visual and verbal areas essential to reading. *Used with permission of* Science *and Oxford University Press.*

across different cultures. This conclusion should, however, be tempered, since the Chinese dyslexics also displayed a drop in brain activity in the left occipito-temporal cortex, within only eight millimeters of the anomaly peak common to Italian, French, and English dyslexics. The hypothesis of a universal reading mechanism thus seems to be reinforced rather than weakened by these results. Inexplicably, the authors played down this converging result, preferring to stress the difference in frontal lobe activity. But why would this region be different in Chinese dyslexics? The anomaly falls close to Exner's area, a region whose lesions in English readers lead to agraphia, or the inability to trace well-formed letters and words. Apparently, what is most difficult for a learner of Chinese characters is to commit three thousand of them to memory. To ease this burden Chinese children may rely in part on a motor memory of how strokes are drawn. My hypothesis is that early impairment of Exner's area may interfere with the memory for writing gestures, and perhaps lead to pronounced reading difficulties in Chinese.

In summary, although reading relies on globally similar anatomical pathways in all cultures, sensitivity to brain pathologies may be different. Phonological impairments are predominant in dyslexics who are taught an alphabetical writing system, while a form of "graphomotor" dyslexia may prevail in Asian writing systems—even if the two subtypes exist in all countries.[292]

A Prime Suspect: The Left Temporal Lobe

Generally speaking, the left temporal lobe remains the prime suspect in our quest for the biological origins of dyslexia. Essentially all brain imaging studies of dyslexia find a reduction of brain activity in this area when it is compared to that of normal readers.[293] In Paulesu's original study, this decrease in temporal lobe activity was found in adults who had suffered from lifelong reading deficits. But reduced activity can also be seen in young dyslexic children aged from eight to twelve years old. Crucially, at this age, the extent of the brain dysfunction correlates with the severity of the reading deficit. Imaging techniques have become so powerful that measurement of the amount of underactivation in the letterbox area allows for a prediction of the severity of a child's reading impairment.[294]

Another anomaly is also frequent in dyslexia. The left inferior frontal cortex, which hosts Broca's area, a region critical for syntax and articulation, is often *hyper*active when dyslexics attempt to read or perform a variety of phonological tasks.[295] This finding can be interpreted as a compensatory strategy. To counter insufficient activation of their posterior regions for automatic decoding, dyslexics probably engage in explicit speech production strategies that mobilize the frontal lobe—a brave but often fruitless endeavor.

Upon close examination, Paulesu's images suggest that two nearby areas are insufficiently active in dyslexic brains (figure 6.1): the left lateral temporal cortex and a region just below it, which happens to coincide rather nicely with the letterbox area. Several researchers speculate that the first area plays a causal role in dyslexia, while the second is underactivated as a result of the first, because the children never make the transition to expert reading. Indeed, one of the functions associated with the lateral temporal cortex is the processing of phonological information in speech (see, for instance, figure 2.19). Assuming that this area is disorganized at birth, a simple cascading scenario can be proposed for dyslexia. Early damage to the brain's speech processing networks explains why phonological impairments are so frequent in dyslexics, even before the onset of reading. The disproportionate difficulties in phonemic awareness complicate the acquisition of the alphabetic principle. These problems, in turn, impact the left letterbox area, which is unable to acquire visual expertise for written words. All this explains why a second, derivative drop in activation is observed in this region.

This scenario, however plausible, is debatable. The possibility remains that the occipito-temporal cortex itself, the future site of "the brain's letterbox," is directly impaired in some children. A functional imaging study strongly suggests that this region may not work properly in dyslexics.[296] Eamon McCrory and his collaborators presented written words and line drawings to normal and dyslexic adults and simply requested that they name them out loud. The left occipito-temporal region, at the precise location of the letterbox area, was the only part of the brain to be strongly underactivated in dyslexics. It showed reduced activity, not only for words, but, more surprisingly, for images. In brief, it seemed to be spectacularly dysfunctional.

Again, however, we do not know if this profound damage is a consequence or a cause of dyslexia. Expert readers develop detectors for T-, L-, or Y-junctions, and since these shapes are commonly found in line drawings it is possible that they enhance cortical responsiveness to drawings as well as to words. The authors of the study favor a more direct explanation. They claim that the cortical location that habitually becomes the letterbox area may be precociously disorganized in many dyslexics. Its impairment might play a causal role in the reading deficit. This site is clearly involved in linking visual shapes with speech sounds and meanings, and its early disruption could have a dramatic impact on reading performance. Because image processing is better distributed on the surface of the cortex than letter-string recognition, this focal anomaly would only have a minor effect on image recognition and naming.

The emerging scenario of a double deficit, both visual and phonological, is strengthened by studies of the temporal sequence of brain activation in dyslexics. Magnetoencephalography, a technique that measures the time course of brain activity on the surface of the cortex, has revealed two successive anomalies in the cascade of processes that lead from vision to word recognition:

- Early image processing appears normal (around 100 milliseconds), but a split second later, a massive anomaly is already visible. Dyslexics do not show the strong left occipito-temporal waveform, at around 150 to 200 milliseconds, that indicates access to "the brain's letterbox" and signals the invariant visual recognition of letter strings.[297] In dyslexics, the left occipito-temporal region does not seem able to simultaneously recognize all the letters that constitute a word—an anomaly that readily explains their slow reading and the persistence of the influence of number of letters on reading time. This length effect often remains present in dyslexics when it has vanished in normal readers.[298]

- During a second stage, at around 200 milliseconds and beyond, while normal readers rapidly activate the left lateral temporal cortex, dyslexics demonstrate weak lateralization of activation to the left hemisphere. Their brain activity, moreover, is much greater than normal in the *right* temporo-parietal region.[299] This anomaly probably reflects the lack of rapid access to word phonology, together with a compensatory reliance on right-hemisphere pathways that are not typically seen in normal readers.

In summary, brain imaging markers readily expose the dyslexic brain. Several key areas are insufficiently active, at both the stage of visual analysis and that of phonological decoding.

Neuronal Migrations

That we can see reduced activation in the dyslexic brain does not mean we understand its causes. Why doesn't the temporal lobe activate to normal levels in dyslexic children? Are the neurons damaged, or perhaps abnormally connected? Does the deficit reside in the macroscopic organization of major fiber tracts? At a more microscopic level, is there an anomaly in the molecules that line up neuron and synapse membranes? In fact all of these hypotheses hold kernels of truth.[300]

Even at a macroscopic level, the basic layout of the cortex and its connections appears disorganized in dyslexics. Anatomical MRIs of dyslexic brains were once viewed as normal, but fine statistical techniques now allow us to detect subtle impairments. One technique, called voxel-based morphometry, consists in quantifying the amount of gray matter at any point in our cortex—the layer that contains the neurons' cell bodies. With MRI, it is possible to roughly estimate the thickness and folds of the cortex at any one point. When this technique was used on the Italian, French, and English dyslexics in Eraldo Paulesu's study, it uncovered profound disorganization in the left temporal cortex. The points where this disruption was found coincided with the locations where brain activity was reduced (figure 6.1). Surprisingly, two types of anomalies appeared: gray matter was rarefied at some places and abnormally dense at others. In fact, the most significant finding was that dyslexics had *more* gray matter than non-dyslexics in the left middle temporal gyrus. When each individual subject was measured, the size of this anomaly neatly predicted the severity of the reading speed deficit.[301]

Why is an excess of gray matter a problem? MRI is not precise enough to reveal the underlying biological mechanisms, but postmortem autopsy studies of adults who happened to be dyslexics shed light on this matter. As early as 1979, the American neurologist Albert Galaburda, at Harvard Medical School, examined the brain of a twenty-year-old dyslexic, followed by the brains of a few other patients.[302] He discovered that their cortex contained numerous "ectopias"—a technical word of Greek origin that means the neurons were not placed correctly. During pregnancy, neurons have to travel considerable distances in the fetal brain. Herds of neurons migrate from the germinal zone around the ventricles, where they are formed by cell division, to their final position in the different layers of the cortex. Neuron division and migration are critical steps in normal brain development, and it is at this very moment that the fetus's brain is most sensitive to pathogens like alcohol. In many dyslexics, neuronal migration seems to have gone awry. With autopsy Galaburda discovered messy piles of neurons on the surface of the cortex. It was as though they had traveled beyond their normal position and crashed. At other places, the six layers of cortical cells were ill-shaped ("dysplasia") or formed miniature folds ("microgyri"), suggesting that the neurons had never reached their proper targets. Finally, some of the brain's fissures had lost their normal asymmetry—a feature that presumably plays an essential role in the functional specialization of the left hemisphere for language.

Galaburda claims that anomalies in neuronal migration are frequent in the dyslexic brain. Their "peppered" effect—too many neurons here, too few there— could explain the macroscopic patchwork of increased or decreased gray matter density found by Paulesu and his colleagues in their anatomical MRIs. For unknown reasons, the misplaced neurons tend to cluster around brain areas involved in speech processing. They also gather around the left occipito-temporal area, "the brain's letterbox" that plays such an essential role in visual word recognition (figure 6.1). A coherent scenario emerges from these elements: these areas, sprinkled with badly positioned neurons, cannot function optimally and cause the subtle phonological and visual deficits associated with full-blown dyslexia.

Naturally, one would expect the disorganization of the basic layout of cortical neurons to also impact on their connections. This is exactly what MRI studies have revealed. All studies of cortical connectivity, coming from different laboratories and using a variety of methods, point to a profound alteration. A narrow set of fiber bundles, located under the parieto-temporal region of the left hemisphere, is always impaired in dyslexics (figure 6.2).[303] The amount of disorganization of cortical connections at this brain location predicts reading scores not only in dyslexics, but even in a population of normal readers. Unfortunately, the current limited spatial resolution of brain images does not allow us to pinpoint the exact cortical areas that are linked by these defective bundles. Nevertheless, that they lie precisely below the underactivated temporal regions strengthens the hypothesis that in dyslexics the left temporal region is partially disconnected from the rest of the brain. By failing to dispatch linguistic information to the rest of the brain, and particularly the left inferior frontal region, such a disconnection causes a major disruption in the flow of information.[304]

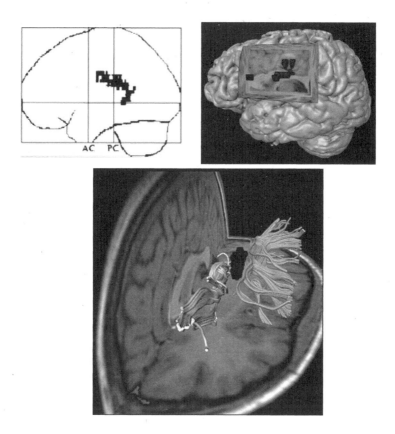

Figure 6.2 Long-distance cortical connections are altered in dyslexia. Several independent studies have revealed disorganization of the fiber bundles located in the depth of the left parieto-temporal region (left, Klingberg et al., 2000; right and bottom, Beaulieu et al., 2005). *Used with permission of* Neuron *and* Neuroimage.

The Dyslexic Mouse

The logical next step would be to dissect the fine-grained organization of left temporal neurons under a microscope, or to record their activity with microelectrodes—but of course, invasive research of this sort is virtually impossible to conduct on humans. To get around this obstacle, Albert Galaburda decided to turn to the rodent's brain.

The idea of looking for the causes of dyslexia in rats and mice seemed at first to be utterly absurd. Indeed, there was no dearth of jokes about Galaburda's studies—"brain scientist discovers reading deficit in mice"! Nonetheless, this innovative research strategy led to some of the most significant recent advances in the neurobiology of dyslexia.

Galaburda's idea was to reproduce in animals the neuronal migration anomalies that had been seen in humans. He correctly foresaw that this would shed considerable light on the mechanisms and consequences of dyslexia. In order to provoke migration defects similar to those of the dyslexic brain, Galaburda and his colleagues developed an original method—freezing small regions of the young rat's cortex. This intrusion perturbed the scaffolding of support cells called glial cells that guide and curb neurons as they migrate. Disorganized piles of neurons (ectopias) appeared in the frozen rats at the place where the neurons had migrated beyond their normal cortical location. As he had hoped, Galaburda now had an animal model, not of dyslexia itself, but of one of its possible underlying causes.

The disruption in young rats' brains produced a chain of unexpected consequences. Locally, cortical disorganization induced abnormal neuronal discharges. These perturbed brain waves sometimes turned into full-blown epilepsy. More surprisingly, effects were found at a great distance from the original lesions. Where the initial lesion occurred had little consequence: all of the cortical lesions produced a backlash at the center of the brain, in the sensory nuclei of the thalamus. This area contains a great many circuits, one of which consists of sensory neurons that relay visual and auditory inputs. In the rats, the largest neurons of the thalamus, which belonged to the sensory circuit known as the "magno-cellular" pathway, died off at a rate faster than normal. In a nutshell, dyslexia-like cortical impairments accelerated cell death in the thalamus.

In a clever move back and forth between their animal model and the human brain, Galaburda and his colleagues returned to their bank of brains taken from deceased human dyslexics. They placed the thalamus rather than the cortex under a microscope and found disorganization comparable to that of the rat. In dyslexics, as compared to normal human adults, the auditory nucleus of the left thalamus contains too many small neurons and too few large cellular bodies.[305] The researchers wondered if these anomalies could explain the sensory deficits frequently found in dyslexics. They returned to animals to design refined behavioral tests that unveiled further parallels between humans and rats. Compared to normal rats, the cooled rats were insensitive to the temporal order of two brief sounds. They were also unable to detect a short pause within a series of sounds—basic auditory deficits very similar to those reported in dyslexics.

An even bigger surprise awaited Galaburda and his team. The anomalies they uncovered were mostly in the male rats. In females, the same initial cortical lesion caused neither cell death in the thalamus nor sensory deficits. Hormonal manipulations pointed to a possible explanation: testosterone, which is more concentrated in males, amplified the distant effects of cortical lesions on the thalamus. Here again, a potential analogy exists between rats and the human species. Although the point remains debated, dyslexia seems more frequent in males than in females. Galaburda's testosterone effect on the death of thalamic neurons could be the deciding factor.

With the animal model as a guide, an even more subtle prediction for human dyslexics can be made. The phonological deficits themselves, which arise from cortical lesions, should be equally frequent in males and females— but in males alone they should be compounded by auditory and visual sensory deficits linked to the thalamus.

The animal model provides a compelling explanatory chain for dyslexia. The account is hypothetical but can explain, in a single sweep, the main anatomical and cognitive features of the disability.[306] This emerging scenario holds that disruptions in the migration of cortical neurons create numerous anomalies such as ectopias and minuscule cortical folds at around the sixth month of pregnancy. The peppering of cortical malformations mostly concentrates in language areas, and these anomalies undermine the phonological representations that will be indispensable to the acquisition of the alphabet principle at around the age of six. In parallel, particularly in males, a cascade of secondary anomalies occurs in sensory circuits of the thalamus, thus further reducing the precision with which auditory and visual inputs are coded.

The Genetics of Dyslexia

Considerable headway has been made in the description of the mechanisms underlying human dyslexia. Nonetheless we still need to understand what causes the initial anomalies in neuronal migration that are thought to bring on this handicap. More than ten years ago Albert Galaburda artificially disrupted the cortex by cooling it. In humans, genetic anomalies perhaps play the same inducing role.

In the late 1990s, Galaburda and his colleagues achieved a genetic break-through. By selective breeding, they created a lineage of mice that spontane-ously developed piles of disorganized cortical neurons (ectopias). This was the first step toward the elucidation of the genetic mechanisms that control neuronal migration.

In parallel, human genome research led to the creation of vast genetic databases on dyslexics. By crossing genetic information with the results of a great many cognitive tests, six vast regions of the human genome, on chro-mosomes 1, 2, 3, 6, 15, and 18, were "linked" to dyslexia. What does this link-age mean? Very roughly speaking, it points to a statistical anomaly in the transmission of chunks of DNA. Some parts of the genome appear as hot spots that tend to be transmitted more frequently in dyslexics than in the ran-dom population. Indirectly, this finding suggests that these DNA sequences contain genes that might induce dyslexia.

The human genome contains three billion base pairs—the "letters" A, T, G, and C, standing for the nucleotides adenine, thymine, guanine, and cyto-sine. Most of the letters are stable in the human species, but millions of them vary and form the genetic signature that makes each of us unique. As a result, the discovery of exactly which genetic variants predispose us to dyslexia is quite a needle-in-a-haystack problem. The DNA of hundreds of families with dyslexic children, over several generations, has contributed toward progres-sively refining the localization of the relevant genes. In 2003, a Finnish team discovered the first serious candidate gene for susceptibility to dyslexia, called DYX1C1, on chromosome 15, and in 2004 and 2005, three other susceptibil-ity genes were identified. They were christened KIAA0319 and DCDC2 (both on chromosome 6) and ROBO1 (on chromosome 3).[307] Many other predispo-sition genes probably remain to be discovered.

If such a vast number of genes are involved, it is probably because they all contribute in part to the complex operations that assemble cortical circuits in the numerous regions essential to reading. The construction of such a compli-cated network can be compared to building a skyscraper. Dozens of architects and contractors work on the project and a mistake by any one of them can jeopardize the whole building. An architect, of course, works on several proj-ects simultaneously. Likewise, no single gene is exclusively dedicated to read-ing, or even to phonology. (Remember that variants of all these genes exist in mice!) Nevertheless, some play a key role in ensuring the solidity of the edifice. If the building's architecture is daring and its structure weak—as is the case for reading in an alphabetic script as complex as English or French—any mistake can make the whole project collapse.

In the case of reading, the early creation of the brain's networks for literacy involves the harmonious migration of cortical neurons toward the left temporal region, and their proper interconnection to visual and language areas. Neurobiology has clarified how this complicated edifice is built during pregnancy. To lead migrating neurons from the ventricular zone to their final destination in the cortex, a scaffolding of cables is first laid down—the radial glial cells, which are cable-like cells pointing toward the cortex. Once this scaffolding is in place, the glial cells divide and each division creates a newborn neuron that literally climbs up its mother's body to reach the heights of the cortex. A variety of signaling and adhesion molecules control the "trafficking" of this literal flow of new neurons into the cortex.

A fascinating recent discovery is that most, if not all, of the susceptibility genes for dyslexia play an essential role on the cortical construction site. In an effort to demonstrate the involvement of these genes, Joe Lo Turco and his collaborators at the University of Connecticut went back to studying rats. They used a genetic technique called RNA interference that momentarily disrupts the action of a selected gene and chooses precisely where and when this disruption will occur.[308] By using this technique in utero, within the ventricular zone where neurons are born at the peak of the migration period, they were able to show that three of the four genes (DYX1C1, DCDC2, and KIAA0319) play a crucial role in neuronal migration. When these genes are blocked, neurons fail to migrate far enough and produce the ectopias and minuscule folds that characterize the human dyslexic brain. Other genes are also involved. The LIS gene, for instance, works hand in hand with DYX1C1. It is so crucial for migration that its disruption causes dramatic mental retardation due to "lissencephaly," with the hemispheres becoming smooth as the cortex is utterly disorganized.

At the time of this writing, the role of the fourth gene associated with dyslexia, ROBO1, is not fully understood. Copies of it can be found in mice, in chickens, and even in drosophila (fruit flies), where they control the growth of dendrites and axons linking the left and right halves of the nervous system. In humans, the corresponding anomaly would be a malformation of the corpus callosum, the bundle of connections that links the two hemispheres. As we will see in the next chapter, such a pathology could be related to the visual symptoms of dyslexia. Genetic research is now rapidly moving ahead, and I am confident that a full causal chain linking genes to dyslexic behavior will be available in the near future.

Overcoming Dyslexia

We are often asked if our growing understanding of the biological mechanisms of dyslexia will result in new treatments. Over the short term, I am sorry to say that no real cure for these brain impairments is anywhere in sight. If our current understanding is correct, dyslexia often relates to neuronal migration anomalies that occur during pregnancy. At this time, treatment using a classic drug approach or the more innovative gene therapy is essentially impossible.

When I lecture on reading and its impairments, I can generally spot the parents of dyslexic children. Many of them look on each scientific advance as a stab in the back bringing them nothing but bad news about what their children are lacking: disorganized gray matter, a temporal cortex that fails to activate, neurons that do not migrate properly, anomalous genes . . . Each of these biological discoveries sounds like a life sentence.

Teachers, on the other hand, often react with a paradoxical mixture of discouragement and relief. When they hear that dyslexia is due to brain anomalies, many of them conclude that they are not competent to cope with the problem. It is outside their sphere of influence. On the face of it, how can a schoolteacher, who already finds it hard to teach reading to normal children, cope with a brain impairment that started before birth? Interfering with the brain does not appear to be any of their business.

Although I can certainly empathize with these feelings of despondency, they are totally misguided. They betray two frequent misconceptions about brain development. The first consists of thinking that biology rhymes with rigidity, as if genes dictated inalterable, ironclad laws that will govern our organisms for the rest of our lives. But are we really defenseless in the face of the almighty gene? This reasoning is deeply flawed. If we think of myopia, another severe genetic condition that afflicts millions of sufferers, it can be erased with the beam of a laser or a pair of glasses. It is not inconceivable that a cognitive equivalent to glasses is within our reach for dyslexia.

The second mistaken belief is more subtle. It betrays a form of "crypto-dualism"—the powerful but erroneous idea that mind and brain belong to two distinct realms. This dualism often crops up on the subject of education. Even the well-informed seem to believe that speech therapy, rehabilitation, computer training, group support, and discussion intervene at a "psychological" level quite different from the stuff that brains are made of. How could any of these forms of therapy constitute a cure for a prenatal anomaly in the brain's makeup?

In reality, a direct one-to-one relation exists between each of our thoughts and the discharge patterns of given groups of neurons in our brains—states of mind *are* states of brain matter. It is impossible to affect the one without also stirring the other. This does not mean that, by thinking hard, we can help our neurons to multiply or migrate! What I mean is that the classic opposition between psychology and the brain sciences is unfounded. Levels of organization in the cortex are so intricate that any psychological interference must produce repercussions in our brain circuits, all the way down to the cellular, synaptic, molecular, and even gene expression levels. The sad fact that a given pathology is caused by microscopic neurobiological anomalies does not imply that psychological treatment cannot help . . . or vice versa. Indeed, the links between the molecular and psychological levels are amazingly potent and direct. Good examples can be found when lithium ions help fight depression, or when molecules of diacetylmorphine hydrochloride, also known as heroin, turn a normal human being into a craving maniac.

Given all of this, I would like to emphasize to families with dyslexic children that genetics is not a life sentence. The brain is a "plastic" organ, which constantly changes and rebuilds itself and for which genes and experience share equal importance. Neuronal migration anomalies, when they are present, affect only very small parts of the cortex. The child's brain contains millions of redundant circuits that can compensate for each other's deficiencies. Each new learning episode modifies our gene expression patterns and alters our neuronal circuits, thus providing the opportunity to overcome dyslexia and other developmental deficits.

Thanks to advances in the psychology of reading, better reading intervention methods are now being designed. With brain imaging, we can trace their impact on the cortex and check whether they do in fact lead to restoring the desired networks for literacy. In the last two decades, several leading researchers have designed efficient intervention strategies for dyslexia.[309] Most of these programs aim at increasing phonemic awareness by helping children manipulate letters and sounds. For instance, some interventions show children pairs of similar words such as "toy" and "boy" and explain how moving one letter turns one word into the other. The next step consists of showing how the very same letter "t," which turns "boy" into "toy," can also be used to write other words such as "train," "ten," or "tug"—and how the letter "b" magically transforms these words into "brain," "Ben," and "bug." Through such games, a dyslexic child can progressively become aware of the phoneme *t* and its correspondence with the letter "t." If he cannot hear the difference between the phonemes *t* and *b*, the speech therapist or the computer will exaggerate it so it becomes very distinct. By enhancing the contrasts between speech sounds and then progressively returning to regular pronunciation, phoneme comprehension can be retrained.

Our knowledge of brain plasticity allows us to understand which ingredients contribute to successful intervention. First, efforts must be intense and prolonged, ideally with short daily training sessions spread over a period of several weeks. A great many studies have shown that brain plasticity is maximized by intense training alternating with sleep. Second, it is vital to solicit the child's motivation, attention, and pleasure networks. Indeed, these attention and reward systems have a massive influence on learning speed. Several neuromodulator systems spread chemical substances such as acetylcholine to a broad range of cortical sites. These chemicals seem to indicate to other circuits the importance of committing a given situation to memory (think, for instance, of those instant memories created by a single strong emotion, like where were you on 9/11?). Experiments on animals show that this modulation effect can be dramatic: when a sound is systematically paired with the activation of acetylcholine networks, the surface of the cortex allocated to this particular sound increases massively until it literally invades nearby cortical territories dedicated to other distracting inputs.[310] In children, maximizing attention and positive emotions can likewise have a very beneficial effect on learning.

An efficient strategy consists of disguising literacy intervention as a computer game.[311] Young children are fascinated by computers. Moreover, rehabilitation software can generate thousands of training situations, at minimal cost and without exhausting the speech therapist. Most crucially, software can adapt to each child. The most impressive programs automatically detect the child's level and propose problems adapted to their abilities. The aim is to target what the Russian psychologist Lev Vygotsky called the "zone of proximal development," where new concepts are maximally learnable because they are just difficult enough to engage the child, yet easy enough to maintain high spirits.[312]

While it is not a miracle cure, literacy intervention can have a very positive impact. After dozens of hours of training, children whose scores were far below their normal age level attain the lower end of normal distribution (or in technical terms, their scores typically move up by one or two standard deviations). The majority of dyslexic children thus end up reading adequately, even if performance continues to lag behind that of their peers. Crucially, the benefits are maintained over the course of several years. In general, word decoding becomes quite efficient, but reading fluency remains imperfect: intervention helps most children learn to read, albeit slowly. Reduced exposure to print may explain this residual lag. Compared to other children, rehabilitated dyslexics have missed out on several years of experience with books. Past the initiation years, children learn to read . . . by reading! It is thus essential that children continue to read so that their literacy skills become automatic and enrich their visual vocabulary of graphemes, morphemes, and words.

Brain imaging demonstrates the positive impact of intensive cognitive intervention. Across studies, two major brain changes are seen: normalization and compensation.[313] Reduced activity in the left temporal region is particularly characteristic of the dyslexic brain. Intervention therapy allows for the partial recovery of this activation. After training, virtually all studies, whether with MRI or with magnetoencephalography, have observed a net gain in activation. Reactivation also occurs in the ventral occipito-temporal "letterbox" region and in the left inferior frontal region associated with articulation (figure 6.3). In each case, fMRI suggests that recovery involves regions close to, but not identical with, those typically seen in normal readers.

Brain imaging also reveals more radical compensation effects. After rehabilitation for dyslexia, brain activity often increases in several regions of the right hemisphere, at locations symmetrical to those of the normal reading circuit. It seems likely that in the presence of left-hemisphere impairment, equivalent regions of the right hemisphere take over. They contain undamaged networks whose initial function is close enough that it also allows recycling for reading.

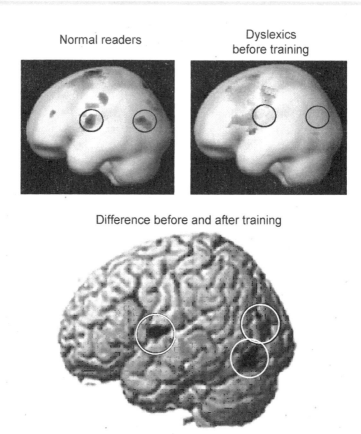

Normal readers

Dyslexics
before training

Difference before and after training

Figure 6.3 Intensive computerized intervention can partially restore a normal pattern of brain activity in dyslexic children. Brain activation images show the regions active when children decided whether two letters rhymed. After training, activity increased in the temporal and parietal regions close to, but not identical with the regions observed in normal children. Other regions of the right hemisphere, not visible in this image, also showed enhanced activity (after Temple et al., 2003). *Used with permission of the Proceedings of the National Academy of Science, USA.*

All the research on dyslexia bears a significant message of hope. In just a few decades, it has clarified the nature of the core deficit—a phonological impairment—its neuronal mechanisms, and how to compensate for it. Nonetheless, several questions remain unanswered. They concern variations between children: Do all dyslexics share the same impairment? Could one diagnose the exact deficit in each child and use the information to fine-tune his treatment? Does current research exclude smaller subgroups of children who might benefit from a wholly different approach?

Although scientific data collected over the past thirty years largely supports the phonological hypothesis, a few dissenting voices remain. It is striking that one of the most sensitive tests for the detection of dyslexia consists of measuring the speed at which children name digits and pictures—a task that does not specifically target phonological processing. Across large groups of children, phonology and fast naming tests explain separate parts of the differences in reading scores—a finding which implies that while a majority of children predominantly suffer from phonological deficits, the difficulty in others comes from another source, perhaps the automatization of the links between vision and language. In the next chapter, I will focus on vision and attempt to unravel why so many children tend to confuse mirror letters such as "p" and "q."

CHAPTER 7

Reading and Symmetry

In everyday language, a dyslexic is someone who confuses left and right and makes mirror errors in reading. Symmetry perception probably plays a significant role in reading, but left-right confusions are not unique to dyslexics. Early in life, virtually all children make mirror errors in reading and writing. Indeed, the ability to generalize across symmetrical views, which facilitates view-invariant object recognition, is one of the essential competences of the visual system. When children learn to read, they must "unlearn" mirror generalization in order to process "b" and "d" as distinct letters. In some children, this unlearning process, which goes against the spontaneous abilities inherited from evolution, seems to present a specific source of impairment.

"Now, if you'll only attend, Kitty, and not talk so much, I'll tell you all my ideas about Looking-glass House. First, there's the room you can see through the glass—that's just the same as our drawing room; only the things go the other way. . . . Well then, the books are something like our books, only the words go the wrong way; I know that, because I've held up one of our books to the glass, and then they hold up one in the other room."

—LEWIS CARROLL, THROUGH THE LOOKING-GLASS

𝒲hen he was five, my son Olivier behaved, for a few months, much like a Lewis Carroll character. His preschool teacher had just taught him to write his first name, and he proudly signed all of his artwork. Roughly half of the time, however, he unwittingly made a rather strange mistake: he wrote his name from right to left! All the letters were there, in the appropriate order, but they were inverted. The result was the exact mirror image of what he should have written: ЯƎIVIJO. Although I felt proud that my child could imitate Leonardo da Vinci, I was nonetheless a little anxious. Was this bizarre behavior a sign of incipient dyslexia?

When I turned to the scientific literature on the topic, I was reassured but also puzzled. It turns out that children all over the world make the same mistake. They suffer from transient difficulties in discriminating letters or words from their mirror images. It is also not unusual for them to spontaneously write backwards (figure 7.1).[314] Mirror writing occurs in all cultures, including China and Japan. It appears for a short period of time at the age when children first begin to write, and then promptly vanishes. Unless this phenomenon extends beyond the ages of eight to ten, there is no cause for alarm. At this late age, mirror errors are indeed more frequent in dyslexic children, though they can disappear later.[315]

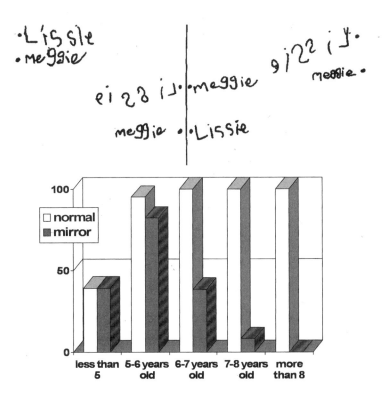

Figure 7.1 Most children go through a "mirror stage" during which they tend to confuse left and right when they read and write. As a simple demonstration, James Cornell asked two sisters to write their names next to a black dot. Whenever there was not enough space for normal left-to-right writing, Lissie, who was five, spontaneously wrote her name in mirror image, from right to left. Meggie however, who was six, always wrote her name in the right direction (after Walsh & Butler, 1996). As the bottom graph shows, almost all children, around the age of five, go through a stage where they seem to be equally competent at writing in both directions (data from Cornell, 1985). *Used with permission of* Behavioral Brain Research.

With my son, mirror writing only lasted a few months. He learned to read and write at a normal speed, but I remained fascinated by the existence of this "mirror stage" in reading acquisition. Where does this mysterious competence come from? As adults, we find it relatively difficult to write our name from right to left. Why do small children who can barely hold a pencil exhibit abilities that exceed those of most educated adults? This, moreover, is an age when children can easily locate anomalies in drawings or missing letters in their names. Why can't they notice that they are writing their names in reverse?

This strange cocktail of talent and blindness gives its unique flavor to the enigma that I now plan to address. Mirror writing, because of its very uniqueness, adds a further solid argument in favor of the neuronal recycling hypothesis. As Charles Darwin and Stephen Jay Gould observed, the best support for evolution lies in nature's imperfections. The perfect design of an eye or the flawless profile of a wing may perhaps be the work of divine genius—but what malicious divinity would have chosen to place one eye of a flatfish facing the ground, thus forcing it to migrate around the head during development? What divine architect would have been twisted enough to let the giraffe's auditory nerve run from its ear to the base of its neck and then back to the brain? If organisms were flawless, I might perhaps have some sympathy for intelligent design. Study of living creatures, however, reveals so many blunders that Darwin's theory comes out as the clear winner: only the blind operation of evolution can explain these stupid quirks.

A parallel argument can be used in psychology. Two competing explanations are always available for the way our behavior fits with our environment. The fit may arise from the effects of evolution on our genetic makeup, or it may come from learning. When behavior seems perfectly adapted, like reading in proficient adults, it is hard to separate nature from nurture. Here too, departures from perfection are particularly interesting. When a child makes a systematic mistake, or on the contrary, displays a competence that goes beyond what he may have learned, we have clear proof that he is relying on tinkered mental mechanisms inherited through evolution.

Thanks to Stephen Jay Gould's eponymous book,[316] the panda's thumb has become the symbol of evolutionary tinkering. This "thumb" is in fact not a true finger but the elongation of a wristbone that, along with appropriated muscle tissue, allows the panda to hold bamboo while eating. Any reasonable engineer would have duplicated one of the existing fingers, but Mother Nature found it easier to co-opt another existing bone, the sesamoid, into this evolved contraption. The very existence of this strange contrivance bears testimony to the fact that evolution, rather than an intelligent designer, shaped the panda's hand.

My argument is that mirror reading is an equally telling anomaly—the brain's panda's thumb. If children spontaneously confuse left and right when reading and writing, it is because their visual system, before formal schooling, already conforms to a strong symmetry constraint. Our visual brain assumes that nature is not concerned with left and right, and therefore obliges children to generalize across mirror orientations. This symmetry constraint, handed down to us by evolution, remains deeply buried in the structure of our cortex and exerts a strong influence on normal and pathological reading.

Overall, the architecture of our visual system is good for reading. It allows us to efficiently recognize the shapes of letters and words regardless of their size and case. Symmetry generalization, however, is a visual property that impedes reading. It slows down learning and leads to systematic confusions between letters such as "p" and "q." Like the panda's thumb, this anomaly makes it clear that our brain was never intended for reading, but converts to it by whatever means it has at its disposal.

When Animals Mix Left and Right

The systematic study of left-right confusion dates back to Pavlov. In the 1920s, his laboratory focused on the limits of animal conditioning. Could a dog learn to salivate for one cue, but not for another? In general, animals excel at discriminating visual stimuli, but Pavlov obtained curious results when he tried to teach dogs to discriminate left and right. When dogs were conditioned to salivate when tapped on their right side, they responded identically to left side stimulation. Mirror discrimination seemed beyond their capacities. Only a surgical cut of the corpus callosum, the fiber tract that connects the two hemispheres, finally enabled them to discriminate symmetrical stimuli. Pavlov concluded that mirror errors arose from an active generalization whereby the knowledge acquired in one hemisphere was transferred to the other across the corpus callosum.

Since Pavlov, numerous animal studies have demonstrated the ubiquity of mirror confusion.[317] Suppose that a pigeon is trained to peck on a key bearing a diagonal line, and is then tested for generalization to new lines with random orientation. Two response peaks can be observed: the pigeon reacts to the learned orientation, but also to the exactly symmetrical one. Although the bird has never seen the new symmetrically oriented line before, it behaves as if this new image is identical to the learned one. Again, this spontaneous competence disappears when interhemispheric connections are severed.[318]

Psychology demonstrates that we behave like pigeons and dogs. Once we have learned a visual shape, we immediately extend the knowledge to its mirror image. To prove this, Michael Tarr and Steven Pinker taught students arbitrary names for novel shapes that looked like trees with a few branches.[319]

After training, the students saw new shapes, some of which were mirror images of the original stimuli. Although they had never seen them before, the students named these novel stimuli at the same speed as the original ones. Thus left-right symmetry does not tax our nervous system—whatever shape we learn is immediately generalized to its mirror orientation.

Here is a simple memory test. Do you remember in which direction Leonardo's Mona Lisa turns her shoulders, or which hand she places on top of the other? Do quarters show George Washington's left or right profile? Does the Statue of Liberty hold a book in the right hand or in the left? Try to make up your mind before checking the answers in figure 7.2—your self-confidence could be shaken.

Experiments by Irving Biederman and his colleagues at the University of Southern California confirm that human memory is marred by mirror confusion. Their studies rely on a perceptual priming effect: we can name an image faster if we have seen it earlier—even several weeks before—than if we see it for the first time. This priming effect indicates that a visual memory trace persists across days or weeks. However, this memory, quite unlike a photograph, does not maintain a perfect image of what we have seen. Visual priming occurs even when the second image is spatially displaced or mirror-reversed.[320] Thus our memory of a visual scene is abstract and withstands image translation or symmetry.

Figure 7.2 Our visual memory is insensitive to orientation. Do these images look familiar? This is probably the first time that you have seen them like this—all are mirror-reversed.

In a nutshell, whenever an image becomes familiar, our memory often fails to specify its spatial orientation. When we see it again, we are unable to tell if it is the same or if it is a mirror image. This mirror generalization has nothing to do with a perceptual limitation. On our retina, two mirror images such as the left and right profiles of the Mona Lisa are different. Our visual system is capable of much finer discriminations, such as the difference between two faces seen from the same angle. If we confuse mirror images, it is not because we don't see them. Rather, errors occur because our visual system bunches together several views under the same label. Left and right angles of the same image are treated like one and the same object.

Evolution and Symmetry

Why is our nervous system so insensitive to the left-right inversion? Probably because we evolved in an environment where this distinction is largely irrelevant.[321] Although the world is three-dimensional, only two of its axes exerted a strong influence on our evolution. The first, which is vertical, is defined by the force of gravity. The second, which only applies to mobile species, goes from front to back. It defines a privileged side of the body, the front, where our sensory organs and feeding apparatus are very logically located. Together, these two axes determine the coordinated frame of the body. Ultimately, they also influence our perception of space. Discriminating vertical from horizontal, far from close, or the front and back of an animal are operations essential to survival. Across millions of years, strong pressures led to the brain's ultimate adoption of these geometrical constraints. Our visual system, like that of all primates, includes complex mechanisms for computing distances, distinguishing impending danger from above or below (an eagle or a snake), and for discriminating faces and their expressions of emotion—our sensitivity to hind parts is less acute!

Crucially, the third parameter of space, the left-right axis, was given less attention by evolution. A tiger's left and right profiles are equally threatening . . . but the beast is less of a threat upside down than right side up![322] Inversion along the left-right axis is of little consequence because it leaves the world essentially unchanged. When we examine a picture of a natural scene, we cannot tell when left and right have been inverted—unless the photograph contains words, cars, street signs, or other man-made objects.

Not only was there no strong evolutionary pressure for distinguishing left from right, but the pressure probably favored confusing them. It can be useful to confuse mirror images if the goal is to generalize rapidly from past knowledge. Imagine, for the sake of argument, that one of our ancestors survived a close encounter with a tiger that had attacked him on the right, and was subsequently approached by the same tiger coming from the left—it would certainly have been helpful for him to immediately recognize the animal.

Across generations, evolution granted us a visual system that could generalize across left and right views. The unfortunate result is that our nervous system runs the risk of confusing objects that are not identical when seen in a mirror, for instance shoes or gloves. The risk, however, is not great. In a natural environment, most animals and plants possess an axis of symmetry or quasi-symmetry, at least superficially—internal organs need not be symmetrical, but external body shape often is. Evolution seems to have calculated that there was more to gain from neglecting left and right if this allowed us to react with greater speed in critical situations.

Symmetry Perception and Brain Symmetry

What biological mechanism causes mirror perception? For the past eighty years, a speculative hypothesis, Orton's theory, has generated much interest and controversy in the psychology community.[323] In the 1920s, Dr. Samuel Orton, an American physician who pioneered the study of reading deficits, attempted to relate them to brain anatomy. His theory was the first systematic attempt to explain dyslexic errors. For this reason it is of special historical interest, even if our present knowledge of brain anatomy renders the hypothesis somewhat far-fetched.

What did Orton propose? He noted that the left hemisphere is almost a mirror image of the right. On this basis, he made the bold speculation that the two hemispheres encoded visual information in mirror-symmetrical ways. Whenever the left hemisphere saw "b," the right hemisphere saw the symmetrical image, "d" (figure 7.3). Orton failed to explain how this hypothetical reversal worked in neural terms—he merely stated that "the groups of cells irradiated by any visual stimulus in the right hemisphere are the exact mirrored counterpart of those in the left."[324]

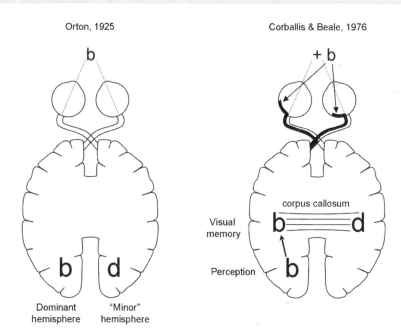

Figure 7.3 Two neurological theories attempted to explain mirror-image confusions. For Orton (1925), symmetry of the brain implied that each visual image was coded twice: once with the appropriate orientation in the dominant hemisphere, and once with the flipped orientation in the other hemisphere. Mirror confusions occurred when a person failed to inhibit the visual code of the minor hemisphere. Corballis and Beale's (1976) alternative theory proposes that the presentation of a letter initially activates visual areas in the opposite hemisphere (in agreement with our modern knowledge of the visual system). However, an active learning mechanism transfers the memorized shape to the other hemisphere across the corpus callosum. During this transfer, the geometry of these connections, which link symmetrical points in the two hemispheres, imposes a left-right inversion. For Corballis and Beale, mirror confusions thus result from an excess of generalization rather than from a genuine mistake. Our visual system, in the course of its evolution, has incorporated the fact that mirror images are often two views of the same object.

If two mental images coexist in our brains, one representing an object and the other its mirror image, it is no wonder, said Orton, that we so frequently confuse left and right. His theory postulated that we perceive the orientation of our surroundings by heeding one hemisphere and silencing the other. Normal subjects usually rely on the left hemisphere, the seat of language, when they have to name letters and words. For Orton, when we read words, we let the left hemisphere dominate and learn to neglect visual information coming from the right hemisphere. By relying on only the left hemisphere, we manage to discriminate "b" from "d"—otherwise they would simply be indistinguishable.

There was one last element in Orton's theory. He assumed that left-right errors occur in some people because of poor hemispheric differentiation. This impairment prevents them from selectively attending to a single hemisphere at a time. By randomly relying on one hemisphere or the other, they inevitably mix up left and right and fail to discriminate mirror images. Orton termed this hypothetical pathology "strephosymbolia," a Greek root for "symbol reversal."

Eighty years later and with the wisdom of hindsight, Orton's theories about vision seem simplistic. He was right, nonetheless, about one important point: the visual maps that occupy the occipital region of the left and right hemispheres are indeed organized as mirror images of each other. However, in the primary visual area, contrary to what Orton believed, incoming images are *not* duplicated in the two hemispheres. Visual projections are much simpler: the left half of the visual field projects to the right hemisphere, and the right half to the left hemisphere. Thus when we see a "p" on the right-hand side of the visual field, this shape is encoded by the primary visual area of the left hemisphere, but the corresponding area of the right hemisphere, or homolog, codes for . . . nothing, since it does not receive any visual information from that part of space.

As incoming information progresses into the hierarchy of visual areas, Orton's duplication problem becomes more real. The secondary and tertiary visual areas, which are functionally more distant from the retina, respond to increasingly larger sectors of the visual field. As a result, a single visual object begins to activate groups of neurons in both hemispheres. However, the neural code also becomes increasingly abstract. The spatial distribution of neurons no longer faithfully reflects the shape of the object on the retina. Rather, as we saw in chapter 3, a combinatorial code for line orientation, curvature, color, and other shape identifiers begins to emerge. Thus, higher up in the visual hierarchy, the two sets of neurons that code for the letter "p" in the left and right hemispheres are *not* shaped like the visual stimulus. Contrary to Orton's belief, there is no way in which they can be "exactly symmetrical" to each other.

Another problem with Orton's theory is that even if the neuronal circuits coding for the letter "p" in the two hemispheres were mirror images of each other, it is unclear why this neuronal symmetry necessarily entails mirror perception errors. On this point, Orton's hypothesis seems misguided. It is reminiscent of the naïve idea that visual processing first turns the image of the visual world upright because the lenses in our eyes start out by inverting it. We are reminded of the theory of the homunculus, the imaginary man who supposedly controls our brains, scrutinizes visual areas as if they were computer screens, and gets fooled if one of these reverses left and right. This simplistic view of the brain is untenable. There is no hidden observer in our cerebrum. Our visual system does not need to display its results for somebody else to read them—it merely extracts information and encodes it in various formats that will help our behavior.

As opposed to what Orton believed, it does not matter that the letter "p" is represented by one group of neurons in the left hemisphere and by one exactly symmetrical to it in the right hemisphere—either of these groups could be a perfectly appropriate abstract code for the letter "p." To assert that one of these groups is correctly oriented and represents the true image, while the other is mirror-reversed, is a fundamental error—as if the word "red" had to be encoded by red neurons, or visual motion by moving neurons!

Orton's vision of hemispheric lateralization seems equally simplistic—if our left hemisphere always encodes the orientation of objects correctly, and the right hemisphere erroneously, we should make mirror errors in all sorts of tasks that primarily call upon the right hemisphere, such as spatial memory or visual attention.

In brief, Orton's hypothesis mixes up the code and what it represents. In the 1920s, it was hard to think of the neural code for written words as anything but a kind of "image" on the cortical surface. The emergence of information theory and computer science, in the 1950s, has given us a better understanding of the nature of a code. We now know that any system of marks can serve as a vehicle for information, providing that it is not ambiguous and comes with adequate encoding and decoding routines. There is no need for the neural code to be shaped like the object it is supposed to represent.

In spite of this avalanche of criticisms, Orton's hypothesis remains strangely appealing. Above all, to refute his theory does not remove the facts that were behind it. A great many children do confuse left and right and make mirror errors in reading and writing. Moreover, it is not unusual for them to be left-handed or to suffer from speech impairments that suggest abnormal hemispheric lateralization. Can the key ingredients of Orton's theory be salvaged and its loopholes discarded?

Dr. Orton's Modern Followers

In the 1970s, Michael Corballis and Ivan Beale, from the University of Auckland, proposed a modern rereading of Orton's hypothesis.[325] They contended, in agreement with modern neuroscientific data, that the visual areas of both hemispheres start out by processing independent parts of the incoming image (see figure 7.3). The anatomical symmetry of visual areas only comes into play when the data are transferred from one hemisphere to the other. Corballis and Beale postulated that each time one hemisphere acquires new visual information, the memory trace is immediately transferred to the other one, in order to maintain coherence across the two hemispheres. Naturally, this transmission relies primarily on the corpus callosum, the large set of fibers that links symmetrical areas of the two hemispheres. Since these fibers establish a one-to-one mapping between symmetrical visual areas, Corballis and Beale note that the transfer should reverse left and right. In this way, our visual memories are immediately mirrored—just a trivial consequence of the geometry of our interhemispheric connections.

An interesting aspect of Corballis and Beale's model is that it does not imply that perception itself is misguided. In the first stages of visual processing, we correctly perceive a properly oriented visual world. This information can be used to guide our actions efficiently—we can, for instance, trace the contour of the letter "b" without inverting its two sides. The acquired visual knowledge is only generalized at a second stage, when we memorize the letter. This is why we cannot remember which way Mona Lisa is facing.

A concrete example of how Corballis and Beale's theory accounts for mirror reversals goes like this. During his first reading lesson, a child sees the letter "b" and learns that it sounds like *bee*. If this image first appears to the child's left, it is initially coded in the right hemisphere, where neuronal populations learn to link this shape with its name. Once memorized, the information is transferred to the other hemisphere, where the "b" shape becomes a "d" shape. This shape thus also becomes associated with the sound *bee*. The next morning, the child sees the shape "d" in his right visual field and confidently asserts that it is a *bee*. To his great surprise, the teacher tells him that the letter is a *dee*. This new information is memorized . . . but immediately destabilizes previous knowledge of letter "b."

If Corballis and Beale are right, the child must turn off the automatic mirroring in order to learn to read. He must cease to see the letters "b" and "d" as the same object, seen from different angles. This unlearning process is difficult for all children, and raises insurmountable problems for some dyslexic children.

The problems with Orton's earlier proposal do not apply to Corballis and Beale's model. It does not maintain that the neuronal code has the proper orientation in one hemisphere and the wrong orientation in the other. It merely states that interhemispheric transfer extends our visual memory to mirror images of the objects we perceived earlier. This transfer of information across hemispheres implies that a letter "b" in the left visual field activates the same visual neurons as a letter "d" in the right visual field. As a result, brain areas farther down the line have no means to differentiate between these two images. Information about letter orientation is jettisoned without ever being committed to memory.

Why did evolution select a symmetrical learning mechanism that actually discards information rather than storing it? No doubt because it provides a simple way to take advantage of the symmetry often present in the natural world. An organism equipped with this mechanism can just as easily recognize a tiger's left or right profiles.

In Corballis and Beale's model, the active maintenance of a symmetrical layout of the visual brain underlies our mirroring abilities. Their model assumes that we are born with a one-to-one array of visual connections linking the two hemispheres, and that learning mechanisms work constantly to preserve this initial symmetry. With the help of a few genes and spatial gradients, it is relatively easy to imagine how the connections of the corpus callosum innately establish the point-to-point mapping between symmetrical brain areas—much as the wings of a butterfly unfold with perfect symmetry.[326] Prior to birth, and before any interaction with the visual world, the brain integrates left-right symmetry as an essential property that it can expect to find in its surroundings. We are probably all prewired for symmetry generalization, a built-in mechanism that gives us an edge in our interactions with the environment. To maintain this advantage, however, our brain must struggle to preserve its initial symmetry. Here lies the key to Corballis and Beale's mechanism: every time learning modifies the neural networks in one hemisphere, callosal transfer ensures that similar learning also occurs in the other hemisphere, thus restoring symmetry.

The Pros and Cons of a Symmetrical Brain

Let us imagine, for the sake of hypothesis, that we are perfect "Corballis machines"—ideally symmetrical organisms whose neurons and connections are mirrored, down to the minutest detail, across the two hemispheres. How would we behave? Could we still move around in space, distinguish left from right, or name visual objects? What types of errors would we make? Could they be related to young children's mirror errors?

Here we must correct a mistaken idea. Even if an organism were perfectly symmetrical, down to the deepest recesses of its nervous system, it would not necessarily confuse left and right in all situations. In fact, such an organism would be able to perform spatially oriented actions in response to asymmetrical stimulation. Imagine, for instance, that it sees a tiger coming from the right and is wired to respond by running to the left. The symmetry of its body plan implies that it should take symmetrical action (run toward the right) in response to the symmetrical scene (a tiger coming from the left). Anyone observing this scene would thus conclude that this organism, although perfectly symmetrical, accurately distinguishes left from right.

Our hypothetical symmetrical organism only responds identically to two mirror images when the response is conventional and nonspatial. If a symmetrical organism, for instance, learns to say the word "tiger" when it sees a tiger's right profile, the symmetry of its body plan will oblige it to give the same answer to the tiger's left profile. In other words, it can recognize and even name a tiger regardless of its orientation in space. This useful feature is true even the first time the organism sees the tiger from a new angle.

In brief, the possession of a symmetrical brain and its preservation over the course of learning has two advantages:

- The symmetry of the body plan enables the organism to store visual objects invariantly, independent of their left-right orientation.
- Yet it does not prevent the emission of appropriate and spatially oriented gestures.

There are, however, two kinds of operations that a symmetrical organism cannot perform. The first one consists in emitting an arbitrary response to an asymmetrical object, but not to its mirror image. This type of operation is useful in real life. For instance, we name the shape "b" as *bee*, but not the shape "d." We also can say "left" at the sight of the left hand, but not at the sight of the right hand. A perfectly symmetrical organism would not be able to perform these tasks, simple as they may seem.

The second operation that a mirrored brain cannot perform consists in making asymmetrical gestures in response to a command that contains no spatial information. A real-life example might be to raise the right hand alone in response to the verbal command "right." A perfectly symmetrical organism would not be able to do this—the connections linking auditory areas to motor commands should also produce a left arm movement. One would therefore expect such an organism to perform symmetrical gestures with its two limbs—or, if some degree of noise creeps in, randomly raise the left or right hand in response to a verbal command.

Amazingly, the theoretical limits of symmetrical organisms coincide with experimental observations of animals and young children. No child has any difficulty in naming a familiar object, even seen from a new angle. When he holds a ball in the right hand no child ever moves his left arm in order to throw it! However, all children initially experience difficulties in distinguishing the letters "b" and "d," and in labeling their left and right hands.

In summary, although it seems somewhat artificial, the concept of a symmetrical brain that actively maintains its symmetry in the course of learning deserves attention. It seems to capture some of our cognitive limits in dealing with left and right. Is there, however, any proof that interhemispheric transfer imposes the left-right reversal postulated by Corballis and Beale? The preceding discussion is not as naïvely geometrical and unrelated to biology as it seems. Results from several experiments link symmetry perception to brain symmetry.

Single-Neuron Symmetry

Neurophysiology is only just beginning to study the neural basis of symmetry perception, but the available data already point to mirror symmetry as one of the primary invariants extracted by neurons in the inferior temporal cortex—

which in humans is, of course, the seat of "the brain's letterbox." As I pointed out earlier, neurons in the inferior temporal cortex tend to prefer certain visual objects, irrespective of changes in viewpoint. A neuron that prefers, say, Jennifer Aniston's face keeps to this preference regardless of its size and position on the retina. Recently, neurophysiologists have reported that some of these neurons are also invariant for mirror symmetry:[327] their response profile, across a range of images, remains identical even when they undergo a left-right inversion.

Crucially, these neurons often also demonstrate greater invariance to horizontal than to vertical reversals. For instance, if a neuron fires for the shape "p," it also discharges strongly for its mirror image on the left-right axis, or the shape "q," but the same neuron does not necessarily respond identically to shapes that are symmetrical along the vertical axis. In other words, this neuron distinguishes "p" from "b," but not "p" from "q."

Collectively, inferior temporal neurons implement perceptual invariance: they allow us to recognize an object regardless of its size and location. The new electrophysiological data imply that left versus right also figures among the irrelevant distinctions that are systematically neglected by our ventral visual system.

If Corballis and Beale's theory is correct, mirror invariance does not need to be learned. It belongs to the structural invariants that are inherent in the geometry of our interhemispheric connections. The work of the neurophysiologist Nikos Logothetis[328] strongly supports this hypothesis. Logothetis trained macaque monkeys to recognize unusual three-dimensional shapes that look like twisted paper clips—a jumble of line segments folded at random angles (figure 7.4). What he wanted to ascertain was whether the monkeys would continue to recognize them later, when they were seen from other angles. During training, Logothetis was careful to ensure that the monkeys only saw each object from an essentially fixed angle—only a very minimal rotation around the vertical axis created a sense of depth. But was this minimum exposure enough to allow for invariant recognition? To answer this question, Logothetis showed the animal all the possible views of the object, mixed with other distracter images of unrelated shapes.

As it turned out, when the new orientation departed from the one learned earlier by 40 degrees or more, the monkey failed to recognize the trained object. However, when rotation reached 180 degrees, recognition suddenly

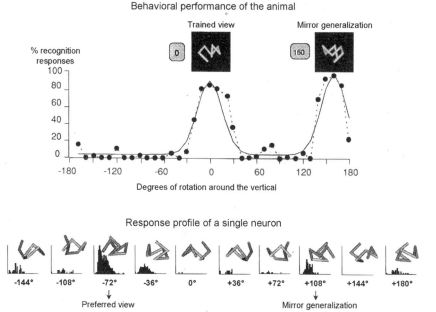

Figure 7.4 Following training with a specific shape, the visual system of primates generalizes to its mirror image. In this experiment, a monkey was taught to recognize a wire-frame object from a specific viewpoint. It was then tested with the same object as seen from various angles (top graph). The animal recognized the object when it was presented at the trained angle, but also after a rotation by 180 degrees, which corresponded to the mirror image of the original view. After learning, neurons responded both to the original view and to its mirror image (bottom graph) (after Logothetis, Pauls, & Poggio, 1995). *Used with permission of Oxford University Press.*

soared. Similar behavior was observed for the firing of inferior temporal neurons: most neurons discharged maximally when an object was presented at the orientation they had learned, stopped firing when the object was turned by about 40 or 50 degrees, but returned to discharging vigorously when the angle reached 180 degrees. Why 180 degrees? This angle corresponds to a half turn, and for wire-frame objects such as paper clips the ensuing image looks almost exactly like the mirror image of the initial object—if you twist a paper clip into the letter "b," you can change it into a "d" by turning it by 180 degrees.

In brief, without any additional learning, neurons in the inferior temporal cortex can recognize the mirror image of a known shape, even when they see it for the first time. The study of face-specific neurons, which are also numerous in this area, confirms this conclusion. Many neurons discharge identically to two mirror-symmetrical views of the same face, for instance its left and right profiles.[329] Amazingly, this mirror invariance is already present in newborn monkeys.[330] In our species too, four-month-old babies recognize the mirror image of a familiar object.[331] From birth on, we all seem to possess an uncanny sense of mirror symmetry.

Logothetis's research points to an interesting contrast between rotation and symmetry. Rotation invariance does not seem to belong to the initial competence of our visual system. We have to learn what an object looks like when it is turned around. Our visual system, at this stage, seems to lack any deep knowledge of 3-D shapes: it merely relies on the 2-D views that it has seen. It is all the more intriguing that symmetry escapes this rule.

Brain imaging experiments with both macaques and humans have recently revealed a remarkable sensitivity to symmetry in the visual cortex at an early stage. The addition of just a few symmetrically distributed dots in the midst of a cloud of random dots dramatically increases activation of the lateral regions of the occipito-temporal cortex. The ability to detect symmetry is directly linked to this enhanced firing.[332] Although evolution never prepared us to see symmetrical dot patterns, they resonate in our visual circuitry, which fires promptly at any hint of symmetry.

Symmetrical Connections

The geometry of the neural connections that underlie symmetry perception has not yet been the object of much research. Perhaps Beale and Corballis are right and the corpus callosum, with its array of symmetrical connections that link the two hemispheres, holds the keys to the perception of symmetry. In the absence of strong data, however, we are currently reduced to speculating on the basis of fragmented but suggestive evidence. I have already cited Pavlov's and Beale's pioneering experiments on dogs and pigeons. Both animals experience great difficulty in learning to discriminate pairs of mirror images. A pigeon, for instance, easily learns the difference between vertical and horizontal bars (— versus |), but performs miserably when the same bars are turned 45 degrees into a mirror-image pair (╱ versus ╲). However, when the corpus callosum is severed, as both Pavlov and Beale have noted, the lesion actually makes the task *easier*, suggesting that interhemispheric connections do in fact contribute to symmetry perception.

But what exact role do they play? Other experiments confirm that the transfer of information across hemispheres, via the corpus callosum, turns over visual space like a glove, thus inverting left and right. In the 1960s, Nancy Mello, at Harvard, published a series of articles, now long forgotten, on pigeons' sense of symmetry.[333] This species presents a useful anatomical peculiarity: unlike what happens in mammals, the pigeon's right eye projects solely to visual areas in the left hemisphere, while the left eye symmetrically projects exclusively toward the right hemisphere. This means that by masking one eye, one can ensure that visual information only enters one hemisphere.

Mello trained pigeons to discriminate, using only one eye, between two visual images like Λ versus V, or ◁ versus ▷. They were rewarded when they pecked at one of the shapes but not the other. Once discrimination had been established, Nancy Mello examined how it generalized to the other eye. Surprise! Pigeons still performed well when the pair of images had vertical symmetry (Λ versus V), but systematically failed with left-right symmetry (◁ versus ▷). They even performed worse than random! A pigeon trained to peck at the shape ◁ seen with the right eye systematically pecked at the symmetrical shape ▷ when it was presented in the left eye. Thus the learned visual information did indeed travel across the two hemispheres—but the transfer systematically reversed left and right, just as Corballis and Beale's model predicts.

One year later, similar research was performed with a species closer to ours, the macaque monkey. John Noble, from University College London, first severed the optical chiasma, thus forcing each eye to project to the opposite hemisphere alone, as in pigeons. He then trained animals to discriminate two symmetrical objects with one eye. Finally, he tested for generalization to the other eye. Noble, too, observed a puzzling mirror-image generalization. After a short period of random responding, the animals began to perform *worse* than chance. Like the pigeons, they systematically chose the wrong shape, the mirror image of the one with which they had been trained.

In later experiments, Noble showed that the transfer of visual knowledge passed either through the corpus callosum or through another bundle of fibers called the anterior commissure. Mirror transfer disappeared when both fiber bundles were cut, but persisted when one of them was left intact. In one final experiment, Noble removed the interhemispheric connections after the learning period, but before testing the other eye. Mirror reversal was still observed, suggesting that it relied on the corpus callosum at the time of learning, but not during the test period.

Noble concluded that, in the monkey, every learning episode is accompanied by mirror learning in the other hemisphere. He noted how this mechanism might account for the peculiar difficulties that animals and humans experience with left-right discrimination. In binocular vision, mirror learning ensures that each hemisphere receives both a primary visual input and a secondary mirror image of the outside world. When the scene that is observed is symmetrical, both representations are identical and enhance one another. However when we have to discriminate asymmetrical inputs such as "b" versus "d," the two internal representations compete with each other. This cerebral conflict slows down learning, even if the primary representation ends up winning.

If this reasoning is correct, a symmetry principle should govern the fine anatomical layout of interhemispheric connections. One-to-one projections should link symmetrical sectors of the visual areas directly. Such is the case,[334] and it constitutes further evidence in favor of Corballis and Beale's model. The geometry of interhemispheric connections is indeed as beautiful as a butterfly's wings. A neuron located in the left visual cortex sends an axon that first goes in a direction perpendicular to the cortical surface, then joins the bundle of callosal fibers and reaches the opposite hemisphere. Then, suddenly, it heads toward a location in the cortex exactly symmetrical to its starting point (figure 7.5). Such neurons have been seen in the primary visual cortex of rats and mice, and in the secondary visual area of macaque monkeys. Their function is unknown, but their geometry seems ideally suited to the creation of mirror images of visual inputs, thereby easing the detection of symmetrical objects.

If these symmetrical connections have occasionally been neglected, it is because they only constitute a fraction of the connections that link the two hemispheres. The major bundle of interhemispheric fibers is not symmetrical, but serves to stitch together the two cortical representations of the visual world. It answers a well-known enigma: why do we perceive one single integrated visual world when, objectively speaking, it is split into two halves, each coded by

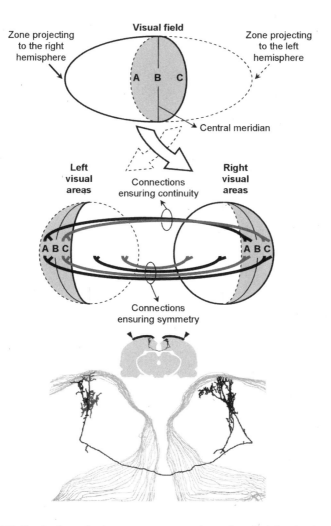

Figure 7.5 Continuity and mirror symmetry are the main principles that govern the architecture of visual connections across the two hemispheres. The top diagram recalls the well-known fact that the left half of the visual field projects to the right hemisphere, and the right half to the left hemisphere, with only minor overlap in the center (after Olavarria & Hiroi, 2003). If we do not perceive any discontinuity in the visual scene, it is thanks to interhemispheric connections that ensure the functional continuity of the visual field by linking cortical regions that code for the same point. These "stitching" connections only concern the central region of the retina, and they are not symmetrical, but shifted by translation. Other interhemispheric connections exist that link symmetrical points of the visual field, even on the periphery. They probably play a special role in symmetry perception and mirror generalization. The bottom figure illustrates their beautiful geometry: a single neuron in the rat's primary visual cortex projects to symmetrical regions of the left and right hemispheres (after Houzel, Carvalho, & Lent, 2002). *Used with permission of* Brazilian Journal of Medical and Biological Research.

a different hemisphere (see figure 7.5)? Interhemispheric connections play an essential role in restoring the functional continuity of the world in spite of this anatomical segregation. The stitching connections concentrate upon the area of discontinuity (the vertical midline) and link the sectors of the left and right hemispheres that code for similar visual locations, point by point.

Research by Jaime Olavarria and his colleagues in Seattle has revealed a competition that takes place during development between the principle of visual continuity, which calls upon stitching connections linking the two hemispheres, and the symmetry principle, which requires larger-scale symmetrical projections.[335] Symmetrical connections are abundant at birth, probably laid down by gradients of chemical attraction. They are, however, considerably pruned in the first days of life. Thanks to this pruning process, the stitching connections that ensure the continuity of the visual field tend to end up dominating the situation.

Interestingly, exposure to the environment may not be needed for this pruning process to take place. Even in utero, spontaneous waves of neuronal activity that travel across the retina tell neurons in the left and right hemispheres that they code for the same retinal location. In rats, the stitching connections are established several days before their eyes open. Furthermore, they become disorganized when the spontaneous retinal activity is perturbed. In this case, the stitching axons lose the race and an excess of symmetrizing connections can be observed.

If I have gone into so much detail over these biological observations, it is because they may turn out to be very relevant to developmental dyslexia. Indeed, they suggest that early impairments in synaptic pruning, in the last months of pregnancy or the first few months of life, may produce too many symmetrical visual connections. Could this excessive symmetry explain why some children experience difficulties in telling left from right, or in distinguishing the letters "b" and "d"? Before discussing this issue in depth, I will first examine how the brain is finally able to make these distinctions.

Dormant Symmetry

If our visual system made everything it learned symmetrical, we would constantly make mirror errors—we would find it impossible to distinguish "p" and "q" or our left shoe from our right. Fortunately, part of our visual system preserves the distinction between left and right.

A number of authors have pointed out[336] that our visual cortex is divided into two major functional streams, almost as though we possessed two distinct visual systems in the same brain. The ventral pathway, on which I have concentrated so far, focuses on invariant object recognition. It is highly sensitive to image identity, shape, and color, but ignores size and spatial orientation—including left versus right. There is, however, a second visual pathway, the dorsal route, that goes through the parietal cortex and is primarily concerned with space and action. In this route, color or the exact nature of a given object is not nearly as important as its distance, position, speed, and orientation in space—all the parameters that determine how we should interact with it. David Milner and Mel Goodale summarize this fundamental distinction in two words: the ventral route cares about the "what," the dorsal route about the "how."

"In the beginning was the deed!" states Goethe's Faust. If we manage to distinguish left from right, we probably owe it to the "how" pathway. As children, we learn that our right hand is used for drawing, writing . . . and, in Roman Catholic families, to make the sign of the cross! The "how" system must be able to tell right from left, because this information is crucial for motor gestures. We do not use the same gesture to grasp a saucepan if it is pointed towards our left or toward our right: our dorsal pathway automatically adapts our gestures to object orientation, even if our ventral system, which is insensitive to orientation, tells us that in both cases this is the same kitchen utensil. It is also the dorsal system that allows us to imagine purely virtual gestures or movements. We rely on it whenever we turn around objects in our mind's eye, read a map, or judge whether two images show the same object from different angles.

In summary, our brain behaves as if it were inhabited by two visual experts. The first, which I will call the "collector," recognizes and labels objects, but pays little or no attention to location and orientation in space. The second, the "manual worker," acts, compares, manipulates . . . but fails to identify the objects it handles.

Our two visual systems usually collaborate so closely that we are unaware of the division of labor. Brain injuries, however, occasionally expose these internal divisions. Scottish researcher Oliver Turnbull has studied dozens of patients in whom brain injury induced dissociations, sometimes as clear-cut as a razor's edge, between the two visual systems. Consider two patients, Mr. D.M., who suffered a head trauma, and Mrs. L.G., who endured multiple strokes in the right temporo-parietal region.[337] Turnbull showed them pictures of familiar objects oriented in a number of different directions: an upside-down helicopter, an upright kangaroo, a telephone rotated by 90 degrees, and so on. When asked to name them, D.M. failed miserably while L.G. performed rather well. When asked to judge if the objects were in their proper positions the reverse occurred: D.M. performed flawlessly, even for objects that he no longer recognized, while L.G. did not. In D.M., the ventral "what" pathway had been seriously damaged, while in L.G. it was the dorsal "how" pathway that no longer functioned.

The same dissociation between object identity and orientation occurs in the macaque monkey. Lesions of the ventral temporal cortex prevent the animal from distinguishing shapes as simple as a circle and a cross but have no influence on the discrimination of shape orientation (for instance "6" versus "9," or "b" versus "d"). After a lesion of the dorsal parietal route, the opposite occurs: the animal, much like a young child, can discriminate "a" from "b," but not "b" from "d."[338]

The dorsal "how" pathway can be broken down into multiple circuits whose organization is only beginning to be understood. Some regions are concerned with hand gestures, others with programming eye movements, and still others code for different types of spatial relations. Parameters of distance, size, number, shape, movement, and orientation each call upon partially distinct subregions of the parietal cortex. In particular, after a right parietal lesion, it is not unusual for a patient to selectively lose the ability to tell left from right. This type of lesion reveals the dormant symmetry computed by our ventral occipito-temporal "what" system.

Oliver Turnbull and Rosaleen McCarthy have described one of the first convincing cases of this "mirror blindness" syndrome in R.J., a sixty-one-year-old man who suffered bilateral parietal lesions due to a coronary thrombosis.[339] R.J. was adamant about the fact that he could see *no* difference between an object and its mirror image. If the two pictures were superimposed, he admitted that the images were not perfectly aligned. But as soon as they were drawn apart, he no longer saw any difference. "I know from what you did that they should be different," he said, "but when I look at one and then the other, they look just the same to me."

To explore this strange pathology, Turnbull and McCarthy presented R.J. with three almost identical images, from which he had to spot the one that did not belong with the others. R.J. had no difficulty in detecting even subtle changes in object identity, for instance a rabbit with four ears. He could, similarly, locate an upside-down rabbit. Amazingly, however, he was unable to spot a rabbit looking right in a group of several other rabbits looking left (figure 7.6). Other cases of "mirror blindness" have since been reported.[340] Whether patients are asked to deal with drawings of animals, tools, or completely novel wire-frame objects, they are helpless if they have to distinguish

Figure 7.6 Following a brain lesion, some patients no longer can discriminate mirror images. To probe their perception of spatial orientation, Oliver Turnbull and Rosaleen McCarthy (1996) designed an odd-one-out test, in which patients must detect, in each row, the image that differs from the others. Patient R.J. systematically failed when the difference lay in the left-right orientation of the images (top row), but succeeded in all other tests, including those that bore upon word or letter orientation (bottom rows). *Used with permission of* Neuropsychologia.

an object from its mirror image. Their dorsal parietal lesion deprives them of any information about left and right, and their intact ventral visual system seems blind to orientation.

Breaking the Mirror

There is, however, a notable exception to mirror blindness, which brings us back to the heart of our subject: reading. When patients with mirror blindness read, they have no difficulty with left and right. They all manage to detect whether letters and words are written normally or in mirror image. They experience no difficulty in spotting the odd word in a list when the intruder reverses individual letters (REASON-ЯƎAƧOИ), the whole word (SMILE-ELIMS), or both (FEVER-ЯƎVƎꟻ). All of these differences are as obvious to them as they are to us—although because of their lesion, they do not see *any* difference between nonverbal mirror shapes such as ꙰ and ꙰.

This simple fact confirms that reading relies on special rules. Part of the ventral pathway in expert readers unlearns mirror-image generalization. In the early stages of reading, we initially need our dorsal system to distinguish letters such as "b" and "d." Only this structure can tell us if a letter "points" left or right, while our native ventral system tells us that both images represent the same shape. Progressively, however, our ventral system learns to break with symmetry. It stops considering "b" and "d" as two views of the same object. Ultimately, it assigns them distinct neuronal populations that cease to generalize across mirror reversals. These asymmetric letter detectors serve as entry points for the extraction of detailed statistics about reading, which only apply to properly oriented words. In this way, the occipito-temporal cortex acquires an asymmetrical neuronal hierarchy for visual word recognition. Unlike its neighboring cortical regions for object or face recognition, which continue to generalize over left-right changes, our reading architecture ceases to confuse mirror images.

This symmetry-breaking process seems to occur for written characters alone. Even then, it is only useful if the writing system itself is strongly asymmetrical. This is not always the case. Egyptian hieroglyphs, for instance, could be written in both directions, from right to left or from left to right. In the latter case, all of the characters were reversed. Human figures and animals indicated the direction of reading: their head was always turned toward the beginning of the line. Similarly, ancient Greek was written in "boustrophedon," literally "as the ox plows": one line was written from left to right, and the next one from right to left, in mirror image. Thus neither the Egyptians nor the Greeks needed to break the symmetry of their ventral visual system. Indeed, their curious bidirectional writing systems would probably never have been adopted if the human visual system did not spontaneously treat shapes like 🦅 and 🦅 as identical. To this day, some writing systems do not impose strict symmetry-breaking. Chinese has a handful of characters in mirror pairs comparable to our "b" and "d," such as 信 and 訃, or 亻 and 卜. Other writing systems, like Tamil, rely solely on elegant cursive characters like ௫, ௷, or ௸, which do not allow for mirror confusion. This cultural difference leads to diminished perceptual competence: readers of Tamil easily confuse mirror images and think, for instance, that the triangle ◢ is present in the image ◥● .[341]

Broken Symmetry . . . or Hidden Symmetry?

Even in readers of the Latin alphabet, who can distinguish "b" from "d," it is unclear if symmetry perception ever totally disappears. Does our visual system really cease to generalize across mirror images of words? Or does mirroring still occur although we are not aware of it? Samuel Orton postulated that the right hemisphere of even a proficient reader might continue to represent letters in reverse. As far as Orton was concerned, learning to read simply required focusing on the correct orientation of words. Their mirror image, although unheeded, was nonetheless present in the nondominant hemisphere. The mirror was not broken, but merely hidden.

The idea that we are unconsciously haunted by the mirror images of the words we read may not be as absurd as it might first appear. In the first stages of visual processing, letters are meaningless shapes. Before our brain recognizes them, they are probably processed like any other visual image, perhaps even while we read. In this case, interhemispheric transfer should still turn them over like a glove. Indeed, we are not helpless when faced with mirror words. With some effort, we can manage to decipher them one letter at a time. This suggests that recognition of single letters is possible in mirror images, even in the absence of access to the deeper visual word form system.

At the motor level, many anecdotes suggest that we host dormant programs that are adapted to mirror writing. If you try to mirror write, you will most probably fail—but try writing on a piece of paper placed under a table, as though you could see through the top, and you will see that the appropriate mirror gestures come quite naturally. It seems almost too easy to even count as genuine mirror writing, but if you filmed your hand and replayed it, you would see that it moves exactly as needed to write from right to left. Likewise, your right hand can write in the correct direction while the left hand writes simultaneously in mirror form. Mirror writing is usually easier with the left hand (for right-handed people), thus substantiating the hypothesis that interhemispheric transfer plays a positive role.

The most convincing support for the dormant representation of mirror reading and writing, however, comes, again, from brain-lesioned patients. Approximately one hemiplegic patient in forty shows transient evidence of mirror writing. This curious symptom almost always appears when hemiplegia obliges a right-handed patient to write with his left hand. In most cases, the deficit only concerns writing, not reading: the patient usually remains unable to read the mirror words that he has written. However, there are a few remarkable cases where both mirror reading and writing appear after a brain injury.

Such is the case with Mrs. H.N., a fifty-one-year-old American woman who was the victim of a car crash.[342] After her accident, Mrs. H.N. suffered from moderate inattention, planning, and memory deficits linked to a bilateral decrease in frontal lobe metabolism. Her main complaint, however, was that she could no longer read or write. One day, as she drove her car, she noticed that the stop signs said "POTS." She suddenly became aware that it was easier for her to mirror read and write rather than write normally. This amazing claim was confirmed by laboratory experiments. Regardless of the hand she used, Mrs. H.N.'s writing was riddled with mistakes when she wrote from left to right. When she wrote in the opposite direction, however, her writing became better, if still a little clumsy (figure 7.7).

Even more surprisingly, Mrs. H.N.'s reading was faster and more accurate when the words were mirror-reversed. When she was asked to recognize a word written from right to left, it took her one second, or essentially the normal time. However, when the word was written properly, the same task took

(A) Before the accident

(B) After the accident, normal writing

(C) After the accident, mirror writing

D) Mirror image of text C

Figure 7.7 Some patients with brain lesions suddenly start to read and write in mirror image. The figure shows writing samples of patient H.N., before and after her car accident (after Gottfried, Sancar, & Chatterjee, 2003). *Used with permission of Neuropsychologia.*

her over two and a half seconds, which is too slow by a factor of four. The same finding was replicated with words flashed on a computer screen. Here, when she saw a mirror letter string flashed for a quarter of a second, she had no trouble deciding if it was an English word or not. As you can see for yourself if you try, this is amazingly fast.

⊃ɿⱯqiɹɔ

ɈɥǫiloɈ

ɿoɈƨɐq

Strangely enough, when the words were written properly, from left to right, her responses remained random until the duration exceeded half a second. Mrs. H.N.'s accident had severely perturbed her reading and writing and had also revealed a dormant representation of mirrored words of which she was initially unaware.

Mrs. H.N's case is by no means an isolated one. A handful of similar cases have been reported in neurological journals—not enough, unfortunately, to draw strong conclusions about the underlying brain mechanisms.[343] The brain lesions are often diffuse and poorly localized, frequently due to a head injury—several brain regions probably need to be simultaneously affected for this exceptional mirror-reading syndrome to emerge. It would also be an exaggeration to claim that mirror-image performance approaches that of normal readers. In fact, the patients' mirror writing is often clumsy, and all tend to be slow at reading mirror words, rather like the letter-by-letter decoding seen in young readers.

Mrs. H.N's behavior points to a third interesting clue: during mirror reading, although she read relatively well, she often failed on irregular words like

woman

colonel

choir

She only appeared able to read with the slow grapheme-to-phoneme route.

Overall, these observations suggest that mirror readers have lost part of their normal reading circuitry. I speculate that in mirror readers, cerebral lesions eradicate the parallel mechanisms of written word recognition. Hence patients rely on intact letter-recognition units to painstakingly decipher words. In the brains of the well-read, this level of letter recognition is probably the only one that has not unlearned how to recognize the mirror images of words. That patients have preserved this level would explain why they can still read reversed words, even if their pace is slow and regularization errors betray a letter-by-letter decoding strategy.

This explanation, however, still fails to explain why the vast majority of brain-lesioned patients who read letter-by-letter do not have a gift for mirror reading. Another ingredient is needed—I suspect that it has to do with a second deficit, an inability to perceive the spatial organization of stimuli. Most mirror readers demonstrate spatial orientation deficits other than reading. One patient confused left and right and drew mirror-image maps of Australia and the United Kingdom.[344] Another could not remember how to draw the face of a clock and appeared to have lost the notion of time and number: she claimed that Thursday came after Friday, and that the birthday following her forty-fourth would be her forty-third![345] Mrs. H.N., as mentioned earlier, had no obvious spatial deficits, but had trouble in planning a strategy and sticking to it.

Such spatial and strategy deficits may play a role in mirror reading. When we read, the parietal and frontal regions of the brain perform crucial spatial computations, first to orient our eyes, then to send top-down attention signals to our visual areas so that they can select a target region for reading. I speculate that damage to one of these areas may perturb reading by altering its direction. As a result, patients adopt anomalous strategies, either by starting with the right-hand side of words, or by only paying attention to the mirror representations that are present but concealed in most of us. This interpretation is reminiscent of Orton's suggestion that expert readers use selective attention to neglect the representations of words that are inappropriately oriented. Patients with mirror reading have lost this basic spatial attention mechanism.

Even if we cannot fully explain the phenomenon, mirror reading reveals new information on a well-concealed aspect of reading. A mirror reader may well be lying dormant in each of us—but this quiescent competence only manifests itself when certain, multiple, and as yet undefined brain injuries bring it into action.

Symmetry, Reading, and Neuronal Recycling

To summarize this chapter, the visual circuits of the child's brain possess many properties that make them useful for reading, but one of them is detrimental: generalization across mirror images. This ancient property, already present in monkeys and pigeons, explains why children make mirror errors when they start to read and write. They see the letters "b" and "d" as though they were the same object viewed from two different angles.

I speculate that one must go beyond the mirror stage to become an expert reader—we have to "unlearn" our spontaneous competence for mirror-image generalization. But how? The distinction between left and right probably begins in the dorsal visual pathway that programs gestures in space. When children trace letters, they associate each one with a distinct gesture. Progressively, this spatial and motor learning is transferred to the ventral visual path-

way for object recognition. Young readers learn to attend to shape orientation. They begin to see letters as two-dimensional curves rather than as
three-dimensional shapes that can rotate in space. Progressively, they become
proficient at spotting groups of letters. This representation of bigrams, graphemes, and morphemes develops exclusively in the left hemisphere, and solely
for letters appearing in proper, left-to-right orientation.

At this point, symmetry vanishes. The expert reader possesses a rich
knowledge of the statistics of letters in normal writing, but is essentially
unaware of mirror writing. The only remnant of his initial competence is a
dormant, unconscious representation of the mirror images of letters no longer exploited by the reading system.

That it takes a normal child several months to unlearn mirror-image generalization strongly supports the neuronal recycling hypothesis. If the brain
were a blank slate, devoid of any structure, but able to absorb everything in its
cultural environment, there would be no reason to learn mirror reading. It is
hard to imagine how the brain could learn mirror representations that it never
sees in its surroundings and that are useless for normal reading. Recycling, on
the other hand, implies that before cortical regions convert to other uses, they
already possess prior structural properties inherited from evolution. Each
cortical region starts out with a portfolio of assets and liabilities that are only
partially rearranged by learning. Human culture works on the basis of existing
biological biases and must discover how they can be adapted to a new use.

A Surprising Case of Mirror Dyslexia

This vision of reading can, to some extent, shed new light on children's learning disabilities. In the majority of dyslexics, as I pointed out in chapter 6, reading deficits come from a core impairment in the processing of speech sounds.
I also pointed out, however, that another, smaller number of children probably suffers from a primarily visual impairment. In these cases their problem
could be due, at least in part, to the unlearning of symmetry.

The general public generally believes that a dyslexic child not only has a
reading deficit, but also fails to distinguish left from right. As a matter of fact,
letter orientation does pose specific difficulties for large numbers of dyslexic
children, who often confuse "p" with "q," or "b" with "d."[346] However, these are

problems that are present in all children and not just dyslexics. Furthermore, they are only transient. Mirror errors peak at between seven and ten years of age, when children learn to recognize letters, and then vanish. As a group, dyslexic children are not particularly gifted at mirror reading. Visual difficulties do not seem to play a dominant role in most of them.[347] The transient visual impairments that do occur may simply be the result of a core phonological deficit: if I can't figure out the difference between phonemes *b* and *d*, it will take me longer to figure out that the mirror shapes "b" and "d" refer to distinct letters. By sheer bad luck, the pronunciation of mirror letters also tends to be confusing, because they all refer to similar phonemes (all are "plosive" consonants that are produced by suddenly stopping the airflow in the vocal tract). It is not surprising that this conjunction of difficulties poses specific problems for children with phonological impairments.

In a few rare cases, however, left-right confusion does seem to be the true cause of dyslexia. A striking example of this can be found in the case of a young woman known as A.H., who was studied in great detail by the American psychologist Michael McCloskey.[348] A.H.'s disability was diagnosed quite late, when she was in her twenties. By that time, she was a brilliant psychology undergraduate at Johns Hopkins University. At the end of one of McCloskey's lectures on developmental dyslexia, she told him that her own childhood had been marred by severe difficulties in reading and arithmetic. McCloskey decided to submit her to a few cognitive tests. He was indeed well inspired! This brilliant student turned out to be suffering from a rare and fascinating form of dyslexia.

A.H.'s reading problems seemed to arise from spatial confusion at essentially all levels of the reading process. When she read isolated letters, her errors consisted almost exclusively of spatial reversals. For instance, she mistook a "p" for a "q," and an "m" for a "w." Interestingly, however, she almost never erred on other letters whose identity did not depend on their spatial orientation, such as "g" or "k." When she read words, Miss A.H. again reversed individual letters, but she also mixed up their position within the string (figure 7.8): "snail" became "nails." Finally, at the sentence level, she frequently inverted word order, especially when it was not tightly constrained by context.

In short, A.H. seemed to live in an unstable visual world where spatial relations never ceased to fluctuate. Letters seemed to float around, turning many words into ambiguous strings. Even a simple word like "brain" became an array of possible transpositions and anagrams such as "drain" or "brian." She could only resolve these ambiguities by relying on context. To prove the extent of her mental confusion, McCloskey made her read sentences in which the letters and words had been scrambled, such as "The stock market is storngly affected dy dercebtions the the of pulbic." Not only could A.H. read these sentences easily, but she spontaneously corrected them and announced without hesitation that 60 percent of the sentences were perfectly normal! Having lived with this deficit all her life, she was unaware of the dramatic instability of her spatial vision.

A.H.'s deficit was not unique to reading. She was unable to make a properly oriented copy of even a simple drawing. When she copied an image, there was a 50 percent chance that the result would be mirror-reversed (figure 7.8). Her school records already showed many such reversals. On a school trip to the Phillips Collection in Washington, D.C., with her art class, she sketched Renoir's *The Luncheon of the Boating Party*—a lighthearted summer painting showing flirtatious young men and women enjoying themselves at a lovely riverside *guinguette* (open-air café). Her sketch concentrated mostly on the hats, but she had managed systematically to invert their locations: those appearing on the left side of the painting ended up on the right side of her sketch, and vice versa.

Even more surprisingly, A.H. occasionally made mirror errors with some of the most commonplace gestures of everyday life. A simple grasping test required her to close her eyes while someone placed a wooden cube in front of her. Upon hearing a signal, she was supposed to open her eyes and grasp the cube as fast as possible. A video camera, filming the scene, captured a fascinating phenomenon (figure 7.8): on two-thirds of the trials, her hand initially reached toward the opposite side! Only later did it veer in the right direction. A.H. reversed left and right so frequently on this grasping test that her performance was worse than if she had responded at random. Nonetheless, only the initial direction of her gesture was wrong: all other parameters—angle, distance, and speed—were appropriate. A.H. obviously suffered from an extremely selective problem of spatial reversal. Her visual system seemed to flip constantly from left to right, as though an evil spirit had forced her to inhabit Alice's looking-glass world.

In the course of their work with A.H., Michael McCloskey and his colleagues fortuitously discovered how they could help their patient. When the

Stimulus word	Reading response
dog	bog
bone	done
pig	dig
star	tars
rib	rip
sun	nuns
skirt	skit
dust	dusk

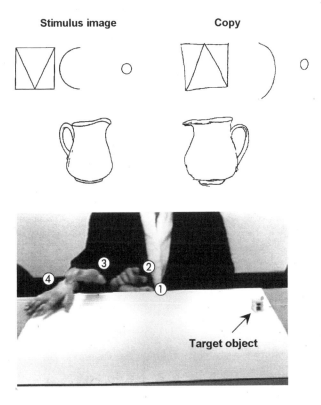

Figure 7.8 In rare cases, dyslexia seems to be caused by an excess of symmetry in the visual system. In a young dyslexic woman, A.H., reading errors mostly concerned the reversible letters "b," "d," "p," and "q," but also affected the order and, more rarely, the identity of other letters (top). A.H. frequently reversed the orientation of the drawings that she attempted to copy, especially along the left-right axis (middle). When she attempted to quickly grasp an object (bottom), her hand often reached to the location exactly symmetrical, as shown in this multiple-exposure photograph (after McCloskey et al., 1995; McCloskey & Rapp, 2000). *Used with permission of* Journal of Memory and Language and Psychological Science.

target object moved, flickered, or was presented for less than one-tenth of a second, her performance improved. Indeed, flickering letters, words, or phrases at ten cycles per second was enough to bring her reading score to a normal level. This phenomenon remains very mysterious—it stands in direct conflict with the predominant view that dyslexia results from an impaired "magnocellular" pathway for fast visual processing.[349] With A.H., flickering was so effective that she began reading under a strobe light!

A.H.'s case is such an exception to everything that has been written in the scientific literature on dyslexia that it raises more questions than it answers. How many children labeled as dyslexics actually suffer, like her, from a primary deficit in spatial perception? Would other dyslexics also benefit from a strobe light? Above all, what impairment mechanism can explain this mysterious mirror reversal of the visual world? Did A.H. suffer from an early lesion of the dorsal pathway that localizes objects in space and distinguishes left from right? Or did her deficit arise from the ventral pathway, which perhaps suffered from an excess of mirror-image connections from the corpus callosum? Or perhaps a putative "unlearning deficit": during reading acquisition, a deficient synaptic elimination mechanism left her with the visual system of a young child, unable to perceive the difference between a word and its mirror image. Brain imaging should no doubt shed some light on these open questions.

In the final analysis, A.H.'s fascinating case reminds us that visuospatial attention is of paramount importance to the normal development of reading.[350] Good decoding skills do not arise from associations between letters and speech sounds alone—letters must also be perceived in their proper orientation, at the appropriate spatial location, and in their correct left-right order. In the young reader's brain, collaboration must take place between the ventral visual pathway, which recognizes the identity of letters and words, and the dorsal pathway, which codes for their location in space and programs eye movements and attention. When any one of these actors stumbles, reading falls flat on its face.

That both the dorsal and ventral pathways need to be trained for reading may explain the remarkable success of teaching methods that emphasize motor gestures. In preschools that follow the method first instigated by the Italian psychologist Maria Montessori, one of the activities that prepares children for reading consists of tracing the contours of large sandpaper cutout letters. Children are taught to do this from left to right, paying careful attention to the order in which the strokes are drawn. This activity brings together gestures, touch, vision, and a sense of space. By imposing an asymmetrical route, always going from left to right, symmetry-breaking in the ventral visual pathway is no doubt simplified. In fact, the French psychologist Edouard Gentaz has demonstrated experimentally that, in normal children, this multisensory method is more efficient than classic phonological and visual training alone.[351]

Multisensory studies open up a whole new line of future research. Brain imaging may perhaps be able to show that the tactile method improves the functional connections linking the dorsal and ventral pathways. Spatial and tactile letter exploration could also help children who suffer from reading deficits of visual origin. A touch-based intervention method might perhaps allow A.H. to improve her reading. As the science of reading progresses, I am confident that we will not just gain a better understanding of how the human brain converts to reading. This new knowledge should, in time, allow for the development of better teaching and rehabilitation techniques that are optimally suited to the child's brain.

CHAPTER 8

Toward a Culture of Neurons

Reading opens up whole new vistas on the nature of the inter-actions between cultural learning and the brain. The neuronal recycling model should extend to cultural inventions other than reading. Mathematics, art, and religion may also be con-strued as constrained devices, adjusted to our primate brains by millennia of cultural evolution. There is, however, a key unresolved question: why are humans the only species to have created a culture and thus to conceive of new uses for their brain circuits? I propose that the expansion of a "conscious neuronal workspace," a vast system of cortical connections, allows for the flexible rearrangement of mental objects for novel purposes.

When it was proclaimed that the Library comprised
all books, the first impression was one of extravagant joy.
All men felt themselves lord of a secret, intact treasure.

—JORGE LUIS BORGES, "THE LIBRARY OF BABEL"

If God existed, he would be a library.   —UMBERTO ECO

*A*s we reach the end of our journey into the reader's brain, it becomes apparent that only a stroke of good fortune allowed us to read. If books and libraries have played a predominant role in the cultural evolution of our species, it is because brain plasticity allowed us to recycle our primate visual system into a language instrument. The invention of reading led to the mutation of our cerebral circuits into a reading device. We gained a new and almost magical ability—the capacity to "listen to the dead with our eyes." But reading was only possible because we inherited cortical areas that could learn to link visual graffiti to speech sounds and meanings.

Unsurprisingly perhaps, our brain circuits turned out to be imperfectly adapted to reading. Poor visual resolution, a steep learning curve, and an annoying propensity to mirror symmetry bear witness to our past evolution. Unfortunately, evolution did not anticipate that our brain circuitry would one day be recycled for word recognition. But our imperfect hardware did not prevent many generations of scribes, from antique Sumer on, from finding ways to take advantage of those circuits. They designed efficient writing systems whose refinement over the ages has allowed the marks on this page to speak to your brain.

Resolving the Reading Paradox

In the first pages of this book, I wondered how the invention of writing could have emerged in a brain that had not originally been designed for that skill. How did the human primate, with an unchanged genome, turn into a bookworm? The mystery, which I called the reading paradox, only thickened when we discovered that although reading has only been around for a few thousand years, the brains of all adult readers are equipped with a finely tuned reading mechanism.

We now know there is no paradox. The human brain never evolved for reading. Biological evolution is blind, and no divine architect ever wired our brain so we could read his book! The only evolution was cultural—reading itself progressively evolved toward a form adapted to our brain circuits. After centuries of trial and error, writing systems the world over converged onto similar solutions. They all rely on a set of shapes that are simple enough to be stored in our ventral visual system and are linked to our language areas. Cultural evolution tuned our writing systems so well that it now takes them only a few years to invade the neuronal circuits of the beginning reader. I introduced the idea of "neuronal recycling" to describe writing's partial or total invasion of cortical areas that were originally devoted to a different function.

Because I believe that details matter, I have delved at length into how our reading circuits work. I hope that the reader will forgive my long-windedness, but I wanted to illustrate two essential aspects of brain function in very concrete terms. My first aim was to demonstrate that cultural acquisition does not rest on general learning mechanisms—it is anchored in specific preexisting neuronal circuits whose function is tightly defined. In the case of reading, we are coming to understand these circuits with ever-increasing accuracy. We now know that they belong to the visual pathways for invariant object recognition present in all primates. Their rich internal structure and their partial capacity for learning new shapes put a limit on acceptable writing systems.

My second point, stemming from the first, is that human cultures are not the vast areas of infinite diversity and arbitrary inventiveness favored by social scientists. Brain structure keeps a tight rein on cultural constructions. The human capacity for invention is not endless but is narrowed by our limited neuronal construction set. If human cultures present an appearance of teeming diversity, it is because an exponential number of cultural forms can arise from the multiple combinations of a restricted selection of fundamental cultural traits.

In the case of reading, the hypothetical cultural invariants are concrete and tangible. From Chinese characters to the alphabet, all writing systems are based on a morpho-phonological principle—they simultaneously represent word roots and phonological structures. They also rely on a small inventory of visual shapes shared throughout the world, and first discovered by Marc Changizi (see chapter 4). A broad range of universal, neurologically constrained features underlies the apparent diversity of writing.

The Universality of Cultural Forms

In this closing chapter, I would like to defend the idea that reading is only one of many examples of how cultural invention is constrained by our neuronal architecture. If we extend the neuronal recycling model to other human activities, we should be able to connect them to their corresponding brain mechanisms. A link of this kind would allow us to understand the neuronal restrictions that define the scope of human cultural invention. At present this prospect seems remote, but we may one day be able to list the essential ingredients of all human culture (including family, society, religion, music, art, and so on) and understand how each of them relates to the vast array of our brain capabilities.

I look forward to the blooming of such a "culture of neurons," whose first stepping stones were laid down over the last thirty years by the French neurobiologist Jean-Pierre Changeux.[352] An accurate account of the diverse dimensions of human culture is now within our grasp. From this "neuro-anthropological" perspective, each cultural feature should ultimately be linked to well-defined neuronal circuits whose collective combinations would explain the multiple forms that cultural representations can take. It should be clear, however, that I am not advocating a direct and naïve reduction of cultural customs to a single gene, molecule, or category of neurons. As with reading, a series of bridging laws must ultimately anchor cultural constructions to their relevant brain networks. These bridging laws must integrate, rather than eliminate, the laws of human psychology. They must also include the historical, political, and economic forces that shaped human societies.

This perspective may sound daunting, but I am not alone in thinking that we can get there. A number of my colleagues have expressed similar arguments in favor of a unity of knowledge, or "consilience,"[353] between the humanities, psychology, and ultimately the brain sciences. The linguist Noam Chomsky was one of the first scientists to forcefully stress the limits of cultural relativism:

[According to a commonly held view,] it is the richness and specificity of instinct of animals that accounts for their remarkable achievements in some domains and lack of ability in others, so the argument runs, whereas humans, lacking such articulated instinctual structure, are free to think, speak, discover and understand without such limits. Both the logic of the problem, and what we are now coming to understand, suggest that this is not the correct way to identify the position of humans in the animal world.[354]

A few years earlier, in 1974, the French anthropologist Dan Sperber had underscored how "the image that emerges from the cumulated ethnographic record is not at all one of indefinite variability, but rather one of extremely elaborate variations within a seemingly arbitrarily restricted set."[355] The hypothesis that cultural diversity hides a restricted array of universal mental structures was, of course, primarily put forward by the founder of structural anthropology, Claude Levi-Strauss. Recently, his theory has been largely confirmed by the research of cognitive anthropologist Donald Brown. In his book *Human Universals*,[356] Brown provides a long list of close to four hundred features shared by all cultures. These universal traits range from the use of color and number terms to territoriality, facial expressions, musical creation, games, legal systems, and many more.

What are the origins of these fundamental cultural features? Sperber holds that they originate from the fundamentally modular structure of the human mind.[357] In an elaboration of Jerry Fodor's hypothesis,[358] Sperber views the mind as a collection of specialized "modules" that evolved in response to a need for the adoption of a specific domain of competence. For Sperber, each module starts out with a narrow domain of application, limited to the situation that motivated its evolution. This domain can, however, be extended to

encompass a greater range of inputs, thus leaving room for the cultural invention of entirely new triggering stimuli.

Face recognition is a good concrete example. A visual competence for
faces emerges in all humans early in infancy. It is associated with a specialized
visual area that some have called the "face module."[359] This module's "proper
domain," or the information that its biological function makes it process, is
supposedly only made up of human faces—the module evolved for this stimulus set because its mastery gave us an evolutionary edge. But our face recognition device also fires for objects that are not a genuine human face, such
as the sight of a statue or a photograph, the vision of Christ's face in patches
of melting snow, or the giant human face that astronomers see on Mars
(actually the shadow of a mountain). Even minimal arrangements of shapes,
such as Internet "smileys" and other "emoticons" (☺), can trigger it. In Sperber's
framework, all of these stimuli, natural or artificial, belong to the module's
"actual domain," or the minimal stimulus that satisfies the hypothetical face
module's input conditions.

For Sperber, it is precisely the discrepancy between the actual domain and
the proper domain of our mental modules that leaves room for human inventiveness and is exploited by human cultures. People invent a variety of cultural
objects to attract the attention of their fellow humans. Their sole purpose is to
excite brain modules in the most unexpected and appealing way. The universal
propensity of human cultures to create portraits, statues, caricatures, masks,
makeup, and tattoos is probably the product of some sort of cultural game
played at the boundaries of our face recognition module. These cultural artifacts do not always imitate the human face—they exaggerate its features into
what ethnologists call a "super-stimulus," or a display that activates the face
module more strongly than a real face. Once optimized to the point of becoming a super-stimulus, a cultural device increases its reproductive efficiency and
can propagate throughout a human group like a mental epidemic.

Neuronal Recycling and Cerebral Modules

An obvious kinship, and even a certain complementarity, links Dan Sperber's
concept of proper versus actual domains to my neuronal recycling hypothesis.
Both views emphasize how a cultural object that could not be anticipated by

evolution finds its place in our brain architecture. It can only do this inasmuch as it belongs to our inventory of learnable mental forms, contained in the normal operating conditions of our cortical circuits.

My only reservation about Sperber's position is that he tends to underestimate the role of learning and brain plasticity in the origins of cultural universals. Although he describes all the modules as "instincts to learn," machines for acquiring new knowledge rather than rigidly wired devices, he emphasizes that they are only capable of learning within a relatively constrained sphere. Indeed, their inflexibility plays an essential role in his argument that a set list of innate modules accounts for the universal features of human cultures: hypothetically, all men and women possess the same fixed array of modules defined by human evolution.

The vision of culture that emerges from Sperber's theory may seem reductive. If we follow this line of thought, each human culture is just a bag of clever tricks to stimulate preestablished cerebral modules. Reading suggests that cultural invention goes far beyond this description. It did not simply emerge thanks to stimulation of our visual system. Writing created the conditions for a proper "cultural revolution" by radically extending our cognitive abilities. Exposure of the child's brain to reading at an age when it is most plastic creates a massive reorientation of human cognition that goes far beyond the mere redirection of a single module's inputs.

I have to add that I am not very fond of the "module" concept. The term locks cognitive functions into fixed cells that supposedly pave the surface of our brains. Cortical reality is vastly more variable, redundant, and plastic. Jean-Pierre Changeux emphasized the explosion of synapses that occurs in the first years of life and spawns diversity and redundancy in the child's cortex. Although brain circuits are already organized at birth, they do not seem to be rigidly allocated to a single domain, but may simply be biased to ease its acquisition. Thanks to their exuberant connections, brain circuits that evolved under the selective pressure of a given cognitive constraint can be converted to very different functions.

This neuronal recycling mechanism significantly enriches the scope of Sperber's modular hypothesis. The contours of our cerebral processors, far from being rigid, contain an element of plasticity. Their function may change, particularly in cultures that enforce intensive education and expose children's brains to cultural objects at a very young age. Plasticity allows for a cultural diversity that goes beyond a fixed list of modules. Revolutionary inventions like the Internet or the computer mouse, moreover, demonstrate that we are probably far from having discovered all of our cerebral toolbox's potential.

Finally, we have to acknowledge that the speed and ease with which cultural inventions emerge varies greatly, probably as a function of the amounts of cortical recycling needed to master them. All objects of thought fit, by definition, into the envelope of learnable representations that our brain circuits can accommodate. However, some may require more complex changes than others. While faces, tattoos, and masks abound in cultures all over the world, reading, mathematics, and music are recent inventions of uneven complexity and variable spread. The intensive and early teaching required to learn them betrays the extent to which they impose synaptic reorganization on our existing brain circuitry. These cultural objects are not all equally accessible—our brains may have to go through a number of intermediate steps on the cultural ladder before certain specific cortical niches become available.

Toward a List of Cultural Invariants

Ambitious contemporary research programs have begun to probe the most salient features of human culture, such as music, religion, art, mathematics, and science. They seek to identify the brain circuits associated with these creations and examine their putative phylogenetic and epigenetic precursors. A quick survey reveals several domains in which significant advances are on the way.

NATURAL SCIENCES

All cultures have an interest in the classification of plants and animals.[360] Scott Atran and Dan Sperber relate this concern for taxonomy to a hypothetical "folk biology" module specialized in the acquisition of knowledge about living things. Possession of such a module may play a key role in a species's survival by affecting its feeding patterns, health, and overall integration into the environment. Developmental psychology, brain imaging, and neuropsychology have indeed begun to reveal brain specialization for the knowledge of plants and animals, which emerges precociously in infants and is eradicated by some brain lesions.[361]

MATHEMATICS

My own laboratory, together with a number of others, has studied how simple mathematical objects are anchored in the brain.[362] All cultures share a minimum set of abstract mathematical concepts, present in both arithmetic and geometry. These concepts emerge in early childhood, even in the absence of any formal education. The prime example is the concept of number. Even an uneducated Amazonian child, whose language has no number words beyond five, understands the difference between sets of twenty versus forty dots, can order them, and can even combine them into approximate additions or subtractions.[363] This core competence for elementary arithmetic is related to the parietal lobes of both hemispheres, where a region sensitive to number of objects can be found in any human baby, child, or adult.[364] We now know, moreover, that even the macaque monkey brain contains neurons tuned to a particular number of objects—at a brain location that seems to be a clear homolog of the human parietal region active during mental arithmetic.[365] It is amusing to think that number theory, which is often touted as the pinnacle of "pure" mathematics, actually has roots in an animal competence for tracking food items and fellow animals, two functions of obvious value for animal survival.

Research into numbers and the brain exemplifies how mathematics rests on basic and universal competences of the human mind. Its ultimate foundations lie in the structured mental representations of space, time, and number that we inherit from our evolutionary past, and that we learn to recombine in novel ways with the help of written and spoken symbols.

ARTS

All cultures produce works of "art," expensive decorative creations whose practical use is far from obvious. In his books and in his course on "neuroesthetics" at the Collège de France (2003–2005), Jean-Pierre Changeux proposed a neurobiological appraisal of art forms, particularly painting and music.[366] As the "synthesis of multiple evolutions," our brain harbors a wide selection of mental representations that can explain the complex emotional reactions brought on by a work of art. In neuronal terms, a painting is broken down into many parts that are conveyed to numerous brain areas. Each of these regions processes one of the artwork's attributes: color, texture, faces, hands, emotional expression. A work of art becomes a masterpiece when it stimulates multiple distributed cerebral processors in a novel, synchronous, and harmonious way.

Ideas similar to these have been proposed by the American neurologist Vilayanur Ramachandran. He holds that "the purpose of art . . . is not merely to depict or represent reality . . . but to enhance, transcend, or indeed even to *distort* [it], . . . to amplify it in order to more powerfully activate the same neural mechanisms that would be activated by the original object."[367] For the British neurophysiologist Semir Zeki, "artists are in some sense neurologists, studying the brain with techniques that are unique to them."[368] Zeki speculates that art forms can be related to partially distinct brain circuits: fauvism stimulates our brain color area, kinetic art pushes the limits of our motion area, and so on . . . Furthermore, the work of gifted artists hides a vast number of interpretations that constantly renew the viewer's pleasure.

Changeux goes one step further. He considers that the mere stimulation of several brain systems is not enough to create a genuine art form. The most admired works of art are not just "super-stimuli" that fascinate viewers by using hypnotic motion or shocking colors—although this description does seem to apply to certain contemporary art trends! Above all, a masterpiece should exude harmony. The neuronal bases for the feeling of coherence between the individual parts and the greater whole, the *consensus partium* that Changeux places at the core of artistic creation, are only now beginning to be studied for music.[369] Here the concept of harmony in musical chords has a strictly physical definition. A basic sensitivity to the musical intervals of octaves and fifths is a cultural universal already present in infants. This feature seems to arise at the level of the primary auditory area, where consonant chords evoke coherent and synchronous neuronal activity while dissonant tones evoke painfully fluttering, beating rhythms. Can harmony in painting or sculpture be explained in much the same way, by some sort of phase synchrony in the multiple distributed brain areas solicited by a masterpiece? At present, we do not have any definitive answers to this question—neuroesthetics remains a fledgling and controversial area of neuroscience.

RELIGION

An even more controversial question is whether neuroscience can shed light on the universal human propensity for religion. Can religious thinking be tracked down to its putative evolutionary and neurobiological roots? Three recent books[370] advance cognitive and even "neurotheological" explanations for the universality of religion and its millennial stability.

Daniel Dennett elaborates on Richard Dawkins's original memetic explanation. Our capacity for cultural transmission through language and imitation, he argues, necessarily leads to the emergence of parasitic mental representations called "memes." These mental objects need not be beneficial to their hosts—the sole requirement is that they operate as efficient "replicators" that facilitate transmission from one mind to another. Religions are nothing more than self-reproducing assemblies of mental representations with one particular characteristic: their very content stipulates to the faithful that it would be sacrilegious not to adhere to them blindly and spread them loyally and diligently.

Pascal Boyer probes more deeply into the cognitive bases that explain the ease with which religious ideas spread. Like Sperber, he assumes that these "mental epidemics" find an echo in our brain's preexisting mental modules. Religion is a kind of parasite that rides piggyback on the cognitive modules destined to social intelligence, moral sense, and causal inference—those that make us infer a causal agent when we see that objects move in a goal-directed fashion. In essence, says Boyer, we are wired to discover an almighty design, intention, and morality in our environment.

Finally, Scott Atran concentrates on explaining the human inclination toward the supernatural. As he sees it, supernatural creatures such as invisible ancestors, thinking trees, living dead, malevolent objects, or enchanted woods present a unique combination that makes them extremely attractive. On the one hand, they fall squarely into the innate mental categories with which we parse the natural world (places, objects, animals, human beings), but on the other hand they violate some of the categories' basic rules. The magnetic hold exerted by this discrepancy explains why we find these purely imaginary concepts so attractive. They are highly efficient super-stimuli for the brains of both children and adults. Like Janus, the two-faced Roman god, one face allows them to penetrate our mental modules, while the other

boosts our attention and memory. Much as kinetic art fascinates us by stim-
ulating the brain's motion area, supernatural thoughts are mesmerizing
decoys for the brain areas that encode places, objects, animals, and human
beings.

All of these evolutionary accounts ring true, but they also seem to fall
short of the complexity of the social and cultural phenomena that they aim
to explain. I can only praise their attempts to relate the mysteries of art and
religion to standard cognitive phenomena that may ultimately be studied
using psychological and brain imaging tools. Nevertheless, they always
remind me of the Rudyard Kipling "just-so story" about how the camel got
his hump or the leopard his spots. Theories of art and religion, like Kipling's
stories, rest on speculative mechanisms that currently remain largely
detached from objective experimentation. In contrast, perhaps the most
interesting aspect of reading is that it readily lends itself to the rigorous dissection of its underlying neuronal mechanisms. Although it is a high-level
and specifically human invention, it ultimately relates to well-identified and
reproducible brain circuits. I look forward to future experimental advances
that will ultimately bring neuroesthetics and neurotheology up to the same
level of scientific rigor.

Why Are We the Only Cultural Species?

The hypothesis that every major feature of human cultures relates to given
cerebral modules or processors inherited from our evolutionary past meets
with one major objection. If we share most, if not all of these processors with
other primates, why are we the only species to have generated immense and
well-developed cultures?

In spite of recent debates about the extent of animal cultures, *Homo sapiens* is a truly singular species in the cultural sphere. Decades of patient etho-
logical observation have only turned up a list of thirty-nine authentic "cultural" traits in our closest cousins, the chimpanzees. Culture is defined as a
behavior that varies across groups of animals of the same species, is transmitted across generations, and is not just a simple consequence of variations in
local environment.[371] The supporters of chimpanzee cultures triumphantly

tout as "cultural" traits the fact that the chimpanzees from Mahale, Tanzania, pick their noses with a stick or that hand-clapping is used at four of seven African sites to attract another chimp's attention. I do not deny the importance of these observations—they serve to qualify the uniqueness of the human species and attempt to relate it to classic mechanisms of unconscious imitation and social transmission. However, the scarcity of animal cultures and the paucity of their contents stand in sharp contrast to the immense list of cultural traditions that even the smallest human group develops spontaneously.

The case of graphic arts is particularly edifying. There are almost no human societies that do not practice some form of drawing or engraving, be it on rock, mud, or the human body. These art forms were already amazingly well mastered by our ancestors in the upper Paleolithic age, as attested for instance by the rock paintings in the Chauvet cave (32,000 years before the present). In a great many cultures, including those of Inuits and Indians, drawings quickly took on a symbolic meaning, particularly to denote number. Full-blown writing emerged three thousand years before our time in at least four distinct sites (Sumeria, Egypt, China, and Mesoamerica). In contrast, no species of monkey or great ape ever created a single genuine symbol system.

The surprising lack of graphic invention in nonhuman primates is not due to any trivial visual or motor limitation. When provided with a drawing pad, a one-year-old chimp manages to produce dots, lines, and curves.[372] According to Japanese ape researcher Tetsuro Matsuzawa, chimpanzees "seem to possess an intrinsic motivation to draw." When the pad is programmed to leave no trace on a screen, they cease to use it, suggesting that "visible traces have some form of reinforcing value for the subjects." Why then, when given drawing and painting kits, don't chimpanzees ever create more than a few haphazard patches of color, seemingly devoid of any communicative or referential goal (see figure 8.1)? If cultural invention stems from the recycling of the brain mechanisms that humans share with other primates, the immense discrepancy between the cultural skills of human beings and chimpanzees needs to be explained.

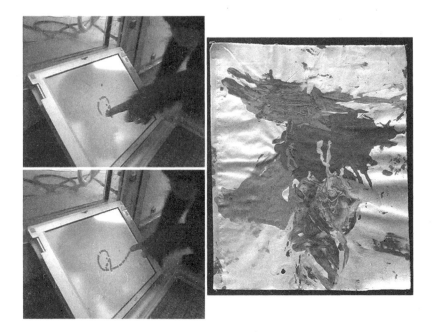

Figure 8.1 In nonhuman primates, graphic production is minimal. A thirteen-month-old chimpanzee learned to control a graphic tablet and trace elementary curves, but never developed any ability to communicate ideas through drawings (left, after Tanaka, Tomonaga, & Matsuzawa, 2003). The "composition" at right was produced by an adult chimpanzee who was living semi-independently in the Mefou Forest Reserve in Cameroon and was provided watercolors and a canvas (© Canadian Ape Alliance). *Used with permission of* Animal Cognition.

Uniquely Human Plasticity?

Several solutions to this conundrum have been proposed. One of them stresses that the human brain is endowed with greater plasticity than that of our cousins the chimpanzees. Undeniably one of *Homo sapiens*'s most salient features, which sets us apart from other primates and even from our ancestor *Homo erectus*, is that we are born with an immature and highly plastic brain.[373] Compared with other primates, delivery of the human newborn takes place at a time when its brain development is far from complete. The cortex is still neurologically immature and its synaptic maturation will continue for a number of years.

While prolonged plasticity probably contributes to our species's capacity to adapt to a broad variety of environmental conditions, I doubt that it plays a dominant role in our bent for cultural invention. Learning does not seem to be the limiting factor that prevents chimpanzees from developing cultural artifacts. A great many experiments, in fact, demonstrate that the primate brain is quite capable of abstract learning and even of radical conversion to new activities. Even macaque monkeys, as shown by Nikos Logothetis and Yasushi Miyashita, learn to recognize the arbitrary shapes of curves and fractals that they never encounter in their natural setting. Such training, as we saw in chapter 3, radically changes the tuning of many neurons in the macaque's visual cortex, which become selective for these new shapes.[374] Other experimenters even taught chimpanzees and macaque monkeys to recognize the shapes of Arabic numbers, order them rapidly, and associate them with their corresponding quantities.[375]

Monkeys can also learn to manipulate tools. Atsushi Iriki and his colleagues at the University of Tokyo had no trouble teaching monkeys to use a long rake to reach for objects. The monkeys became so good at this task that they managed to grasp hidden objects with the help of video feedback on a computer screen.[376] Tool learning led to massive changes in a small region of the anterior parietal cortex: neuronal receptive fields expanded, neurotrophic factors were expressed, and connections with distant posterior cortical areas increased. Similarities with the human brain suggest that this region may well be a forerunner of the human network for learned manual gestures, including writing.

In brief, the rudiments of a humanlike ability for symbol learning are obviously present in other primate species—yet not one of them ever invented cultural symbols on its own. Since the monkey brain can acquire symbols and tools, reduced brain plasticity cannot be responsible for the lack of cultural innovation in nonhuman primates. What is missing is not the capacity to learn, but rather the ability to invent and to transmit cultural objects.

Reading Other Minds

It has been proposed that the human brain is specifically "preadapted" for cultural transmission. Michael Tomasello explicitly defends this position:

Human beings are biologically adapted for culture in ways
that other primates are not. The difference can be clearly seen
when the social learning skills of humans and their nearest
primate relatives are systematically compared. The human
adaptation for culture begins to make itself manifest in human
ontogeny at around one year of age as human infants come to
understand other persons as intentional agents like the self
and so engage in joint attentional interactions with them. This
understanding then enables young children to employ some
uniquely powerful forms of cultural learning to acquire the
accumulated wisdom of their cultures.[377]

For Tomasello, the singularity of our species rests upon a unique capacity
for cultural transmission, due to the recent expansion of a cerebral module for
"theory of mind"—the mental representation of the intentions and beliefs of
others. This brain system impacts on cultural propagation in at least three
distinct ways. First, it enables adults to represent the extent and limits of their
children's knowledge, thus motivating them to teach and even to design pedagogical strategies. Second, "theory of mind" helps children understand the
communicative and pedagogical intentions of adults—they are not content
with imitating adults slavishly, but do so with a thorough understanding of
adults' goals.[378] Finally, theory of mind grants each member of the human
species the ability to represent itself, to pay attention to its own mental states,
and to manipulate them at will through the design of new cultural inventions.

Dozens of experiments show that very young children are remarkably sensitive to the communicative intentions of others. This sharing of intentions
plays a key role in the acquisition of language.[379] Contrary to a widespread
view, children do not learn words by repeated association with their corresponding objects. When they hear a new word, they track the speaker's gaze
to figure out what he or she is referring to. Only when they have understood
what the speaker has in mind—and have taken into account a variety of cues
about his knowledge and competence—do they assign a meaning to the spoken word. No learning occurs in "mindless" situations, for example if a loud-
speaker merely repeats a word in fixed relation to an object. Cultural transmission requires minding other minds.

According to Tomasello, the uniquely human sensitivity to another's mental states does not result from sudden mutation. A primitive representation of intentions, beliefs, and goals is already present in the pedigree of all great apes. It is amplified, however, in the human species and is accompanied by a peculiar motivation to share emotions and activities with other human beings. For Tomasello, none of these human competences, strictly speaking, predetermines us to develop a culture. However, at some point a critical mass of culture-enhancing devices is reached. A "cultural ratchet effect" emerges whereby each novel invention in turn facilitates the transmission of others to the social group. Intelligent imitation and active pedagogy ultimately stabilize cultural representations across several generations.

Although Tomasello's hypothesis is an interesting one, it only partially accounts for the extraordinary expansion of the human cultural sphere. It explains culture's almost epidemic spread from a single innovation center to a large group of human beings, and accounts for the stable or even irreversible character of our most dominant cultural traits (fire, farming, animal breeding, cities, writing, legal system . . .). It says little, however, about the initial spark that triggers cultural invention. No doubt the human species is particularly gifted at spreading culture—but it is also the only species to *create* culture in the first place. *Homo sapiens* has an imagination unparalleled in the animal world, which is obvious when one observes a young child at play. About cultural creativity itself, Tomasello's social transmission hypothesis remains essentially silent.

A Global Neuronal Workspace

My own view on the uniqueness of human culture is that beyond the development of a "theory of mind," another singular change was needed—the capacity to arrive at new combinations of ideas and the elaboration of a conscious mental synthesis.

When one compares the brain anatomy of today's humans to that of other primates, early hominids, or even Neanderthal man, a salient feature in our species is the disproportionate expansion of our frontal lobe. The prefrontal cortex, which lies in the frontmost part of the cerebral hemispheres, is a key player in the vast networks of associative areas that also include the inferior parietal and anterior temporal regions. These regions are up to forty times larger in humans than in macaques.[380]

Perhaps the most notable transformation that has taken place in the human brain is the expansion of fiber bundles underlying the frontal lobe— they are larger in humans that in any other primate species, even after correction for an overall change in brain or body volume.[381] Some long-distance connections, such as those that link the inferior prefrontal cortex to the occipital pole, may exist only in humans (figure 8.2). Human prefrontal neurons show a clear adaptation to this massive increase in connectivity: their dendritic trees, which receive incoming inputs, are bushier, and synaptic contacts are massively more numerous than those of other primates.[382] One category of neuron, the giant fusiform cell, seems only to exist in *Homo sapiens* and the other great apes.[383] These neurons are located in a part of the frontal lobe, the anterior cingulate cortex, and send extremely long axons to other distant regions of the cortex.

A new functionality must be behind this massive increase in connectivity. Jean-Pierre Changeux and I propose that it serves to convey input coming from several brain areas and to assemble it in a common cortical workspace. Once a mental synthesis is reached, the divergent connections can also broadcast the information to the entire brain.

Most of the primate cortex is largely modular—it is subdivided into specialized territories, each of which has its own specific inputs, internal structure, and outputs. The prefrontal cortex and other associative areas, however, are different—they emit and receive much more diverse cortical signals, and therefore tend to be less specialized. Their transversal connections break the brain's modularity and massively expand the cortex's transmission bandwidth. It is hardly surprising to discover that this system reaches its maximal expansion in the human brain. My proposal is that this evolution results in a large-scale "neuronal workspace" whose main function is to assemble, confront, recombine, and synthesize knowledge. It contributes to preventing the division of data and allows our behavior to be guided by any combination of information from past or present experience. Our prefrontal cortex, thanks to its connections to all the high-level areas, provides a space for internal deliberation fed by a whole set of perceptions and memories. Bluntly put, what we term "conscious thinking" may simply be the manipulation of information within this global neuronal workspace.

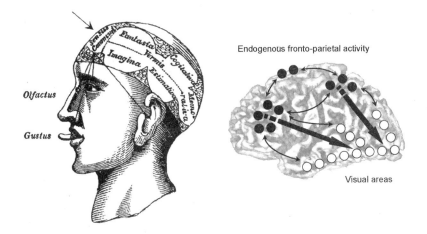

Endogenous fronto-parietal activity

Visual areas

Olfactus

Gustus

Long-distance associative fiber tracts

Macaque monkey   Homo sapiens

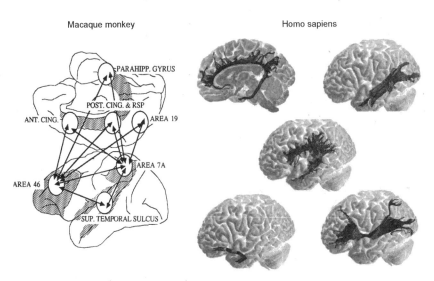

Figure 8.2 Reading is but one example of the extraordinary inventiveness of the human species. Why does *Homo sapiens* alone recycle its cortical networks for novel cultural uses? The Aristotelian notion of a "common sense," a cerebral center where the fusion of the five senses supports imagination and invention, finds an echo in the modern discovery of vast long-distance fiber tracts, particularly developed in humans, that bring together and flexibly recombine the information arising from many brain areas. These networks are the seat of constant spontaneous activity. Jean-Pierre Changeux and I postulate that this top-down activity, as it meets with bottom-up sensory inputs, provides conscious access to the external world and supports an enhanced capacity for mental exploration and invention.
Used with permission of Annual Review of Neuroscience *and Oxford University Press.*

My attempt to draw a link between the frontal lobe and the highest mental faculties is certainly not new. Over the last thirty years, a connection has been proposed by physiologists like Patricia Goldman-Rakic and Joaquim Fuster as well as by neuropsychologists like Alexander Luria, François Lhermitte, and Tim Shallice. All these scientists noticed that although damage to the frontal lobe often leaves elementary and automatized behaviors intact, it leads to difficulties in the organization of conscious reflection and in the planning of complex goal-directed action. At the beginning of the twentieth century, pioneering lesion experiments by the Italian neurologist Leonardo Bianchi already led him to term the frontal lobe the "organ of mental synthesis." He stressed how frontal lobe connectivity, which is particularly exuberant in humans, supported a confrontation of auditory and visual modalities and a rapid connection to speech comprehension and production systems.[384]

Looking even further back in time, we owe Aristotle and Galen the idea that our mind includes a *sensus communis* or "common sense," a private forum of consciousness where we bind inputs from the sense organs into a coherent and intelligible representation. Avicenna, in the tenth century, already associated this capacity with the front of the brain, although he pinpointed the brain's cavities (the ventricles) rather than the cortical mantle itself:

> One of the animal internal faculties of perception is the faculty of fantasy, i.e. *sensus communis*, located in the forepart of the front ventricle of the brain. It receives all the forms which are imprinted on the five senses and transmits to it from them. Next is the faculty of representation located in the rear part of the front ventricle of the brain, which preserves what the *sensus communis* has received from the individual five senses even in the absence of the sensed object. . . . Next is the faculty of "sensitive imagination" in relation to the animal soul, and "rational imagination" in relation to the human soul. This faculty is located in the middle ventricle of the brain.[385]

In this passage, Avicenna grants the frontal lobe three capabilities that he believes exist in animals, but are much expanded in human beings: *synthesis* of the five senses; *memory*, which keeps this synthesis "in mind" even after its object has vanished; and *imagination*, which enriches perception and, coupled to reason, can conceive of new ways to achieve a goal. With intriguing foresight, these ideas anticipated three major discoveries by modern neuroscience about frontal lobe function:

- Prefrontal neurons are massively multimodal. Experiments show that they are capable of gathering all the information relevant to a current goal, whether it arises from hearing, vision, or touch. There is practically no input to the brain that does not make its way to the prefrontal cortex.

- Prefrontal neurons exhibit a remarkable capacity to remain active even after the perceived object has vanished. Their firing can remain intense for several dozens of seconds, and keep the working memory of a past episode for essentially as long as the information will be needed.

- Prefrontal regions participate in a brain-scale network whose activity constantly fluctuates, even in the absence of any sensory inputs. It is tempting to associate this fluctuation with the spontaneous flow of consciousness and imagination.[386] When a volunteer undergoes a brain imaging scan at rest, and is not asked to perform any particular task, the brain is not quiescent, but goes through an unpredictable series of coordinated states.[387] From one moment to the next, activation appears in distinct sectors of the frontal cortex, each linked with particular territories of the distant parietal, temporal, or occipital lobes. The endogenous activation of these networks marks the conscious resting state in humans—it only disappears in deep sleep, anesthesia, or coma.[388]

In summary, a well-developed network of dense long-distance connections is present in the human brain. They form a global workspace that allows for the confrontation, synthesis, and distribution of information arising from other brain processors. This system is further endowed with a fringe of spontaneous fluctuation that allows for the testing of new ideas. I believe that the recent expansion of this global workspace network is closely related to both the emergence of reflexive consciousness and to the human competence for cultural invention.

Indeed, the spontaneous fluctuations that occur in the prefrontal cortex and other workspace areas provide a potential neuronal mechanism for the cultural recycling process that permeates this book. Their stochastic activity may indicate a permanent exploration of the various ways in which the contents of our brain processors can be assembled. With its expanded network of top-down connections, our species, more than any other primate, can select an arbitrary bunch of ideas and play with them. Ultimately these ideas can be put to new, unforeseen uses. Within our global neuronal workspace, we can bring an infinite number of thoughts to consciousness and recombine them at will. I suggest that the secret of our species's peculiar competence for brewing new cultural objects lies in this neuronal melting pot.

Crucially, although conscious brain activity fluctuates stochastically it does not wander at random. Within an individual brain, as well as at the level of a society, selection mechanisms stabilize the combinations of ideas that are most interesting, most useful, or, sometimes, merely most shocking or contagious. The dorsolateral prefrontal cortex is well equipped for this kind of sorting. It is the recipient of privileged neuronal projections coming from the evaluation and reward circuits located in the orbitofrontal and cingulate cortex as well as the subcortical nuclei such as the amygdala and the basal ganglia.

My hypothesis of an exploration of cortical circuits by the prefrontal cortex is no doubt reminiscent of Dan Sperber's idea that cultures play with the fringe of variability in brain modules. The possibility, however, of a synthesis across multiple brain processors, under the aegis of the prefrontal cortex, considerably enriches Sperber's model. Cortical recycling does not merely consist in finding novel inputs for a single module. Above all, the network of long-distance connections that culminates in *Homo sapiens* allows for the confrontation of multiple brain processors that would otherwise fail to communicate. I am convinced that innovation often arises from such links and cross-domain metaphors.[389] The invention of reading, in particular, did not merely consist of the creation of a set of signs that efficiently stimulated our visual cortex. It relied, above all, on an *association* of those signs with auditory, phonological, and lexical representations of spoken language. If other great apes never hit upon this idea, it may be because the architecture of their brains, contrary to ours, did not allow them to test these novel combinations.

A great many researchers have arrived, via different routes, at the conclusion that our species has a gift for putting thoughts together. Other species are unable to integrate all sources of information with the same degree of flexibility. Even in children, integration is limited prior to the full development of the prefrontal cortex. Children's spatial knowledge has been studied by my friend and colleague Elizabeth Spelke, of the Harvard Psychology Department. Her work suggests that, before the age of five or six, children's search behavior is modular and poorly integrated. Younger children can use the color of a wall to guide their search for a hidden object, and their sense of location to tell them on which side of the room to look—but it is only around the age of six that children represent a combinatorial concept such as "left of the green wall." Other species like rats never attain this integration stage, even as adults.[390] Our capacity for putting together two separate representations ("where" and "what") seems to be linked to language and might therefore be uniquely human.[391]

Annette Karmiloff-Smith, from the London Institute of Child Health, has developed a related theory which postulates that children's cognitive development consists of moving "beyond modularity."[392] A fraction of the child's brain, including the prefrontal cortex, learns to redescribe, in an explicit and abstract form, the older implicit knowledge that is buried within specialized modules. The acquisition of the alphabetic principle provides a nice example of internal redescription. At birth, infants already possess, albeit implicitly, some knowledge of phonemes. It is only after explicit learning of the alphabet that this buried knowledge is consciously extracted and becomes full-fledged phonemic awareness—the ability to consciously represent and manipulate phonemes. In this domain, mental synthesis, conscious representation, and cultural invention clearly go hand in hand.

Many other fields of human invention can be submitted to a similar analysis. In the case of mathematics, my previous book, *The Number Sense*, stressed how mathematical invention rests upon drawing new links or metaphors between domains such as number, space, and time.[393] In the field of artistic creation, Jean-Pierre Changeux views the search for a harmonious synthesis between brain processors as the very essence of art. Finally, the archaeologist Steven Mithen reaches a similar conclusion from an entirely different perspective: the study of the vestigial tools left by our ancestors over the last few million years of human evolution. Mithen speculates that the accomplishments of our species result from a mutation that allowed the multiple "chapels" of our mind's specialized intelligence to fuse into a single "cathedral" where thoughts of any kind could fluidly wander.[394]

In brief, a consensus is emerging around a simple idea: our species's brain has a special knack for mental recombination. The evolution of a global work-space allows it to optimally exploit the cultural cognitive niche made possible by neuronal recycling. Only human beings invent radically new ways to use their ancient brain processors and string them together to come up with innovative rules. Our prefrontal cortex functions like a primitive "Turing machine." It operates slowly and makes frequent mistakes, but the novel syntheses it generates can be genuinely creative. Its inventions, accumulated through cultural transmission over many thousands of years, go far beyond the competence that our species inherited through biological evolution. In the midst of many cultural treasures, reading is by far the finest gem—it embodies a second inheritance system that we are duty-bound to transmit to coming generations.

The Future of Reading

Writing—the art of communicating thoughts to the mind, through the eye—is the great invention of the world. Great in the astonishing range of analysis and combination which necessarily underlies the most crude and general conception of it—great, very great in enabling us to converse with the dead, the absent, and the unborn, at all distances of time and of space; and great, not only in its direct benefits, but greatest help, to all other inventions.

—ABRAHAM LINCOLN

*R*eading is both the result of human evolution and a major actor in its cultural explosion. The expansion of this "cathedral" of the mind, our prefrontal cortex, allowed our species to come up with writing. This invention, in turn, sharpened our mind. Its exercise endowed us with additional external memory that allows us, as Francisco de Quevedo put it, to "listen to the dead with our eyes" and share the thoughts of past thinkers. In this respect, reading is the first "prosthesis of the mind"[395]—a prosthesis that successions of ancient scribes adapted to our primate brain.

The neuroscience of reading demonstrates that every child's brain contains neuronal circuits that can be recycled for reading. Because of this scientific advance, much hope remains even for dyslexic children and adult illiterates. Brain plasticity often allows dyslexia to be sidestepped through the use of alternative cerebral pathways. Our growing understanding of how reading develops in young children, the emergence of reading software based on solid cognitive foundations, and its adjustment to each child's brain should bring renewed hope to all of those for whom reading is an ordeal.

One point should be clear: I do not claim that neuroscience will brush away all language learning problems, nor that computers optimized by cognitive scientists will soon replace schoolteachers. My message is more modest: there is nothing that a little bit of science cannot help. Parents and educators must have a better understanding of what reading changes in a child's brain. Children's visual and language areas constitute an extraordinary machine that education recycles into an expert reading device. I am convinced that increased knowledge of these circuits will greatly simplify the teacher's task.

I do not claim either that neuroscience should take over experimental psychology and pedagogical research. In brain imaging laboratories such as mine, it is psychologists who are behind the definition of neuroimaging protocols on reading. In the classroom, teachers are at the helm. They alone are responsible for arousing the child's interest in reading with clever illustrations, exercises, and creative activities. To handle classroom dynamics calls for a pedagogical expertise for which I have the deepest respect. I believe, however, that neither educators nor psychologists can afford to ignore recent scientific findings. Neuroscience today sheds indispensable light on how a reader's brain works, and what makes it more or less receptive to different teaching methods.

My hope is that, in due course, research on teaching, psychology, and neuroscience will merge into a single, unified science of reading. How reading should be taught is the first question that this new science will have to address. The principles that will allow children to learn to read without tears will have to be defined. A new approach to reading instruction could be achieved through the introduction of experimental classrooms and research laboratories in schools. Conditions such as these would allow for proper pedagogical experimentation. Teachers and researchers could work jointly to design future teaching methods. Many new questions need to be asked. Should writing be taught at the same time as letter recognition? Is it useful, even at a very young age, for children to write short sentences and passages? Do we save time by explicitly drawing the child's attention to the traps of symmetrical letters like "b" and "d"? There are no ready-made answers to these questions. Only collective and rigorous exploration will allow us to arrive at constructive solutions and interesting pedagogical advances.

No one can deny, however, that some questions about reading instruction have already been answered. We now know that the whole-language approach is inefficient: all children regardless of their socioeconomic backgrounds benefit from explicit and early teaching of the correspondence between letters and speech sounds. This is a well-established fact, corroborated by a great many classroom experiments. Furthermore, it is coherent with our present understanding of how the reader's brain works. To backtrack on this point, under the pretext of experimentation or pedagogical independence, would be disastrous for reading acquisition.

As a scientific consensus emerges on the mechanisms of literacy, reading instruction could become the prime example of genuine "neuro-psycho-pedagogy"—an integrated and cumulative approach where teacher autonomy is safeguarded but instruction is aimed toward a pragmatic search for efficient education strategies.

Science can contribute to teaching by introducing educators to the demanding concept of *experimentation*. To experiment does not mean to test vague ideas picked up at the last minute. It requires patient and meticulous design. Before generating innovative teaching strategies, all previous sources of knowledge must be tapped. To experiment also implies that any invention be evaluated by comparing it to a control situation (a different day, a different exercise, a different classroom . . .). I am confident that experimentation can significantly improve reading instruction. The achievement of this promise, however, will require rigor and attention, not just to recent cognitive discoveries, but also to the vast experience accumulated by teachers—often the prime experts of reading acquisition.

Unfortunately, we live in a world where education reform is still primarily a useful electoral platform for policymakers and politicians. Education policies swing back and forth on the whims of school boards. Decisions are often grounded in well-meaning ideologies, but in the absence of rational thought, good intentions are often transformed into largely misguided teaching practices. Left-wing progressives supported the whole-language approach under the pretext that it protects children from the tyranny of decoding and spelling instruction, and that children should be free to learn at their own pace. In a similar vein, some teachers still think that the constraints exercised by our genes and brain structure on learning are "right-wing." These attitudes do not have much to do with the hard facts about reading acquisition.

I cannot accept that the intuitions of school administrators should replace carefully accumulated scientific knowledge. A few simple truths should be accepted by all. In spite of variations in the speed at which we learn, all children have similar brains. Their cerebral circuits are well tuned to systematic grapheme-phoneme correspondences and have everything to gain from phonics—the only method that will give them the freedom to read any text. Classroom size is largely irrelevant—rigorously planned teaching methods can be used by all children, even in large classes of twenty pupils or more. However, it is essential to quickly detect at-risk children, and thus to develop efficient and standardized screening tests for dyslexia. Early diagnosis will allow these children to be oriented, at least temporarily, toward special reading classes with reinforced phonological training. Finally, although decoding is essential for beginning readers, vocabulary enrichment is equally important. A child must learn the morphology of English (prefixes, suffixes, and roots of words), particularly when he comes from an underprivileged background or a family where English is a second language.

These ideas are simple, straightforward, and easy to apply. They are the first offshoots of the new science of reading—the most extraordinary gift that we can transmit to future generations:

> Reading that pleases and profits, that together delights
> and instructs, has all that one should desire.
>
> —JACQUES AMYOT, 1513–1593

ACKNOWLEDGMENTS

my research on reading and its cerebral bases has been conducted in tight collaboration with my friend and colleague Laurent Cohen, professor of neurology at the Hôpital de la Salpêtrière, and a major player in the neuropsychology of reading. Heated discussions with him forged many of the ideas developed in this book. I would also like to thank all of my research collaborators, and notably Raphaël Gaillard, Antoinette Jobert, Sid Kouider, Denis Le Bihan, Stéphane Lehéricy, Jean-François Mangin, Nicolas Molko, Lionel Naccache, Jean-Baptiste Poline, Philippe Pinel, Mariano Sigman, Marcin Szwed, and Fabien Vinckier.

I have lost count of the many colleagues who guided me through the vast literature on reading and its impairments. They each contributed to my work in different ways. Some sent me their articles, others corrected my work prior to publication and frequently gave it a useful critical look, and yet others read some of the chapters in this book or authorized me to reproduce figures from their research. It is impossible for me to mention them all, but I am particularly grateful to Irving Biederman, Catherine Billard, Brian Butterworth, Alfonso Caramazza, Jean-Pierre Changeux, Roger Chartier, Joe Devlin, Guinevere Eden, Uta Frith, Albert Galaburda, Jonathan Grainger, Ed Hubbard, Alumit Ishai, Nancy Kanwisher, Régine Kolinsky, Heikki Lyytinen, Christian Marendaz, Bruce McCandliss, Yasushi Miyashita, Jose Morais, John Morton, Kimihiro Nakamura, Tatiana Nazir, Eraldo Paulesu, Monique Plaza, Michael Posner, Cathy Price, Franck Ramus, Sally and Bennett Shaywitz, Dan Sperber, Liliane Sprenger-Charolles, Xiaolin Sun, Ovid Tzeng, and Joe Ziegler.

I gratefully acknowledge the many places that hosted me while I was writing this book: Seattle (with thanks to Ann and Dan Streissguth), Boston (with special thanks to Elizabeth Spelke and Eliot Blass), Piriac (summer home of the Dehaene family), the Fondation des Treilles in Tourtour near Nice, the ski resort La Plagne in the French Alps, the Altiplanico hotel in San Pedro de Atacama (thanks to Marcela Peña), my hometown of Palaiseau, and my laboratory at the Commissariat à l'Energie Atomique in Saclay (France).

The structure of this book greatly benefited from careful readings by Odile Jacob, my French editor, and several rounds of rewriting by Susana Franck.

This book is dedicated to my wife, Ghislaine Dehaene-Lambertz, my longtime collaborator, materfamilias, and a constant source of encouragement and positive criticism. Nothing would have been possible without her help and support.

NOTES

INTRODUCTION: The New Science of Reading

1. Gould, 1992.
2. Dawkins, 1996.
3. Barkow, Cosmides, & Tooby, 1992; Pinker, 2002.
4. Quartz & Sejnowski, 1997.
5. Changeux, 1983.

CHAPTER 1: How Do We Read?

6. Rayner & Bertera, 1979.
7. Rayner, 1998.
8. Sere, Marendaz, & Herault, 2000.
9. Morrison & Rayner, 1981; O'Regan, 1990.
10. McConkie & Rayner, 1975.
11. Rayner, Well, & Pollatsek, 1980; Rayner, Inhoff, Morrison, Slowiaczek, & Bertera, 1981. For a review, see Rayner, 1998.
12. Pollatsek, Bolozky, Well, & Rayner, 1981.
13. Rubin & Turano, 1992.
14. See Rayner & Pollatsek, 1989, pp. 440–449.
15. Paap, Newsome, & Noel, 1984; Besner, 1989; Mayall, Humphreys, & Olson, 1997; Mayall, Humphreys, Mechelli, Olson, & Price, 2001.
16. McConkie & Zola, 1979; Rayner, McConkie, & Zola, 1980.
17. Rastle, Davis, Marslen-Wilson, & Tyler, 2000; Longtin, Segui, & Hallé, 2003.
18. Decomposition into morphemes seems to be a highly systematic process, which can be applied even to words like "repair" where it can lead us astray. However, it is unclear whether it occurs before or after we have accessed a mental lexicon of known words. The processing of morphemes remains hotly debated at both the empirical and theoretical levels. For further details, see, for instance, Caramazza, Laudanna, & Romani, 1988; Taft, 1994; Ferrand, 2001, chapter 5.

19. I shall use quotation marks to refer to the visual form of letters and words, and italics to refer to their pronunciation. Thus the word "women" is pronounced *wimen*.

20. Rey, Jacobs, Schmidt-Weigand, & Ziegler, 1998; Rey, Ziegler, & Jacobs, 2000.

21. Prinzmetal, Treiman, & Rho, 1986; see also Prinzmetal, 1990.

22. Augustine, *Confessions*, book 6, chapter 3. The quotes from Augustine and Isidore of Seville are drawn from Manguel, 1997.

23. Rubenstein, Lewis, & Rubenstein, 1971; Coltheart et al., 1977; Seidenberg et al., 1996; Ferrand, 2001, chapter 4.

24. Van Orden, Johnston, & Hale, 1988; Jared & Seidenberg, 1991.

25. I owe this pun and several other wordplays to Hammond & Hughes, 1978.

26. Perfetti & Bell, 1991; Ferrand & Grainger, 1992, 1993, 1994; Lukatela, Frost, & Turvey, 1998.

27. Coulmas, 1989, p. 251.

28. Coulmas, 1989.

29. Wydell & Butterworth, 1999.

30. Marshall & Newcombe, 1973; Shallice, 1988; McCarthy & Warrington, 1990; Coltheart & Coltheart, 1997.

31. Seidenberg & McClelland, 1989.

32. Seidenberg & McClelland, 1989; Grainger & Jacobs, 1996; Plaut, McClelland, Seidenberg, & Patterson, 1996; Ans, Carbonnel, & Valdois, 1998; Zorzi, Houghton, & Butterworth, 1998; Coltheart et al., 2001; Harm & Seidenberg, 2004. The most recent of these explicit models, called CDP+ (Perry, Ziegler, & Zorzi, 2007), provides an interesting synthesis of earlier theoretical proposals.

33. Selfridge, 1959.

34. Dennett, 1978.

35. McClelland & Rumelhart, 1981; Rumelhart & McClelland, 1982.

36. Weekes, 1997; Lavidor & Ellis, 2002. This is only true if the words are sufficiently familiar and appear in their normal horizontal orientation, within the center of the fovea or slightly to its right (Lavidor, Babkoff, & Faust, 2001; Lavidor & Ellis, 2002 Ellis, Young, & Anderson, 1988; Ellis, 2004).

37. Reicher, 1969; Ferrand, 2001, pp. 58–63.

38. Spoehr & Smith, 1975; Rumelhart & McClelland, 1982.

39. In chapter 3, we will see that recent observations on the neural coding of words qualify this definition of lexical neighbors as words that differ by a single letter. If words are encoded using pairs of letters, or bigrams, as suggested by several studies (Grainger & van Heuven, 2003; Grainger & Whitney, 2004; Dehaene, Cohen, Sigman, & Vinckier, 2005), we should probably adopt a slightly different estimate based on the number or proportion of shared bigrams as a measure of orthographic similarity. See Grainger, Granier, Farioli, Van Assche, & van Heuven, 2006.

40. The scientific literature on neighborhood effects is large and sometimes contradictory, probably because the presence of numerous neighbors may trigger opposite effects. For a recent discussion, see Ferrand, 2001, pp. 96–115.

41. Segui & Grainger, 1990.

42. Rey et al., 1998; Rey, Ziegler, & Jacobs, 2000. For a recent review of competition effects at multiple levels within the reading processing chain, see Grainger & Ziegler, 2007.

43. Swinney et al., 1979; Seidenberg et al., 1982.

CHAPTER 2: The Brain's Letterbox

44. Déjerine, 1892. All subsequent quotations are drawn from this essential source on the neuropsychology of reading.

45. Warrington & Shallice, 1980; Patterson & Kay, 1982; Montant & Behrmann, 2000.

46. Cohen & Dehaene, 1995, 2000; Gaillard, Naccache et al., 2006.

47. Patterson & Kay, 1982; Hanley & Kay, 1996.

48. Miozzo & Caramazza, 1998.

49. A study by Fiset and others (2005) qualifies this statement by suggesting that the similarity of each letter with the remainder of the alphabet contributes to the word length effect. When letter confusability is adequately controlled, the patients do not necessarily take more time to read longer words. Thus the true nature of the patients' deficit could be the loss of an acquired neural code that amplifies the minuscule differences between letters. This loss would increase the similarity between letters (much as Chinese characters look alike to someone who has not learned them). Beyond a certain threshold, this exaggerated resemblance may force the patient to slowly scrutinize each letter, like a beginning reader. In other words, the parallel recognition of words would not be abolished in pure alexia, but merely deteriorated (Arguin, Fiset, & Bub, 2002). The hypothesis of a residual parallel reading ability could explain the phenomenon of "implicit reading," whereby some patients remain able to classify words by meaning, even though they can no longer read them out loud (Coslett & Saffran, 1989; Coslett, Saffran, Greenbaum & Schwartz, 1993). However, this competence could also reflect the partial recovery of reading by the undamaged right hemisphere (Coslett & Monsul, 1994; Cohen et al., 2003).

50. Cohen et al., 2003. Laurent Cohen's book *L'homme thermomètre* (*Thermometer Man*) is an excellent introduction to visual deficits and their interpretation.

51. Damasio & Damasio, 1983; Binder & Mohr, 1992; Leff et al., 2001.

52. Leff et al., 2001.

53. Cohen et al., 2000; Cohen et al., 2002; Dehaene et al., 2002; Gaillard, Naccache et al., 2006. It should be noted that the term "visual word form area" and its connotations remain a matter of debate (see Price & Devlin, 2003; Cohen & Dehaene, 2004).

54. For examples, see Cohen et al., 2003, and its bibliography.

55. Geschwind, 1965.

56. Petersen et al., 1988; see also Posner et al., 1988; Petersen et al., 1989; Petersen et al., 1990.

57. Dehaene et al., 2002.

58. Cohen et al., 2000. In the Montreal Neurological Institute coordinate system, the measurements are −42 mm along the lateral axis, −57 mm along the posterior direction, and −12 mm along the vertical axis.

59. In figure 2.4 one can discern a source of variability: although the left occipito-temporal region activates in all the subjects we have studied, activation in the corresponding region of the right hemisphere is only apparent in some of them. Why this variability exists is not yet fully understood, but it may predict the potential recovery of some capacity for reading following a left-hemisphere infarct (Cohen, Henry et al., 2004).

60. Hasson et al., 2002.

61. Dehaene et al., 2002; Cohen, Jobert et al., 2004.

62. Booth et al., 2002; Cohen, Jobert et al., 2004.

63. Tokunaga et al., 1999; Nakamura et al., 2000.

64. Burton, Small, & Blumstein, 2000.

65. Ishai et al., 1999; Haxby et al., 2000; Ishai et al., 2000; Haxby et al., 2001; Levy et al., 2001; Hasson et al., 2002; Malach, Levy, & Hasson, 2002; Hasson et al., 2003.

66. Puce et al., 1996.

67. Haxby et al., 2001; Grill-Spector, Sayres, & Ress, 2006.

68. Grill-Spector, Sayres, & Ress, 2006.

69. Tsao et al., 2006.

70. Tarkiainen et al., 1999; Tarkiainen, Cornelissen, & Salmelin, 2002.

71. Allison et al., 1994; Nobre, Allison, & McCarthy, 1994; Allison et al., 1999. See also Gaillard, Naccache et al., 2006.

72. Kanwisher, McDermott, & Chun, 1997.

73. Cohen et al., 2000.

74. McCandliss, Curran, & Posner, 1993.

75. Cohen et al., 2000; Molko et al., 2002.

76. Ellis, Young, & Anderson, 1988; Cohen et al., 2000; Ellis, 2004. The transfer of information across hemispheres is just one of the factors that contribute to the advantage of the right visual field for reading (see Brysbaert, 1994). Reading direction, from left to right in English, also plays an important role because it leads to an enhanced training of the right visual field for letter recognition (Nazir et al., 2004).

77. Molko et al., 2002.

78. It is unlikely that a fiber tract this large carries only the identities of letters and words. It most probably also conveys a lot of additional information, for instance about the color, the shape, and the identity of objects. Indeed, further research showed that AC was also unable to transmit such information across the two hemispheres (Intriligator, Hénaff, & Michel, 2000).

79. Le Bihan et al., 2006.

80. Polk & Farah, 2002.

81. Grill-Spector & Malach, 2001; Naccache & Dehaene, 2001.

82. Dehaene et al., 2001.

83. Dehaene et al., 2004.

84. Devlin et al., 2004.

85. For further evidence of subliminal word processing, see, for instance, Naccache et al., 2005; Gaillard, Del Cul et al., 2006; Del Cul, Baillet, & Dehaene, 2007.

86. Dehaene et al., 2004.

87. See, for instance, Price, Wise, & Frackowiak, 1996; Büchel, Price, & Friston, 1998; Cohen et al., 2002; Dehaene et al., 2002; Polk & Farah, 2002. All these studies have found less activation of the ventral visual region by consonant strings than by words or pseudo-words. This result is easily obtained as long as the participants perform a simple and neutral task such as detecting whether the strings contain a descending letter like "g" or "p." It is crucial, however, to avoid difficult tasks such as memorizing the stimuli or detecting repetitions. In the latter cases, performing the task is slower and more difficult for consonant strings than for words, and the results can reverse (see, for instance, Tagamets et al., 2000).

88. Binder et al., 2006; Vinckier et al., 2007.

89. Baker et al., 2007.

90. Polk et al., 2002.

91. Dehaene, 1995; Bentin et al., 1999; Tarkiainen et al., 1999; Pinel et al., 2001.

92. See Tan et al., 2000; Kuo et al., 2001; Kuo et al., 2003; Lee et al., 2004. However, some areas are more active when reading Chinese characters than English words, including the right parietal and middle frontal regions (Tan et al., 2001; Siok et al., 2004). These regions may play a role in visuospatial attention and motor coding, thus helping Chinese readers memorize the several thousand characters of their language. Indeed, as a memory aid, Chinese readers often mentally retrace the spatial sequence of strokes that compose a character.

93. Chen et al., 2002; Fu et al., 2002. In some studies, Chinese characters induce somewhat greater activity than Pinyin strings, presumably because they have been overlearned and have become more familiar than the Pinyin script.

94. Kuo et al., 2001.

95. Ding, Peng, & Taft, 2004.

96. Nakamura et al., 2005. For similar results, see Ha Duy Thuy et al., 2004, and, with magnetoencephalography, Koyama et al., 1998. Korean reading has also been studied, with similar results (Lee, 2004).

97. Sakurai, Takeuchi et al., 2000.

98. Sasanuma, 1975; Sakurai, Momose et al., 2000; Sakurai, Takeuchi et al., 2000; Sakurai, Ichikawa, & Mannen, 2001.

99. Catani et al., 2003.

100. Di Virgilio & Clarke, 1997.

101. Marinkovic et al., 2003. See also Pammer et al., 2004, for globally similar images, with an additional suggestion of surprisingly early activation of the left inferior frontal region.

102. Fiez et al., 1999; Hagoort et al., 1999; Tagamets et al., 2000; Bokde et al., 2001; Xu et al., 2001; Cohen et al., 2002; Fiebach et al., 2002; Simos, Breier et al., 2002; Binder et al., 2003.

103. Pugh et al., 1996; Price et al., 1997; Friederici, Opitz, & von Cramon, 2000; Booth et al., 2002; Cohen, Jobert et al., 2004.

104. For recent reviews of research in this field, see Fiez & Petersen, 1998; Price, 1998; Jobard, Crivello, & Tzourio-Mazoyer, 2003; Mechelli, Gorno-Tempini, & Price, 2003; Price et al., 2003.

105. van Atteveldt, Formisano, Goebel, & Blomert, 2004.

106. Raij, Uutela, & Hari, 2000.

107. Dehaene-Lambertz & Dehaene, 1994; Dehaene-Lambertz & Baillet, 1998; Dehaene-Lambertz, Dehaene, & Hertz-Pannier, 2002.

108. Dehaene-Lambertz, 1997; Naatanen et al., 1997; Cheour et al., 1998; Jacquemot et al., 2003.

109. Simon et al., 2002.

110. Paulesu, Frith, & Frackowiak, 1993.

111. Binder et al., 1999, 2000; Kotz et al., 2002.

112. Vandenberghe et al., 1996.

113. Binder et al., 1999; Cohen et al., 2003.

114. Devlin et al., 2004.

115. Nakamura et al., 2005.

116. Rissman, Eliassen, & Blumstein, 2003.

117. Mazoyer et al., 1993; Price et al., 1997; Vandenberghe, Nobre, & Price, 2002.

118. Thompson-Schill, D'Esposito, & Kan, 1999; Rodd, Davis, & Johnsrude, 2005.

119. My thanks to http://monster-island.org/tinashumor/.

120. Damasio, 1989b, 1989a.

121. Warrington & Shallice, 1984; Caramazza & Hillis, 1991; Caramazza, 1996, 1998; Shelton, Fouch, & Caramazza, 1998.

122. Damasio et al., 1996; Grabowski, Damasio, & Damasio, 1998.

123. Martin et al., 1996; Chao, Haxby, & Martin, 1999; Beauchamp et al., 2004.

124. Pulvermuller, 2005.

125. Dehaene, 1995.

126. Wydell & Butterworth, 1999.

127. Chen et al., 2002; Fu et al., 2002.

128. Paulesu et al., 2000.

CHAPTER 3: The Reading Ape

129. Klüver & Bucy, 1937.

130. Mishkin & Pribram, 1954.

131. Lissauer, 1890; Goodale & Milner, 1992.

132. Humphrey & Weiskrantz, 1969; Weiskrantz & Saunders, 1984.

133. Desimone & Gross, 1979; Schwartz, Desimone, Albright, & Gross, 1983; Perrett, Mistlin, & Chitty, 1989; Tanaka, 1996; Tamura & Tanaka, 2001; Tanaka, 2003.

134. Sary, Vogels, & Orban, 1993; Ito et al., 1995.

135. Vogels & Biederman, 2002.

136. Tanaka, 2003.

137. Logothetis, Pauls, & Poggio, 1995; Booth & Rolls, 1998.

138. Quiroga et al., 2005.

139. Rolls, 2000.

140. Grill-Spector et al., 1998; Lerner et al., 2001; Vuilleumier et al., 2002.

141. Thorpe, Fize, & Marlot, 1996.

142. Riesenhuber & Poggio, 1999; VanRullen & Thorpe, 2002.

143. Tanaka, 2003.

144. Tsunoda et al., 2001.

145. Ito & Komatsu, 2004.

146. Brincat & Connor, 2004.

147. Biederman, 1987.

148. Ibid.

149. Biederman & Bar, 1999.

150. Vogels et al., 2001.

151. Miyashita, 1988; Logothetis, Pauls, & Poggio, 1995.

152. Baker, Behrmann, & Olson, 2002.

153. Gould & Marler, 1987.

154. Miyashita, 1988; Sakai & Miyashita, 1991.

155. Gould & Vrba, 1982.

156. Jacob, 1977.

157. Dehaene, 2005; Dehaene & Cohen, 2007.

158. Dawkins, 1989.

159. Blackmore, 1999.

160. Sperber, 1996; Sperber & Hirschfeld, 2004.

161. Dehaene et al., 2005. This proposal, called the LCD (Local Combination Detectors) model, extends an earlier proposal by Mozer, 1987.

162. Sigman & Gilbert, 2000; Sigman et al., 2005.

163. A region close to the V4 area, particularly in the left hemisphere, is activated by letter strings more than by checkerboards, faces, or houses when words are presented in the right visual field. See Cohen et al., 2000, 2002; Hasson et al., 2002; and figure 3.10.

164. Whitney, 2001; Grainger & Whitney, 2004; Schoonbaert & Grainger, 2004.

165. Humphreys, Evett, & Quinlan, 1990; Peressotti & Grainger, 1999.

166. Perea & Lupker, 2003; Schoonbaert & Grainger, 2004; Grainger et al., 2006.

167. McClelland & Rumelhart, 1981; Grainger & Jacobs, 1996; Ans, Carbonnel, & Valdois, 1998; Zorzi, Houghton, & Butterworth, 1998; Harm & Seidenberg, 1999; Coltheart et al., 2001; Perry, Ziegler, & Zorzi, 2007.

168. For an in-depth discussion of the science behind the raednig of mixed-up wrods, see the Web site of Cambridge psychologist Matt Davis: http://www.mrc-cbu.cam. ac.uk/~mattd/Cmabrigde/.

169. Vinckier et al., 2006.

170. Rastle et al., 2000; Longtin, Segui, & Hallé, 2003.

171. Reicher, 1969; Spoehr & Smith, 1975; Rumelhart & McClelland, 1982.

172. Binder et al., 2006.

173. Cohen et al., 2002.

174. Bouma, 1973; Lavidor, Babkoff, & Faust, 2001; Nazir et al., 2004; Vinckier et al., 2006.

175. Dehaene et al., 2004; Vinckier et al., 2007.

176. Allison et al., 1994, 1999; Haxby et al., 2001; Grill-Spector, Sayres, & Ress, 2006.

177. McCrory et al., 2005.

178. Ding, Peng, & Taft, 2004.

179. Ziegler & Goswami, 2005.

180. Paulesu et al., 2000.

181. Ha Duy Thuy et al., 2004; Nakamura et al., 2005.

182. Quartz & Sejnowski, 1997.

183. Levy et al., 2001; Hasson et al., 2002; Malach, Levy, & Hasson, 2002.

184. Turing, 1952.

185. Grill-Spector et al., 1998; Lerner et al., 2001.

186. Dehaene et al., 2004.

187. See, for instance, Kitterle & Selig, 1991; Robertson & Lamb, 1991.

188. Cohen, Lehéricy et al., 2004.

189. Turkeltaub et al., 2004.

190. Cohen et al., 2003.

CHAPTER 4: Inventing Reading

191. Levi-Strauss, 1958; Sperber, 1974; Chomsky, 1988; Brown, 1991.

192. Changizi & Shimojo, 2005.

193. Changizi, Zhang, Ye, & Shimojo, 2006.

194. There is no doubt that motor constraints also exerted a strong selective pressure on the evolution of writing. However, several arguments suggest that the regularities discovered by Marc Changizi are more tightly linked to the ease of reading than of writing (Changizi & Shimojo, 2005; Changizi et al., 2006). In particular, the same regularities are also found in computer fonts and in the logos of major companies, two categories of symbols where ease of writing cannot possibly be an issue. Conversely, they are not found in the shapes of handwritten characters, whose curves obey qualitatively different statistics.

195. Leroi-Gourhan, 1983; Calvet, 1998.

196. Leroi-Gourhan, 1993, pp. 190–191, 202.

197. Schmandt-Besserat, 1996.

198. Brain imaging has recently demonstrated that the sight of bodies and their configurations activates a specific region of the occipital cortex (Downing et al., 2001).
199. DeFrancis, 1989.
200. Coulmas, 1989, p. 78.
201. Zali & Berthier, 1997, p. 38.
202. A similar phenomenon occurred in Korea when King Sejong expressed his dissatisfaction with Chinese characters, which he judged unfit for the Korean language. "Because our language differs from the Chinese language," he said, "my poor people cannot express their thoughts in Chinese writing. In my pity for them, I create 28 letters which all can easily learn and use in their daily lives." This is how the Hangul writing was born in 1446. Nowadays, close to six centuries later, twenty-four of its twenty-eight characters remain in use. Its rational combination of phonetic, alphabetic, and syllabic principles makes it one of the most elegant writing systems on earth (Coulmas, 1989, pp. 118–122).
203. Coulmas, 1989, p. 146.

CHAPTER 5: Learning to Read

204. For excellent reviews of reading development and its implications for teaching, see Rayner & Pollatsek, 1989; Ehri, Nunes, Stahl, & Willows, 2001; Rayner et al., 2001.
205. Eimas et al., 1971; Werker & Tees, 1984; Kuhl, 2004.
206. Mehler et al., 1988.
207. Dehaene-Lambertz & Dehaene, 1994; Cheour et al., 1998; Dehaene-Lambertz, Dehaene, & Hertz-Pannier, 2002; Pena et al., 2003.
208. Kuhl, 2004.
209. Saffran, Aslin, & Newport, 1996; Marcus et al., 1999; Altmann, 2002; Marcus & Berent, 2003.
210. Chomsky, 1980.
211. Kellman & Spelke, 1983; for a recent overview, see Wang & Baillargeon, 2008.
212. Bhatt et al., 2006.
213. Kraebel, West, & Gerhardstein, 2007.
214. Shuwairi, Albert, & Johnson, 2007.
215. Smith, 2003; Son, Smith, & Goldstone, 2008.
216. Pascalis & de Schonen, 1994; Tzourio-Mazoyer et al., 2002; de Haan, Johnson, & Halit, 2003.
217. Pascalis, de Haan, & Nelson, 2002; Pascalis et al., 2005.
218. Robinson & Pascalis, 2004.
219. Gathers et al., 2004.
220. Southgate et al., 2008.
221. Frith, 1985.
222. Seidenberg & McClelland, 1989; Hutzler et al., 2004.
223. Goswami 1986 has shown that some five-year-olds spontaneously discover analo-

gies in the internal structure of words. For instance, after reading the word "beak," they can exploit this competence to read other words such as "bean" or "peak." Yet the importance of those results remains debated (Nation, Allen, & Hulme, 2001). It seems that only the explicit teaching of grapheme-phoneme correspondences allows children to acquire a genuine mastery of the alphabetic writing system.

224. Morais et al., 1979; Morais et al., 1986.

225. Mann, 1986; Read et al., 1986; Cheung & Chen, 2004.

226. Rayner & Pollatsek, 1989; Rayner et al., 2001.

227. Castles & Coltheart, 2004.

228. Ehri & Wilce, 1980.

229. Stuart, 1990.

230. Or, conversely, he might write "kee" when hearing the word *key*. The development of orthography generally parallels reading acquisition, although some languages, particularly French, can be more irregular in writing (speech-to-letter conversion) than in reading (letter-to-speech conversion). See Pacton et al., 2001; Martinet, Valdois, & Fayol, 2004.

231. Sprenger-Charolles & Siegel, 1997; Sprenger-Charolles, Siegel, & Bonnet, 1998.

232. Aghababian & Nazir, 2000; Zoccolotti et al., 2005.

233. See, for instance, the Web site www.mrisafety.com.

234. Gaillard et al., 2003.

235. Shaywitz et al., 2002.

236. Turkeltaub et al., 2003.

237. Orton, 1925, p. 608.

238. Simos et al., 2001; U. Maurer et al., 2005.

239. U. Maurer et al., 2006.

240. Parviainen et al., 2006.

241. Posner & McCandliss, 1999; McCandliss, Cohen, & Dehaene, 2003.

242. Turkeltaub et al., 2003.

243. Castro-Caldas et al., 1998.

244. Castro-Caldas et al., 1999; see also Petersson et al., 2007.

245. Plato, *Phaedrus* (translated by B. Jowett). Available online at http://oll.libertyfund .org/.

246. Nieder, Diester, & Tudusciuc, 2006.

247. Kolinsky et al., 1987; Kolinsky, Morais, & Verhaeghe, 1994.

248. Gauthier et al., 1999; Gauthier et al., 2000; Wong et al., 2005.

249. Michael Tarr and Isabel Gauthier (2000) even proposed to replace the traditional name of the region that responds to faces (the FFA, for *fusiform face area*, which refers to its localization in the fusiform gyrus of the occipito-temporal cortex) with *flexible fusiform area*.

250. Gauthier, Curran, Curby, & Collins, 2003; Rossion, Kung, & Tarr, 2004.

251. Marks, 1978; Cytowic, 1998; Ramachandran & Hubbard, 2001b.

252. Dehaene, 1997.

253. Ramachandran & Hubbard, 2001a; Hubbard et al., 2005.
254. Dehaene, 1997, chapter 3.
255. Hubbard et al., 2005; see also Sperling et al., 2006.
256. Ramachandran & Hubbard, 2001b.
257. Mondloch & Maurer, 2004; Maurer, Pathman & Mondloch, 2006.
258. National Institute of Child Health and Human Development, 2000; Rayner et al., 2001.
259. National Institute of Child Health and Human Development, 2000; Ehri, Nunes, Stahl et al., 2001; Ehri, Nunes, Willows et al., 2001.
260. Bellenger, 1980.
261. Larson, 2004.
262. Ibid.
263. Pelli, Farell & Moore, 2003.
264. Ahissar & Hochstein, 2004.
265. Paap, Newsome & Noel, 1984.
266. Yoncheva, Blau, Maurer & McCandliss, 2006.
267. Rayner et al., 2001; Bitan & Karni, 2003.
268. Share, 1995, 1999.
269. Braibant & Gérard, 1996; Goigoux, 2000; Ehri, Nunes, Stahl et al., 2001; Ehri, Nunes, Willows et al., 2001.
270. Seymour, Aro & Erskine, 2003.
271. See chapter 2, p. 97.

CHAPTER 6: The Dyslexic Brain

272. Shaywitz, 2003.
273. Shaywitz et al., 1992.
274. Ramus, 2003; Vellutino, Fletcher, Snowling, & Scanlon, 2004.
275. For a review of the visual attention deficit hypothesis, see Valdois, Bosse, & Tainturier, 2004; Bosse, Tainturier, & Valdois, 2007.
276. Zoccolotti et al., 2005.
277. This discovery, which dates from the late 1970s (see, for instance, Liberman et al., 1971; Bradley & Bryant, 1978; Fischer, Liberman, & Shankweiler, 1978; Bradley & Bryant, 1983), has been frequently replicated and extended. For detailed references, see the bibliography.
278. Castles & Coltheart, 2004.
279. Leppanen et al., 2002; Richardson, Leppanen, Leiwo, & Lyytinen, 2003. See also Benasich & Tallal, 2002; Maurer, Bucher, Brem, & Brandeis, 2003.
280. Ahissar et al., 2000; Temple et al., 2000; Breier et al., 2001; Cestnick, 2001; Breier, Gray, Fletcher, Foorman, & Klaas, 2002.
281. Eden et al., 1996; Demb, Boynton, & Heeger, 1997; Demb, Boynton, Best, & Heeger, 1998; Demb, Boynton, & Heeger, 1998.

282. Kujala et al., 2001.

283. Unfortunately, the control group does not seem to have received any training. Thus the children in the trained group might have simply benefited from the nonspecific attention and computer training they received. To assess whether nonverbal audio-visual pairing is really useful to dyslexic children, this study is in great need of replication with an improved design.

284. See the recent review by Tallal & Gaab, 2006.

285. For two recent snapshots on this debate, see Ramus, 2003; Tallal & Gaab, 2006.

286. Nicolson, Fawcett, & Dean, 2001. Anatomical anomalies of the cerebellum have indeed been observed in some group studies of dyslexic children (Eckert et al., 2003; Brambati et al., 2004). It must be noted, however, that the cerebellar theory need not conflict with a phonological origin for dyslexia. The cerebellum is, literally, a "small brain" within the brain, and is connected to virtually all major cortical regions. It thus contributes to practically all cognitive functions, including speech, phonological, and orthographic processing.

287. Galaburda & Livingstone, 1993; Demb et al., 1997; Demb, Boynton, Best & Heeger, 1998; Demb, Boynton, & Heeger, 1998; Stein, 2001.

288. Ramus, Pidgeon, & Frith, 2003.

289. Ramus, Rosen et al., 2003; White et al., 2006. For a clear exposition of a visual theory of dyslexia, see Valdois, Bosse, & Tainturier, 2004.

290. Paulesu et al., 2001.

291. Siok et al., 2004.

292. In agreement with these hypotheses, there is at least one well-documented case of a bilingual child, born in Japan of English-speaking parents, who learned to read Japanese without any difficulty (in both Kanji and Kana), but is severely dyslexic in English (Wydell & Butterworth, 1999).

293. Shaywitz et al., 1998; Brunswick et al., 1999; Georgiewa et al., 1999; Temple et al., 2001; Georgiewa et al., 2002; Shaywitz et al., 2002; McCrory et al., 2005.

294. Shaywitz et al., 2002.

295. Shaywitz et al., 1998; Georgiewa et al., 1999; Georgiewa et al., 2002.

296. McCrory et al., 2005.

297. Salmelin, Service, Kiesila, Uutela, & Salonen, 1996; Helenius, Tarkiainen, Cornelissen, Hansen, & Salmelin, 1999.

298. Zoccolotti et al., 2005

299. Simos et al., 2000; Simos, Fletcher et al., 2002.

300. For recent reviews Habib, 2000; Ramus, 2004; Fisher & Francks, 2006.

301. Silani et al., 2005. For partially similar results, see Vinckenbosch, Robichon, & Eliez, 2005; Brown et al., 2001; Brambati et al., 2004. Even though I focus primarily on temporal lobe anomalies, the literature reveals anatomical anomalies that are vastly more complex. A great diversity of impairments have been reported, although unfortunately not always replicated: asymmetries in the temporal lobe and the cerebellum, reduced gray matter in the left inferior frontal region, the inferior parietal

lobule, the right cerebellum . . . (see, for instance, Brown et al., 2001; Eckert et al., 2003; Eckert, 2004). I share Eraldo Paulesu's opinion when he states, with G. Silani, that these observations "may be relevant for dyslexia as a neurological syndrome and the distribution of the underlying pathology but perhaps less informative about which morphometric abnormality is relevant to the core neuropsychological syndrome of dyslexia." See also Vinckenbosch, Robichet & Eliez, 2005.

302. Galaburda et al., 1985.

303. Klingberg et al., 2000; Beaulieu et al., 2005; Deutsch et al., 2005; Silani et al., 2005; Niogi & McCandliss, 2006.

304. Paulesu et al., 1996. Alternatively, the affected fiber bundle might be a vertical projection exiting from the cortex (Beaulieu et al., 2005; Niogi & McCandliss, 2006), perhaps related to motor articulation deficits or to a reduced competence for fine handwriting movements.

305. Galaburda, Menard, & Rosen, 1994.

306. Ramus, 2004.

307. For a survey of this fascinating genetic hunt, see Grigorenko, 2003; Fisher & Francks, 2006; Galaburda, Lo Turco, Ramus, Fitch, & Rosen, 2006.

308. Meng et al., 2005; Paracchini et al., 2006.

309. See, for example, Torgesen, 2005.

310. Kilgard & Merzenich, 1998.

311. Merzenich et al., 1996.

312. Wilson et al., 2006.

313. Kujala et al., 2001; Simos, Fletcher et al., 2002; Temple et al., 2003; Eden et al., 2004.

CHAPTER 7: Reading and Symmetry

314. Cornell, 1985; McMonnies, 1992; Wolff & Melngailis, 1996.

315. Corballis, Macadie, Crotty, & Beale, 1985; Wolff & Melngailis, 1996; Terepocki, Kruk, & Willows, 2002; Lachmann & Geyer, 2003.

316. Gould, 1992.

317. Corballis & Beale, 1976.

318. Beale, Williams, Webster, & Corballis, 1972.

319. Tarr & Pinker, 1989.

320. Biederman & Cooper, 1991; Fiser & Biederman, 2001.

321. Corballis & Beale, 1976.

322. Rollenhagen & Olson, 2000.

323. Orton, 1925, 1937.

324. Orton, 1925, p. 607.

325. Corballis & Beale, 1976, 1993.

326. It is tempting to speculate that this might be the role of the gene ROBO1, whose mutation has been associated with dyslexia (Hannula-Jouppi et al., 2005) and which

is known from animal research to regulate the journey of growing axons across the brain's midline.

327. Rollenhagen & Olson, 2000; Baylis & Driver, 2001.

328. Logothetis, Pauls, & Poggio, 1995.

329. Perrett, Mistlin, & Chitty, 1989.

330. Rodman, Scalaidhe, & Gross, 1993.

331. Bornstein, Gross, & Wolf, 1978.

332. Sasaki et al., 2005.

333. Mello, 1965, 1966, 1967.

334. Abel, O'Brien, & Olavarria, 2000; Houzel, Carvalho, & Lent, 2002; Olavarria & Hiroi, 2003.

335. See Olavarria & Hiroi, 2003, and the references therein.

336. See notably Ungerleider & Mishkin, 1982; Goodale & Milner, 1992.

337. Turnbull, 1997; Turnbull, Beschin, & Della Sala, 1997.

338. Walsh & Butler, 1996.

339. Turnbull & McCarthy, 1996.

340. Davidoff & Warrington, 2001; Priftis, Rusconi, Umilta, & Zorzi, 2003. See also Feinberg & Jones, 1985; Riddoch & Humphreys, 1988; Davidoff & Warrington, 1999; Warrington & Davidoff, 2000.

341. Danziger & Pederson, 1998.

342. Gottfried, Sancar, & Chatterjee, 2003.

343. Streifler & Hofman, 1976; Heilman, Howell, Valenstein, & Rothi, 1980; Wade & Hart, 1991; Lambon-Ralph, Jarvis, & Ellis, 1997; Pflugshaupt et al., 2007.

344. Lambon-Ralph, Jarvis, & Ellis, 1997.

345. Streifler & Hofman, 1976.

346. Wolff & Melngailis, 1996.

347. Liberman et al., 1971; Fischer, Liberman, & Shankweiler, 1978; Wolff & Melngailis, 1996.

348. McCloskey et al., 1995; McCloskey & Rapp, 2000.

349. Galaburda & Livingstone, 1993; Demb, Boynton, & Heeger, 1997; Demb, Boynton, Best, & Heeger, 1998; Demb, Boynton, & Heeger, 1998; Stein, 2001.

350. Valdois, Bosse, & Tainturier, 2004; Bosse, Tainturier, & Valdois, 2007.

351. Gentaz, Colé, & Bara, 2003.

CHAPTER 8: Toward a Culture of Neurons

352. Changeux & Danchin, 1976; Changeux, 1983; Changeux & Connes, 1989; Changeux, 2002.

353. Wilson, 1998.

354. Chomsky, 1986.

355. Sperber, 1974.

356. Brown, 1991.

357. Sperber & Hirschfeld, 2004; See also Atran, Medin & Ross, 2005.

358. Fodor, 1983.

359. Kanwisher, McDermott, & Chun, 1997.

360. Atran, 1990; Berlin, 1992.

361. Caramazza & Shelton, 1998; Chao, Haxby, & Martin, 1999; Pouratian et al., 2003.

362. Dehaene, 1997; Butterworth, 1999.

363. Pica, Lemer, Izard, & Dehaene, 2004; Dehaene, Izard, Pica, & Spelke, 2006.

364. Dehaene, Piazza, Pinel, & Cohen, 2003; Cantlon, Brannon, Carter, & Pelphrey, 2006.

365. Nieder & Miller, 2004.

366. Changeux, 1994.

367. Ramachandran, 2005.

368. Zeki, 2000.

369. Wallin, Merker, & Brown, 2000.

370. Atran, 2002; Boyer, 2002; Dennett, 2006.

371. Whiten et al., 1999; Byrne et al., 2004.

372. Boysen, Berntson, & Prentice, 1987; Tanaka, Tomonaga & Matsuzawa, 2003.

373. Coqueugniot et al., 2004.

374. Logothetis, Pauls, & Poggio, 1995.

375. Matsuzawa, 1985; Washburn & Rumbaugh, 1991; Kawai & Matsuzawa, 2000.

376. Iriki, 2005.

377. Tomasello, 2000b, p. 37; see also Tomasello, 2000a; Tomasello et al., 2005.

378. Gergely, Bekkering, & Kiraly, 2002.

379. Tomasello, Strosberg, & Akhtar, 1996.

380. Van Essen et al., 2001.

381. Van Essen et al., 2001; Schoenemann, Sheehan, & Glotzer, 2005.

382. Elston, Benavides-Piccione, & DeFelipe, 2001.

383. Nimchinsky et al., 1999.

384. Bianchi, 1921.

385. From Finger, 1994, p. 19.

386. Dehaene & Changeux, 2005.

387. Raichle et al., 2001; Laufs et al., 2003; Fox et al., 2006.

388. Laureys, 2005.

389. For a similar hypothesis in the mathematical domain, see Nunez & Lakoff, 2000.

390. Cheng & Gallistel, 1986; Hermer & Spelke, 1996.

391. See, however, Gouteux, Thinus-Blanc & Vauclair, 2001, for conflicting data.

392. Karmiloff-Smith, 1992.

393. Dehaene, 1997.

394. Mithen, 1996.

CONCLUSION: The Future of Reading

395. Anne Fagot-Largeault, Conference at Collège de France, October 13, 2006.

BIBLIOGRAPHY

USEFUL GENERAL SOURCES

Cavallo, G., & Chartier, R. (1999). *A history of reading in the West.* Boston: University of Massachusetts Press.

Changeux, J. P. (2004). *The physiology of truth: Neuroscience and human knowledge.* Cambridge, MA: Belknap Press.

Coulmas, F. (1989). *The writing systems of the world.* Oxford: Blackwell.

Corballis, M. C., & Beale, I. L. (1976). *The psychology of left and right.* New York: Erlbaum.

DeFrancis, J. (1989). *Visible speech: The diverse oneness of writing systems.* Honolulu: University of Hawaii Press.

Ellis, A. W. (1984). *Reading, writing and dyslexia: A cognitive analysis.* Hillsdale, NJ: Lawrence Erlbaum Associates.

Jean, G. (1992). *Writing: The story of alphabets and scripts.* London: Thames & Hudson.

Manguel, A. (1997). *A history of reading.* New York: Penguin.

Miles, T. R., & Miles, E. (1999). *Dyslexia: A hundred years on* (2nd edition). Buckingham, UK: Open University Press.

Mithen, S. (1996). *The prehistory of the mind: The cognitive origins of art, religion and science.* London: Thames & Hudson.

Posner, M. I., & Raichle, M. E. (1994). *Images of mind.* New York: Scientific American Library.

Rayner, K., Foorman, B. R., Perfetti, C. A., Pesetsky, D., & Seidenberg, M. S. (2001). How psychological science informs the teaching of reading. *Psychological Science in the Public Interest* 2:31–74.

Rayner, K., & Pollatsek, A. (1989). *The psychology of reading.* Englewood Cliffs, NJ: Prentice Hall.

Robinson, A. (1995). *The story of writing: Alphabets, hieroglyphs and pictograms.* London: Thames & Hudson.

Shaywitz, S. (2003). *Overcoming dyslexia.* New York: Random House.

Snowling, M. (2000). *Dyslexia.* Oxford: Blackwell.

Snowling, M. J., & Hulme, C. (Eds.). (2005). *The science of reading: A handbook.* Oxford: Blackwell.

DETAILED REFERENCES

Abel, P. L., O'Brien, B. J., & Olavarria, J. F. (2000). Organization of callosal linkages in visual area V2 of macaque monkey. *Journal of Comparative Neurology* 428:278–293.

Aghababian, V., & Nazir, T. A. (2000). Developing normal reading skills: Aspects of the visual processes underlying word recognition. *Journal of Experimental Child Psychology* 76(2):123–150.

Ahissar, M., & Hochstein, S. (2004). The reverse hierarchy theory of visual perceptual learning. *Trends in Cognitive Sciences* (10):457–464.

Ahissar, M., Protopapas, A., Reid, M., & Merzenich, M. M. (2000). Auditory processing parallels reading abilities in adults. *Proceedings of the National Academy of Sciences* 97(12):6832–6837.

Allison, T., McCarthy, G., Nobre, A. C., Puce, A., & Belger, A. (1994). Human extrastriate visual cortex and the perception of faces, words, numbers and colors. *Cerebral Cortex* 5:544–554.

Allison, T., Puce, A., Spencer, D. D., & McCarthy, G. (1999). Electrophysiological studies of human face perception. I: Potentials generated in occipitotemporal cortex by face and non-face stimuli. *Cerebral Cortex* 9(5):415–430.

Altmann, G. T. (2002). Statistical learning in infants. *Proceedings of the National Academy of Sciences* 99(24):15250–15251.

Ans, B., Carbonnel, S., & Valdois, S. (1998). A connectionist multiple-trace memory model for polysyllabic word reading. *Psychological Review* 105(4):678–723.

Arguin, M., Fiset, S., & Bub, D. (2002). Sequential and parallel letter processing in letter-by-letter dyslexia. *Cognitive Neuropsychology* 19:535–555.

Atran, S. (1990). *Cognitive foundations of natural history.* New York and Cambridge, UK: Cambridge University Press.

———. (2002). *In gods we trust: The evolutionary landscape of religion.* New York: Oxford University Press.

Atran, S., Medin, D. L., & Ross, N. O. (2005). The cultural mind: Environmental decision making and cultural modeling within and across populations. *Psychological Review* 112(4):744–776.

Baker, C., Behrmann, M., & Olson, C. (2002). Impact of learning on representation of parts and wholes in monkey inferotemporal cortex. *Nature Neuroscience* 5(11):1210–1216.

Baker, C. I., Liu, J., Wald, L. L., Kwong, K. K., Benner, T., & Kanwisher, N. (2007). Visual word processing and experiential origins of functional selectivity in human extrastriate cortex. *Proceedings of the National Academy of Sciences* 104(21):9087–9092.

Barkow, J. H., Cosmides, L., & Tooby, J. (Eds.). (1992). *The adapted mind: Evolutionary psychology and the generation of culture.* New York: Oxford University Press.

Baylis, G. C., & Driver, J. (2001). Shape-coding in IT cells generalizes over contrast and mirror reversal, but not figure-ground reversal. *Nature Neuroscience* 4(9):937–942.

Beale, I. L., Williams, R. J., Webster, D. M., & Corballis, M. C. (1972). Confusion of mirror images by pigeons and interhemispheric commissures. *Nature* 238(5363):348–349.

Beauchamp, M. S., Lee, K. E., Argall, B. D., & Martin, A. (2004). Integration of auditory and visual information about objects in superior temporal sulcus. *Neuron* 41(5):809–823.

Beaulieu, C., Plewes, C., Paulson, L. A., Roy, D., Snook, L., Concha, L., & Phillips, L. (2005). Imaging brain connectivity in children with diverse reading ability. *Neuroimage* 25(4):1266–1271.

Bellenger, L. (1980). *Les méthodes de lecture* (2nd edition). Paris: Presses Universitaires de France.

Benasich, A. A., & Tallal, P. (2002). Infant discrimination of rapid auditory cues predicts later language impairment. *Behavioral and Brain Research* 136:31–49.

Bentin, S., Mouchetant-Rostaing, Y., Giard, M. H., Echallier, J. F., & Pernier, J. (1999). ERP manifestations of processing printed words at different psycholinguistic levels: time course and scalp distribution. *Journal of Cognitive Neuroscience* 11:235–260.

Berlin, B. (1992). *Ethnobiological classification: Principles of categorization of plants and animals in traditional societies.* Princeton, NJ: Princeton University Press.

Besner, D. (1989). On the role of outline shape and word-specific visual pattern in the identification of function words: NONE. *Quarterly Journal of Experimental Psychology A* 41:91–105.

Bhatt, R. S., Hayden, A., Reed, A., Bertin, E., & Joseph, J. (2006). Infants' perception of information along object boundaries: Concavities versus convexities. *Journal of Experimental Child Psychology* 94(2):91–113.

Bianchi, L. (1921). *La mécanique du cerveau et la fonction des lobes frontaux.* Paris: Louis Arnette.

Biederman, I. (1987). Recognition-by-components: A theory of human image understanding. *Psychological Review* 94(2):115–147.

Biederman, I., & Bar, M. (1999). One-shot viewpoint invariance in matching novel objects. *Vision Research* 39(17):2885–2899.

Biederman, I., & Cooper, E. E. (1991). Evidence for complete translational and reflectional invariance in visual object priming. *Perception* 20(5):585–593.

Binder, J. R., Frost, J. A., Hammeke, T. A., Bellgowan, P. S., Rao, S. M., & Cox, R. W. (1999). Conceptual processing during the conscious resting state: A functional MRI study. *Journal of Cognitive Neuroscience* 11(1):80–95.

Binder, J. R., Frost, J. A., Hammeke, T. A., Bellgowan, P. S., Springer, J. A., Kaufman, J. N., & Possing, E. T. (2000). Human temporal lobe activation by speech and nonspeech sounds. *Cerebral Cortex* 10(5):512–528.

Binder, J. R., McKiernan, K. A., Parsons, M. E., Westbury, C. F., Possing, E. T., Kaufman, J. N., & Buchanan, L. (2003). Neural correlates of lexical access during visual word recognition. *Journal of Cognitive Neuroscience* 15(3):372–393.

Binder, J. R., Medler, D. A., Westbury, C. F., Liebenthal, E., & Buchanan, L. (2006). Tuning of the human left fusiform gyrus to sublexical orthographic structure. *Neuroimage* 33(2):739–748.

Binder, J. R., & Mohr, J. P. (1992). The topography of callosal reading pathways. A case-control analysis. *Brain* 115:1807–1826.

Bitan, T., & Karni, A. (2003). Alphabetical knowledge from whole words training: Effects of explicit instruction and implicit experience on learning script segmentation. *Brain Research Cognitive Brain Research* 16(3):323–337.

Blackmore, S. J. (1999). *The meme machine.* Oxford: Oxford University Press.

Bokde, A. L., Tagamets, M. A., Friedman, R. B., & Horwitz, B. (2001). Functional interactions of the inferior frontal cortex during the processing of words and word-like stimuli. *Neuron* 30(2):609–617.

Booth, J. R., Burman, D. D., Meyer, J. R., Gitelman, D. R., Parrish, T. B., & Mesulam, M.

M. (2002). Functional anatomy of intra- and cross-modal lexical tasks. *Neuroimage* 16(1):7–22.

Booth, M., & Rolls, E. (1998). View-invariant representations of familiar objects by neurons in the inferior temporal visual cortex. *Cerebral Cortex* 8(6):510–523.

Bornstein, M. H., Gross, C. G., & Wolf, J. Z. (1978). Perceptual similarity of mirror images in infancy. *Cognition* 6(2):89–116.

Bosse, M. L., Tainturier, M. J., & Valdois, S. (2007). Developmental dyslexia: the visual attention span deficit hypothesis. *Cognition* 104(2):198–230.

Bouma, H. (1973). Visual interference in the parafoveal recognition of initial and final letters of words. *Vision Research* 13(4):767–782.

Boyer, P. (2002). *Religion explained: The evolutionary origins of religious thought.* New York: Basic Books.

Boysen, S. T., Berntson, G. G., & Prentice, J. (1987). Simian scribbles: A reappraisal of drawing in the chimpanzee *(Pan troglodytes)*. *Journal of Comparative Psychology* 101(1):82–89.

Bradley, L., & Bryant, P. (1983). Categorizing sounds and learning to read: A causal connection. *Nature* 30:419–421.

Bradley, L., & Bryant, P. E. (1978). Difficulties in auditory organization as a possible cause of reading backwardness. *Nature* 271:746–747.

Braibant, J.-M., & Gérard, F.-M. (1996). Savoir lire: Une question de méthodes? *Bulletin de psychologie scolaire et d'orientation* 1:7–45.

Brambati, S. M., Termine, C., Ruffino, M., Stella, G., Fazio, F., Cappa, S. F., & Perani, D. (2004). Regional reductions of gray matter volume in familial dyslexia. *Neurology* 63(4):742–745.

Breier, J. I., Gray, L., Fletcher, J. M., Diehl, R. L., Klaas, P., Foorman, B. R., & Molis, M. R. (2001). Perception of voice and tone onset time continua in children with dyslexia with and without attention deficit/hyperactivity disorder. *Journal of Experimental Child Psychology* 80(3):245–270.

Breier, J. I., Gray, L. C., Fletcher, J. M., Foorman, B., & Klaas, P. (2002). Perception of speech and nonspeech stimuli by children with and without reading disability and attention deficit hyperactivity disorder. *Journal of Experimental Child Psychology* 82(3):226–250.

Brincat, S. L., & Connor, C. E. (2004). Underlying principles of visual shape selectivity in posterior inferotemporal cortex. *Nature Neuroscience* 7(8):880–886.

Brown, D. (1991). *Human universals.* New York: McGraw-Hill.

Brown, W. E., Eliez, S., Menon, V., Rumsey, J. M., White, C. D., & Reiss, A. L. (2001). Preliminary evidence of widespread morphological variations of the brain in dyslexia. *Neurology* 56(6):781–783.

Brunswick, N., McCrory, E., Price, C. J., Frith, C. D., & Frith, U. (1999). Explicit and implicit processing of words and pseudowords by adult developmental dyslexics: A search for Wernicke's Wortschatz? *Brain* 122:1901–1917.

Brysbaert, M. (1994). Interhemispheric transfer and the processing of foveally presented stimuli. *Behavioural Brain Research* 64(1–2):151–161.

Büchel, C., Price, C. J., & Friston, K. (1998). A multimodal language region in the ventral visual pathway. *Nature* 394(6690):274–277.

Burton, M. W., Small, S. L., & Blumstein, S. E. (2000). The role of segmentation in phonological processing: an fMRI investigation. *Journal of Cognitive Neuroscience* 12(4):679–690.

Butterworth, B. (1999). *The mathematical brain.* London: Macmillan.

Byrne, R. W., Barnard, P. J., Davidson, I., Janik, V. M., McGrew, W. C., Miklosi, A., & Wiessner, P. (2004). Understanding culture across species. *Trends in Cognitive Sciences* 8(8):341–346.

Calvet, L.-J. (1998). *Histoire de l'écriture.* Paris: Hachette.

Cantlon, J. F., Brannon, E. M., Carter, E. J., & Pelphrey, K. A. (2006). Functional imaging of numerical processing in adults and 4-y-old children. *PLoS Biology* 4(5):e125.

Caramazza, A. (1996). The brain's dictionary. *Nature* 380:485–486.

———. (1998). The interpretation of semantic category-specific deficits: What do they reveal about the organization of conceptual knowledge in the brain? *Neurocase* 4:265–272.

Caramazza, A., & Hillis, A. E. (1991). Lexical representation of nouns and verbs in the brain. *Nature* 349:788–790.

Caramazza, A., Laudanna, A., & Romani, C. (1988). Lexical access and inflectional morphology. *Cognition* 28:297–332.

Caramazza, A., & Shelton, J. R. (1998). Domain-specific knowledge systems in the brain: The animate-inanimate distinction. *Journal of Cognitive Neuroscience* 10:1–34.

Castles, A., & Coltheart, M. (2004). Is there a causal link from phonological awareness to success in learning to read? *Cognition* 91(1):77–111.

Castro-Caldas, A., Miranda, P. C., Carmo, I., Reis, A., Leote, F., Ribeiro, C., & Ducla-Soares, E. (1999). Influence of learning to read and write on the morphology of the corpus callosum. *European Journal of Neurology* 6(1):23–28.

Castro-Caldas, A., Petersson, K. M., Reis, A., Stone-Elander, S., & Ingvar, M. (1998). The illiterate brain: Learning to read and write during childhood influences the functional organization of the adult brain. *Brain* 121(Pt 6):1053–1063.

Catani, M., Jones, D. K., Donato, R., & Ffytche, D. H. (2003). Occipito-temporal connections in the human brain. *Brain* 126(Pt 9):2093–2107.

Cestnick, L. (2001). Cross-modality temporal processing deficits in developmental phonological dyslexics. *Brain and Cognition* 46(3):319–325.

Changeux, J. P. (1983). *L'homme neuronal.* Paris: Fayard.

———. (1994). *Raison et plaisir.* Paris: Odile Jacob.

———. (2002). *L'homme de vérité.* Paris: Odile Jacob.

Changeux, J. P., & Connes, A. (1989). *Matière à pensée.* Paris: Odile Jacob.

Changeux, J. P., & Danchin, A. (1976). Selective stabilization of developing synapses as a mechanism for the specification of neuronal networks. *Nature* 264:705–712.

Changizi, M. A., & Shimojo, S. (2005). Character complexity and redundancy in writing systems over human history. *Proceedings of the Royal Society B: Biological Sciences* 272(1560):267–275.

Changizi, M. A., Zhang, Q., Ye, H., & Shimojo, S. (2006). The structures of letters and symbols throughout human history are selected to match those found in objects in natural scenes. *American Naturalist* 167(5):E117–139.

Chao, L. L., Haxby, J. V., & Martin, A. (1999). Attribute-based neural substrates in temporal cortex for perceiving and knowing about objects. *Nature Neuroscience* 2(10):913–919.

Chen, Y., Fu, S., Iversen, S. D., Smith, S. M., & Matthews, P. M. (2002). Testing for dual brain processing routes in reading: a direct contrast of Chinese Character and Pinyin reading using fMRI. *Journal of Cognitive Neuroscience* 14(7):1088–1098.

Cheng, K., & Gallistel, C. R. (1986). A purely geometric module in the rat's spatial representation. *Cognition* 23:149–178.

Cheour, M., Ceponiene, R., Lehtokoski, A., Luuk, A., Allik, J., Alho, K., & Naatanen, R. (1998). Development of language-specific phoneme representations in the infant brain. *Nature Neuroscience* 1(5):351–353.

Cheung, H., & Chen, H. C. (2004). Early orthographic experience modifies both phonological awareness and on-line speech processing. *Language and Cognitive Processes* 19:1–28.

Chomsky, N. (1980). *Rules and representations.* Oxford: Basil Blackwell.

———. (1986). *Knowledge of language: Its nature, origins, and use.* Westport, CT: Praeger.

———. (1988). *Language and the problems of knowledge.* Cambridge: MIT Press.

Cohen, L., & Dehaene, S. (1995). Number processing in pure alexia: the effect of hemispheric asymmetries and task demands. *Neurocase* 1:121–137.

———. (2000). Calculating without reading: Unsuspected residual abilities in pure alexia. *Cognitive Neuropsychology* 17(6):563–583.

———. (2004). Specialization within the ventral stream: The case for the visual word form area. *Neuroimage* 22(1):466–476.

Cohen, L., Dehaene, S., Naccache, L., Lehéricy, S., Dehaene-Lambertz, G., Hénaff, M. A., & Michel, F. (2000). The visual word form area: Spatial and temporal characterization of an initial stage of reading in normal subjects and posterior split-brain patients. *Brain* 123:291–307.

Cohen, L., Henry, C., Dehaene, S., Martinaud, O., Lehéricy, S., Lemer, C., & Ferrieux, S. (2004). The pathophysiology of letter-by-letter reading. *Neuropsychologia* 42(13):1768–1780.

Cohen, L., Jobert, A., Le Bihan, D., & Dehaene, S. (2004). Distinct unimodal and multimodal regions for word processing in the left temporal cortex. *Neuroimage* 23(4):1256–1270.

Cohen, L., Lehéricy, S., Chochon, F., Lemer, C., Rivaud, S., & Dehaene, S. (2002). Language-specifiċ tuning of visual cortex? Functional properties of the visual word form area. *Brain* 125(Pt 5):1054–1069.

Cohen, L., Lehéricy, S., Henry, C., Bourgeois, M., Larroque, C., Sainte-Rose, C., Dehaene, S., & Hertz-Pannier, L. (2004). Learning to read without a left occipital lobe: Right-hemispheric shift of visual word form area. *Annals of Neurology* 56(6):890–894.

Cohen, L., Martinaud, O., Lemer, C., Lehéricy, S., Samson, Y., Obadia, M., Slachevsky, A., & Dehaene, S. (2003). Visual word recognition in the left and right hemispheres: Anatomical and functional correlates of peripheral alexias. *Cerebral Cortex* 13:1313–1333.

Coltheart, M., & Coltheart, V. (1997). Reading comprehension is not exclusively reliant upon phonological representation. *Cognitive Neuropsychology* 14:167–175.

Coltheart, M., Davelaar, E., Jonasson, J. T., & Besner, D. (1977). Access to the internal lexicon. In Dornic, S. (Ed.), *Attention and performance* VI (pp. 535–555). London: Academic Press.

Coltheart, M., Rastle, K., Perry, C., Langdon, R., & Ziegler, J. (2001). DRC: A dual route cascaded model of visual word recognition and reading aloud. *Psychological Review* 108(1):204–256.

Coqueugniot, H., Hublin, J. J., Veillon, F., Houet, F., & Jacob, T. (2004). Early brain growth in *Homo erectus* and implications for cognitive ability. *Nature* 431(7006):299–302.

Corballis, M. C., & Beale, I. L. (1976). *The psychology of left and right.* New York: Erlbaum.

———. (1993). Orton revisited: Dyslexia, laterality, and left-right confusion. In Willows, D., Kruk, R. S. & Corsos, E. (Eds.), *Visual processes in reading and reading disabilities.* Hillsdale, NJ: Erlbaum.

Corballis, M. C., Macadie, L., Crotty, A., & Beale, I. L. (1985). The naming of disoriented letters by normal and reading-disabled children. *Journal of Child Psychology and Psychiatry* 26(6):929–938.

Cornell, J. (1985). Spontaneous mirror-writing in children. *Canadian Journal of Experimental Psychology* 39:174–179.

Coslett, H. B., & Monsul, N. (1994). Reading with the right hemisphere: Evidence from transcranial magnetic stimulation. *Brain and Language* 46:198–211.

Coslett, H. B., & Saffran, E. M. (1989). Evidence for preserved reading in "pure alexia." *Brain* 112:327–359.

Coslett, H. B., Saffran, E. M., Greenbaum, S., & Schwartz, H. (1993). Reading in pure alexia: The effect of strategy. *Brain* 116:21–37.

Coulmas, F. (1989). *The writing systems of the world.* Oxford: Blackwell.

Cytowic, R. E. (1998). *The man who tasted shapes.* Cambridge, MA: MIT Press.

Damasio, A. R. (1989a). The brain binds entities and events by multiregional activation from convergence zones. *Neural Computation* 1:123–132.

———. (1989b). Time-locked multiregional retroactivation: A systems-level proposal for the neural substrates of recall and recognition. *Cognition* 33:25–62.

Damasio, A. R., & Damasio, H. (1983). The anatomic basis of pure alexia. *Neurology* 33:1573–1583.

Damasio, H., Grabowski, T. J., Tranel, D., Hichwa, R. D., & Damasio, A. R. (1996). A neural basis for lexical retrieval. *Nature* 380(11 April 1996):499–505.

Danziger, E., & Pederson, E. (1998). Through the looking glass: Literacy, writing systems and mirror-image discrimination. *Written Language and Literacy* 1:153–167.

Davidoff, J., & Warrington, E. K. (1999). The bare bones of object recognition: Implications from a case of object recognition impairment. *Neuropsychologia* 37(3):279–292.

———. (2001). A particular difficulty in discriminating between mirror images. *Neuropsychologia* 39(10):1022–1036.

Dawkins, R. (1989). *The selfish gene.* Oxford: Oxford University Press.

———. (1996). *The blind watchmaker: Why the evidence of evolution reveals a universe without design.* New York: W. W. Norton.

de Haan, M., Johnson, M. H., & Halit, H. (2003). Development of face-sensitive event-related potentials during infancy: a review. *International Journal of Psychophysiology* 51(1):45–58.

DeFrancis, J. (1989). *Visible speech: The diverse oneness of writing systems.* Honolulu: University of Hawaii Press.

Dehaene, S. (1995). Electrophysiological evidence for category-specific word processing in the normal human brain. *NeuroReport* 6:2153–2157.

———. (1997). *The number sense.* New York: Oxford University Press.

———. (2005). Evolution of human cortical circuits for reading and arithmetic: The "neuronal recycling" hypothesis. In Dehaene, S., Duhamel, J. R., Hauser, M., & Rizzolatti, G. (Eds.). *From monkey brain to human brain* (pp. 133–157). Cambridge, MA: MIT Press.

Dehaene, S., & Changeux, J. P. (2005). Ongoing spontaneous activity controls access to consciousness: A neuronal model for inattentional blindness. *PLoS Biology* 3(5):e141.

Dehaene, S., & Cohen, L. (2007). Cultural recycling of cortical maps. *Neuron* 56(2):384–398.

Dehaene, S., Cohen, L., Sigman, M., & Vinckier, F. (2005). The neural code for written words: A proposal. *Trends in Cognitive Sciences* 9(7):335–341.

Dehaene, S., Izard, V., Pica, P., & Spelke, E. (2006). Core knowledge of geometry in an Amazonian indigene group. *Science* 311:381–384.

Dehaene, S., Jobert, A., Naccache, L., Ciuciu, P., Poline, J. B., Le Bihan, D., & Cohen, L. (2004). Letter binding and invariant recognition of masked words: Behavioral and neuroimaging evidence. *Psychological Science* 15(5):307–313.

Dehaene, S., Le Clec'H, G., Poline, J. B., Le Bihan, D., & Cohen, L. (2002). The visual word form area: A prelexical representation of visual words in the fusiform gyrus. *NeuroReport* 13(3):321–325.

Dehaene, S., Naccache, L., Cohen, L., Le Bihan, D., Mangin, J. F., Poline, J. B., & Riviere, D. (2001). Cerebral mechanisms of word masking and unconscious repetition priming. *Nature Neuroscience* 4(7):752–758.

Dehaene, S., Piazza, M., Pinel, P., & Cohen, L. (2003). Three parietal circuits for number processing. *Cognitive Neuropsychology* 20:487–506.

Dehaene-Lambertz, G. (1997). Electrophysiological correlates of categorical phoneme perception in adults. *NeuroReport* 8(4):919–924.

Dehaene-Lambertz, G., & Baillet, S. (1998). A phonological representation in the infant brain. *NeuroReport* 9(8):1885–1888.

Dehaene-Lambertz, G., & Dehaene, S. (1994). Speed and cerebral correlates of syllable discrimination in infants. *Nature* 370:292–295.

Dehaene-Lambertz, G., Dehaene, S., & Hertz-Pannier, L. (2002). Functional neuroimaging of speech perception in infants. *Science* 298(5600):2013–2015.

Déjerine, J. (1892). Contribution à l'étude anatomo-pathologique et clinique des différentes variétés de cécité verbale. *Mémoires de la Société de Biologie* 4:61–90.

Del Cul, A., Baillet, S., & Dehaene, S. (2007). Brain dynamics underlying the nonlinear threshold for access to consciousness. *PLoS Biology* 5(10):e260.

Demb, J. B., Boynton, G. M., Best, M., & Heeger, D. J. (1998). Psychophysical evidence for a magnocellular pathway deficit in dyslexia. *Vision Research* 38(11):1555–1559.

Demb, J. B., Boynton, G. M., & Heeger, D. J. (1997). Brain activity in visual cortex predicts individual differences in reading performance. *Proceedings of the National Academy of Sciences* 94(24):13363–13366.

———. (1998). Functional magnetic resonance imaging of early visual pathways in dyslexia. *Journal of Neuroscience* 18(17):6939–6951.

Dennett, D. (1978). *Brainstorms.* Cambridge, MA: MIT Press.

———. (2006). *Breaking the spell: Religion as a natural phenomenon.* New York: Viking.

Desimone, R., & Gross, C. G. (1979). Visual areas in the temporal cortex of the macaque. *Brain Research* 178(2–3):363–380.

Deutsch, G. K., Dougherty, R. F., Bammer, R., Siok, W. T., Gabrieli, J. D., & Wandell, B. (2005). Children's reading performance is correlated with white matter structure measured by diffusion tensor imaging. *Cortex* 41(3):354–363.

Devlin, J. T., Jamison, H. L., Matthews, P. M., & Gonnerman, L. M. (2004). Morphology and the internal structure of words. *Proceedings of the National Academy of Sciences* 101(41):14984–14988.

Di Virgilio, G., & Clarke, S. (1997). Direct interhemispheric visual input to human speech areas. *Human Brain Mapping* 5:347–354.

Ding, G., Peng, D., & Taft, M. (2004). The nature of the mental representation of radicals in Chinese: A priming study. *Journal of Experimental Psychology: Learning, Memory, and Cognition* 30(2):530–539.

Downing, P. E., Jiang, Y., Shuman, M., & Kanwisher, N. (2001). A cortical area selective for visual processing of the human body. *Science* 293(5539):2470–2473.

Eckert, M. (2004). Neuroanatomical markers for dyslexia: a review of dyslexia structural imaging studies. *Neuroscientist* 10(4):362–371.

Eckert, M. A., Leonard, C. M., Richards, T. L., Aylward, E. H., Thomson, J., & Berninger, V. W. (2003). Anatomical correlates of dyslexia: Frontal and cerebellar findings. *Brain* 126(Pt 2):482–494.

Eden, G. F., Jones, K. M., Cappell, K., Gareau, L., Wood, F. B., Zeffiro, T. A., Dietz, N. A., Agnew, J. A., & Flowers, D. L. (2004). Neural changes following remediation in adult developmental dyslexia. *Neuron* 44(3):411–422.

Eden, G. F., VanMeter, J. W., Rumsey, J. M., Maisog, J. M., Woods, R. P., & Zeffiro, T. A. (1996). Abnormal processing of visual motion in dyslexia revealed by functional brain imaging. *Nature* 382(6586):66–69.

Ehri, L. C., Nunes, S. R., Stahl, S. A., & Willows, D. M. M. (2001). Systematic phonics instruction helps students learn to read: Evidence from the National Reading Panel's meta-analysis. *Review of Educational Research* 71:393–447.

Ehri, L. C., Nunes, S. R., Willows, D. M., Schuster, B. V., Yaghoub-Zadeh, Z., & Shanahan, T. (2001). Phonemic awareness instruction helps children learn to read: Evidence from the National Reading Panel's meta-analysis. *Reading Research Quarterly* 36:250–287.

Ehri, L. C., & Wilce, L. S. (1980). The influence of orthography on readers' conceptualisation of the phonemic structure of words. *Applied Psycholinguistics* 1:371–385.

Eimas, P. D., Siqueland, E. R., Jusczyk, P. W., & Vigorito, J. (1971). Speech perception in infants. *Science* 171:303–306.

Ellis, A. W. (2004). Length, formats, neighbours, hemispheres, and the processing of words presented laterally or at fixation. *Brain and Language* 88(3):355–366.

Ellis, A. W., Young, A. W., & Anderson, C. (1988). Modes of word recognition in the left and right cerebral hemispheres. *Brain and Language* 35(2):254–273.

Elston, G. N., Benavides-Piccione, R., & DeFelipe, J. (2001). The pyramidal cell in cognition: A comparative study in human and monkey. *Journal of Neuroscience* 21(17):RC163.

Fagot-Largeault, A. (2006). *Vivre le handicap et ses prothèses.* Conference at Collège de France.

Feinberg, T., & Jones, G. (1985). Object reversals after parietal lobe infarction—a case report. *Cortex* 21:261–271.

Ferrand, L. (2001). *Cognition et Lecture: Processus de base de reconnaissance des mots écrits chez l'adulte.* Bruxelles: DeBoeck université.

Ferrand, L., & Grainger, J. (1992). Phonology and orthography in visual word recognition: Evidence from masked non-word priming. *Quarterly Journal of Experimental Psychology A* 45(3):353–372.

———. (1993). The time course of orthographic and phonological code activation in the early phases of visual word recognition. *Bulletin of the Psychonomic Society* 31:119–122.

———. (1994). Effects of orthography are independent of phonology in masked form priming. *Quarterly Journal of Experimental Psychology A* 47(2):365–382.

Fiebach, C. J., Friederici, A. D., Muller, K., & von Cramon, D. Y. (2002). fMRI evidence for dual routes to the mental lexicon in visual word recognition. *Journal of Cognitive Neuroscience* 14(1):11–23.

Fiez, J. A., Balota, D. A., Raichle, M. E., & Petersen, S. E. (1999). Effects of lexicality, frequency, and spelling-to-sound consistency on the functional anatomy of reading. *Neuron* 24(1):205–218.

Fiez, J. A., & Petersen, S. E. (1998). Neuroimaging studies of word reading. *Proceedings of the National Academy of Sciences* 95(3):914–921.

Finger, S. (1994). *Origins of neuroscience: A history of explorations into brain function.* New York: Oxford University Press.

Fischer, F. W., Liberman, I. Y., & Shankweiler, D. (1978). Reading reversals and developmental dyslexia: A further study. *Cortex* 14(4):496–510.

Fiser, J., & Biederman, I. (2001). Invariance of long-term visual priming to scale, reflection, translation, and hemisphere. *Vision Research* 41(2):221–234.

Fiset, D., Arguin, M., Bub, D., Humphreys, G. W., & Riddoch, M. J. (2005). How to make the word-length effect disappear in letter-by-letter dyslexia. *Psychological Science* 16(7):535–541.

Fisher, S. E., & Francks, C. (2006). Genes, cognition and dyslexia: Learning to read the genome. *Trends in Cognitive Sciences* 10(6): 250–257.

Fodor, J. A. (1983). *The modularity of mind.* Cambridge, MA: MIT Press.

Fox, M. D., Corbetta, M., Snyder, A. Z., Vincent, J. L., & Raichle, M. E. (2006). Spontaneous neuronal activity distinguishes human dorsal and ventral attention systems. *Proceedings of the National Academy of Sciences* 103(26):10046–10051.

Friederici, A. D., Opitz, B., & von Cramon, D. Y. (2000). Segregating semantic and syntactic aspects of processing in the human brain: An fMRI investigation of different word types. *Cerebral Cortex* 10(7):698–705.

Frith, U. (1985). Beneath the surface of developmental dyslexia. In Patterson, K. E., Marshall, J. C., & Coltheart, M. (Eds.), *Surface dyslexia: Cognitive and neuropsychological studies of phonological reading* (pp. 301–330). Hillsdale, NJ: Erlbaum.

Fu, S., Chen, Y., Smith, S., Iversen, S., & Matthews, P. M. (2002). Effects of word form on brain processing of written Chinese. *Neuroimage* 17(3):1538–1548.

Gaillard, R., Del Cul, A., Naccache, L., Vinckier, F., Cohen, L., & Dehaene, S. (2006). Nonconscious semantic processing of emotional words modulates conscious access. *Proceedings of the National Academy of Sciences* 103(19):7524–7529.

Gaillard, R., Naccache, L., Pinel, P., Clemenceau, S., Volle, E., Hasboun, D., Dupont, S., Baulac, M., Dehaene, S., Adam, C., & Cohen, L. (2006). Direct intracranial, fMRI, and lesion evidence for the causal role of left inferotemporal cortex in reading. *Neuron* 50(2):191–204.

Gaillard, W. D., Balsamo, L. M., Ibrahim, Z., Sachs, B. C., & Xu, B. (2003). fMRI identifies regional specialization of neural networks for reading in young children. *Neurology* 60(1):94–100.

Galaburda, A. M., & Livingstone, M. (1993). Evidence for a magnocellular deficit in developmental dyslexia. In Tallal, P., Galaburda, A. M., Llinas, R., & Von Euler, C. (Eds.), *Temporal information processing in the nervous system: Special reference to dyslexia and dysphasia* (pp. 70–82). New York: New York Academy of Sciences.

Galaburda, A. M., LoTurco, J., Ramus, F., Fitch, R. H., & Rosen, G. D. (2006). From genes to behavior in developmental dyslexia. *Nature Neuroscience* 9(10):1213–1217.

Galaburda, A. M., Menard, M. T., & Rosen, G. D. (1994). Evidence for aberrant auditory anatomy in developmental dyslexia. *Proceedings of the National Academy of Sciences* 91(17):8010–8013.

Galaburda, A. M., Sherman, G. F., Rosen, G. D., Aboitiz, F., & Geschwind, N. (1985). Developmental dyslexia: Four consecutive patients with cortical anomalies. *Annals of Neurology* 18(2):222–233.

Gathers, A. D., Bhatt, R., Corbly, C. R., Farley, A. B., & Joseph, J. E. (2004). Developmental shifts in cortical loci for face and object recognition. *NeuroReport* 15(10):1549–1553.

Gauthier, I., Curran, T., Curby, K. M., & Collins, D. (2003). Perceptual interference supports a non-modular account of face processing. *Nature Neuroscience* 6(4):428–432.

Gauthier, I., Skudlarski, P., Gore, J. C., & Anderson, A. W. (2000). Expertise for cars and birds recruits brain areas involved in face recognition. *Nature Neuroscience* 3(2):191–197.

Gauthier, I., Tarr, M. J., Anderson, A. W., Skudlarski, P., & Gore, J. C. (1999). Activation of the middle fusiform "face area" increases with expertise in recognizing novel objects. *Nature Neuroscience* 2(6):568–573.

Gentaz, E., Colé, P., & Bara, F. (2003). Evaluation d'entraînements multisensoriels de préparation à la lecture pour les enfants en grande section de maternelle: Une étude sur la contribution du système haptique manuel. *L'Année Psychologique* 104:561–584.

Georgiewa, P., Rzanny, R., Gaser, C., Gerhard, U. J., Vieweg, U., Freesmeyer, D., Mentzel, H. J., Kaiser, W. A., & Blanz, B. (2002). Phonological processing in dyslexic children: a study combining functional imaging and event related potentials. *Neuroscience Letters* 318(1):5–8.

Georgiewa, P., Rzanny, R., Hopf, J. M., Knab, R., Glauche, V., Kaiser, W. A., & Blanz, B. (1999). fMRI during word processing in dyslexic and normal reading children. *Neuroreport* 10(16):3459–3465.

Gergely, G., Bekkering, H., & Kiraly, I. (2002). Rational imitation in preverbal infants. *Nature* 415(6873):755.

Geschwind, N. (1965). Disconnection syndromes in animals and man. *Brain* 88:237–294.

Goigoux, R. (2000). Apprendre à lire à l'école: Les limites d'une approche idéovisuelle. *Psychologie Française* 45(235–245).

Goodale, M. A., & Milner, A. D. (1992). Separate visual pathways for perception and action. *Trends in Neuroscience* 15:20–25.

Goswami, U. (1986). Children's use of analogy in learning to read: A developmental study. *Journal of Experimental Child Psychology* 42:73–83.

Goswami, U., Gombert, J. E., & de Barrera, L. F. (1998). Children's orthographic representations and linguistic transparency: Nonsense word reading in English, French, and Spanish. *Applied Psycholinguistics* 19:19–52.

Gottfried, J. A., Sancar, F., & Chatterjee, A. (2003). Acquired mirror writing and reading: Evidence for reflected graphemic representations. *Neuropsychologia* 41(1):96–107.

Gould, J. L., & Marler, P. (1987). Learning by instinct. *Scientific American* 256 (January): 74–85.

Gould, S. J. (1992). *The panda's thumb: More reflections in natural history.* New York: W. W. Norton.

Gould, S. J., & Vrba, E. S. (1982). Exaptation: A missing term in the science of form. *Paleobiology* 8(1):4–15.

Gouteux, S., Thinus-Blanc, C., & Vauclair, J. (2001). Rhesus monkeys use geometric and nongeometric information during a reorientation task. *Journal of Experimental Psychology: General* 130(3):505–519.

Grabowski, T. J., Damasio, H., & Damasio, A. R. (1998). Premotor and prefrontal correlates of category-related lexical retrieval. *Neuroimage* 7(3):232–243.

Grainger, J., Granier, J. P., Farioli, F., Van Assche, E., & van Heuven, W. J. (2006). Letter position information and printed word perception: The relative-position priming constraint. *Journal of Experimental Psychology: Human Perception and Performance* 32(4):865–884.

Grainger, J., & Jacobs, A. M. (1996). Orthographic processing in visual word recognition: A multiple read-out model. *Psychological Review* 103(3):518–565.

Grainger, J., & van Heuven, W. (2003). Modeling letter position coding in printed word perception. In Bonin, P. (Ed.), *The mental lexicon* (pp. 1–24). New York: Nova Science Publishers.

Grainger, J., & Whitney, C. (2004). Does the huamn mnid raed wrods as a wlohe? *Trends in Cognitive Sciences* 8(2):58–59.

Grainger, J., & Ziegler, J. (2007). Cross-code consistency effects in visual word recognition. In Grigorenko, E. L., & Naples, A. J. (Eds.), *Single-word reading: Biological and behavioral perspectives* (129–158). New York: Erlbaum.

Grill-Spector, K., Kushnir, T., Hendler, T., Edelman, S., Itzchak, Y., & Malach, R. (1998). A sequence of object-processing stages revealed by fMRI in the human occipital lobe. *Human Brain Mapping* 6(4):316–328.

Grill-Spector, K., & Malach, R. (2001). fMR-adaptation: A tool for studying the functional properties of human cortical neurons. *Acta Psychologica (Amsterdam)* 107(1–3):293–321.

Grill-Spector, K., Sayres, R., & Ress, D. (2006). High-resolution imaging reveals highly selective nonface clusters in the fusiform face area. *Nature Neuroscience* 9(9):1177–1185.

Grigorenko, E. L. (2003). The first candidate gene for dyslexia: Turning the page of a new chapter of research. *Proceedings of the National Academy of Sciences* 100(20):11190–11192.

Ha Duy Thuy, D., Matsuo, K., Nakamura, K., Toma, K., Oga, T., Nakai, T., Shibasaki, H., & Fukuyama, H. (2004). Implicit and explicit processing of Kanji and Kana words and non-words studied with fMRI. *Neuroimage* 23(3):878–889.

Habib, M. (2000). The neurological basis of developmental dyslexia: An overview and working hypothesis. *Brain* 123(Pt 12):2373–2399.

Hagoort, P., Indefrey, P., Brown, C., Herzog, H., Steinmetz, H., & Seitz, R. J. (1999). The neural circuitry involved in the reading of German words and pseudowords: A PET study. *Journal of Cognitive Neuroscience* 11(4):383–398.

Hammond, P., & Hughes, P. (1978). *Upon the pun: Dual meaning in words and pictures.* London: W. H. Allen.

Hanley, J. R., & Kay, J. (1996). Reading speed in pure alexia. *Neuropsychologia* 34(12):1165–1174.

Hannula-Jouppi, K., Kaminen-Ahola, N., Taipale, M., Eklund, R., Nopola-Hemmi, J., Kaariainen, H., & Kere, J. (2005). The axon guidance receptor gene ROBO1 is a candidate gene for developmental dyslexia. *PLoS Genetics* 1(4):e50.

Harm, M. W., & Seidenberg, M. S. (1999). Phonology, reading acquisition, and dyslexia: Insights from connectionist models. *Psychological Review* 106(3):491–528.

———.(2004). Computing the meanings of words in reading: Cooperative division of labor between visual and phonological processes. *Psychological Review* 111(3):662–720.

Hasson, U., Harel, M., Levy, I., & Malach, R. (2003). Large-scale mirror-symmetry organization of human occipito-temporal object areas. *Neuron* 37(6):1027–1041.

Hasson, U., Levy, I., Behrmann, M., Hendler, T., & Malach, R. (2002). Eccentricity bias as an organizing principle for human high-order object areas. *Neuron* 34(3):479–490.

Haxby, J. V., Gobbini, M. I., Furey, M. L., Ishai, A., Schouten, J. L., & Pietrini, P. (2001). Distributed and overlapping representations of faces and objects in ventral temporal cortex. *Science* 293(5539):2425–2430.

Haxby, J. V., Ishai, I. I., Chao, L. L., Ungerleider, L. G., & Martin, I. I. (2000). Object-form topology in the ventral temporal lobe: Response to I. Gauthier (2000). *Trends in Cognitive Sciences* 4(1):3–4.

Heilman, K. M., Howell, G., Valenstein, E., & Rothi, L. (1980). Mirror-reading and writing in association with right-left spatial disorientation. *Journal of Neurology, Neurosurgery, and Psychiatry* 43(9):774–780.

Helenius, P., Tarkiainen, A., Cornelissen, P., Hansen, P. C., & Salmelin, R. (1999). Dissociation of normal feature analysis and deficient processing of letter-strings in dyslexic adults. *Cerebral Cortex* 9(5):476–483.

Hermer, L., & Spelke, E. (1996). Modularity and development: The case of spatial reorientation. *Cognition* 61(3):195–232.

Houzel, J. C., Carvalho, M. L., & Lent, R. (2002). Interhemispheric connections between primary visual areas: Beyond the midline rule. *Brazilian Journal of Medical and Biological Research* 35(12):1441–1453.

Hubbard, E. M., Arman, A. C., Ramachandran, V. S., & Boynton, G. M. (2005). Individual differences among grapheme-color synesthetes: Brain-behavior correlations. *Neuron* 45(6):975–985.

Humphrey, N. K., & Weiskrantz, L. (1969). Size constancy in monkeys with inferotemporal lesions. *Quarterly Journal of Experimental Psychology A* 21(3):225–238.

Humphreys, G. W., Evett, L. J., & Quinlan, P. T. (1990). Orthographic processing in visual word identification. *Cognitive Psychology* 22:517–560.

Hutzler, F., Ziegler, J. C., Perry, C., Wimmer, H., & Zorzi, M. (2004). Do current connectionist learning models account for reading development in different languages? *Cognition* 91(3):273–296.

Intriligator, J., Hénaff, M. A., & Michel, F. (2000). Able to name, unable to compare: The visual abilities of a posterior split-brain patient. *NeuroReport* 11(12):2639–2642.

Iriki, A. (2005). A prototype of Homo-Faber: A silent precursor of human intelligence in the tool-using monkey brain. In Deheane, S., Duhamel, J. R., Hauser, M., & Rizzolatti, G. (Eds.), *From monkey brain to human brain* (pp. 133–157). Cambridge, MA: MIT Press.

Ishai, A., Ungerleider, L. G., Martin, A., & Haxby, J. V. (2000). The representation of objects in the human occipital and temporal cortex. *Journal of Cognitive Neuroscience* 12(Suppl 2):35–51.

Ishai, A., Ungerleider, L. G., Martin, A., Schouten, J. L., & Haxby, J. V. (1999). Distributed representation of objects in the human ventral visual pathway. *Proceedings of the National Academy of Sciences* 96(16):9379–9384.

Ito, M., & Komatsu, H. (2004). Representation of angles embedded within contour stimuli in area V2 of macaque monkeys. *Journal of Neuroscience* 24(13):3313–3324.

Ito, M., Tamura, H., Fujita, I., & Tanaka, K. (1995). Size and position invariance of neuronal responses in monkey inferotemporal cortex. *Journal of Neurophysiology* 73:218–226.

Jacob, F. (1977). Evolution and tinkering. *Science* 196(4295):1161–1166.

Jacquemot, C., Pallier, C., Le Bihan, D., Dehaene, S., & Dupoux, E. (2003). Phonological grammar shapes the auditory cortex: A functional magnetic resonance imaging study. *Journal of Neuroscience* 23(29):9541–9546.

Jared, D., & Seidenberg, M. S. (1991). Does word identification proceed from spelling to sound to meaning? *Journal of Experimental Psychology: General* 120:358–394.

Jobard, G., Crivello, F., & Tzourio-Mazoyer, N. (2003). Evaluation of the dual route theory of reading: A metanalysis of 35 neuroimaging studies. *Neuroimage* 20(2):693–712.

Kanwisher, N., McDermott, J., & Chun, M. M. (1997). The fusiform face area: A module in human extrastriate cortex specialized for face perception. *Journal of Neuroscience* 17:4302–4311.

Karmiloff-Smith, A. (1992). *Beyond modularity*. Cambridge, MA: MIT Press.

Kawai, N., & Matsuzawa, T. (2000). Numerical memory span in a chimpanzee. *Nature* 403(6765):39–40.

Kellman, P. J., & Spelke, E. S. (1983). Perception of partly occluded objects in infancy. *Cognitive Psychology* 15:483–524.

Kilgard, M. P., & Merzenich, M. M. (1998). Cortical map reorganization enabled by nucleus basalis activity. *Science* 279(5357):1714–1718.

Kitterle, F. L., & Selig, L. M. (1991). Visual field effects in the discrimination of sine-wave gratings. *Perception and Psychophysics* 50(1):15–18.

Klingberg, T., Hedehus, M., Temple, E., Salz, T., Gabrieli, J. D., Moseley, M. E., & Poldrack, R. A. (2000). Microstructure of temporo-parietal white matter as a basis for reading ability: evidence from diffusion tensor magnetic resonance imaging [see comments]. *Neuron* 25(2):493–500.

Klüver, H., & Bucy, P. C. (1937). Psychic blindness and other symptoms following bilateral temporal lobectomy in rhesus monkey. *American Journal of Physiology* 119:352–353.

Kolinsky, R., Morais, J., Content, A., & Cary, L. (1987). Finding parts within figures: A developmental study. *Perception* 16(3):399–407.

Kolinsky, R., Morais, J., & Verhaeghe, A. (1994). Visual separability: A study on unschooled adults. *Perception* 23(4):471–486.

Kotz, S. A., Cappa, S. F., von Cramon, D. Y., & Friederici, A. D. (2002). Modulation of the lexical-semantic network by auditory semantic priming: an event-related functional MRI study. *Neuroimage* 17(4):1761–1772.

Koyama, S., Kakigi, R., Hoshiyama, M., & Kitamura, Y. (1998). Reading of Japanese Kanji (morphograms) and Kana (syllabograms): A magnetoencephalographic study. *Neuropsychologia* 36(1):83–98.

Kraebel, K. S., West, R. N., & Gerhardstein, P. (2007). The influence of training views on infants' long-term memory for simple 3D shapes. *Developmental Psychobiology* 49(4):406–420.

Kuhl, P. K. (2004). Early language acquisition: Cracking the speech code. *Nature Reviews Neuroscience* 5(11):831–843.

Kujala, T., Karma, K., Ceponiene, R., Belitz, S., Turkkila, P., Tervaniemi, M., & Naatanen, R. (2001). Plastic neural changes and reading improvement caused by audiovisual training in reading-impaired children. *Proceedings of the National Academy of Sciences* 98(18):10509–10514.

Kuo, W. J., Yeh, T. C., Duann, J. R., Wu, Y. T., Ho, L. T., Hung, D., Tzeng, O. J., & Hsieh, J. C. (2001). A left-lateralized network for reading Chinese words: A 3 T fMRI study. *NeuroReport* 12(18):3997–4001.

Kuo, W. J., Yeh, T. C., Lee, C. Y., Wu, Y. T., Chou, C. C., Ho, L. T., Hung, D. L., Tzeng, O. J., & Hsieh, J. C. (2003). Frequency effects of Chinese character processing in the brain: An event-related fMRI study. *Neuroimage* 18(3):720–730.

Lachmann, T., & Geyer, T. (2003). Letter reversals in dyslexia: Is the case really closed? A critical review and conclusions. *Psychology Science* 45:50–70.

Lambon-Ralph, M. A., Jarvis, C., & Ellis, A. W. (1997). Life in a mirrored world: Report of a case showing mirror reversal in reading and writing and for non-verbal materials. *Neurocase* 3:249–258.

Larson, K. (2004). *The science of word recognition*, http://www.microsoft.com/typography/ctfonts/WordRecognition.aspx.

Laufs, H., Krakow, K., Sterzer, P., Eger, E., Beyerle, A., Salek-Haddadi, A., & Kleinschmidt, A. (2003). Electroencephalographic signatures of attentional and cognitive default modes in spontaneous brain activity fluctuations at rest. *Proceedings of the National Academy of Sciences* 100(19):11053–11058.

Laureys, S. (2005). The neural correlate of (un)awareness: Lessons from the vegetative state. *Trends in Cognitive Sciences* 9:556–559.

Lavidor, M., Babkoff, H., & Faust, M. (2001). Analysis of standard and non-standard visual word format in the two hemispheres. *Neuropsychologia* 39(4):430–439.

Lavidor, M., & Ellis, A. W. (2002). Word length and orthographic neighborhood size effects in the left and right cerebral hemispheres. *Brain and Language* 80:45–62.

Le Bihan, D., Urayama, S., Aso, T., Hanakawa, T., & Fukuyama, H. (2006). Direct and fast detection of neuronal activation in the human brain with diffusion MRI. *Proceedings of the National Academy of Sciences* 103(21):8263–8268.

Lee, C. Y., Tsai, J. L., Kuo, W. J., Yeh, T. C., Wu, Y. T., Ho, L. T., Hung, D. L., Tzeng, O. J., & Hsieh, J. C. (2004). Neuronal correlates of consistency and frequency effects on Chinese character naming: An event-related fMRI study. *Neuroimage* 23(4):1235–1245.

Lee, K. M. (2004). Functional MRI comparison between reading ideographic and phonographic scripts of one language. *Brain and Language* 91(2):245–251.

Leff, A. P., Crewes, H., Plant, G. T., Scott, S. K., Kennard, C., & Wise, R. J. (2001). The functional anatomy of single-word reading in patients with hemianopic and pure alexia. *Brain* 124(Pt 3):510–521.

Leppanen, P. H., Richardson, U., Pihko, E., Eklund, K. M., Guttorm, T. K., Aro, M., & Lyytinen, H. (2002). Brain responses to changes in speech sound durations differ between infants with and without familial risk for dyslexia. *Developmental Neuropsychology* 22(1):407–422.

Lerner, Y., Hendler, T., Ben-Bashat, D., Harel, M., & Malach, R. (2001). A hierarchical axis of object processing stages in the human visual cortex. *Cerebral Cortex* 11(4):287–297.

Leroi-Gourhan, A. (1983). *Le fil du temps*. Paris: Fayard.

———. (1993). *Gesture and speech*. Cambridge, MA: MIT Press.

Levi-Strauss, C. (1958). *Anthropologie structurale*. Paris: Plon.

Levy, I., Hasson, U., Avidan, G., Hendler, T., & Malach, R. (2001). Center-periphery organization of human object areas. *Nature Neuroscience* 4(5):533–539.

Liberman, I. Y., Shankweiler, D., Orlando, C., Harris, K. S., & Berti, F. B. (1971). Letter confusions and reversals of sequence in the beginning reader: Implications for Orton's theory of developmental dyslexia. *Cortex* 7(2):127–142.

Lissauer, H. (1890). Ein fall von seelenblindheit nebst einem beitrage zur theorie derselben. *Archiv für Psychiatrie und Nervenkrankheiten* 21:222–270.

Logothetis, N. K., Pauls, J., & Poggio, T. (1995). Shape representation in the inferior temporal cortex of monkeys. *Current Biology* 5(5):552–563.

Longtin, C.-M., Segui, J., & Hallé, P. A. (2003). Morphological priming without morphological relationship. *Language and Cognitive Processes* 18:313–334.

Lukatela, G., Frost, S. J., & Turvey, M. T. (1998). Phonological priming by masked nonword primes in the lexical decision task. *Journal of Memory and Language* 39:666–683.

McCandliss, B. D., Cohen, L., & Dehaene, S. (2003). The visual word form area: Expertise for reading in the fusiform gyrus. *Trends in Cognitive Sciences* 7:293–299.

McCandliss, B. D., Curran, T., & Posner, M. I. (1993). Repetition effects in processing visual words: A high density ERP study of lateralized stimuli. *Neuroscience Abstracts* 19:1807.

McCarthy, R. A., & Warrington, E. K. (1990). *Cognitive neuropsychology: A clinical introduction.* San Diego: Academic Press.

McClelland, J. L., & Rumelhart, D. E. (1981). An interactive activation model of context effects in letter perception: I. An account of basic findings. *Psychological Review* 88:375–407.

McCloskey, M., & Rapp, B. (2000). A visually based developmental reading deficit. *Journal of Memory and Language* 43:157–181.

McCloskey, M., Rapp, B., Yantis, S., Rubin, G., Bacon, W., Dagnelie, G., Gordon, B., Aliminosa, D., Boatman, D. F., Badecker, W., Johnson, D. N., Tusa, R. J., & Palmer, E. (1995). A developmental deficit in localizing objects from vision. *Psychological Science* 6:112–117.

McConkie, G. W., & Rayner, K. (1975). The span of the effective stimulus during a fixation in reading. *Perception and Psychophysics* 17:578–586.

McConkie, G. W., & Zola, D. (1979). Is visual information integrated across successive fixations in reading? *Perception and Psychophysics* 25(3):221–224.

McCrory, E. J., Mechelli, A., Frith, U., & Price, C. J. (2005). More than words: A common neural basis for reading and naming deficits in developmental dyslexia? *Brain* 128(Pt 2):261–267.

McMonnies, C. W. (1992). Visuo-spatial discrimination and mirror image letter reversals in reading. *Journal of the American Optometric Association* 63(10):698–704.

Malach, R., Levy, I., & Hasson, U. (2002). The topography of high-order human object areas. *Trends in Cognitive Sciences* 6(4):176–184.

Manguel, A. (1997). *A history of reading.* New York: Penguin.

Mann, V. A. (1986). Phonological awareness: The role of reading experience. *Cognition* 24(1–2):65–92.

Marcus, G. F., & Berent, I. (2003). Are there limits to statistical learning? *Science* 300(5616):53–55; author reply 53–55.

Marcus, G. F., Vijayan, S., Bandi Rao, S., & Vishton, P. M. (1999). Rule learning by seven-month-old infants. *Science* 283(5398):77–80.

Marinkovic, K., Dhond, R. P., Dale, A. M., Glessner, M., Carr, V., & Halgren, E. (2003). Spatiotemporal dynamics of modality-specific and supramodal word processing. *Neuron* 38(3):487–497.

Marks, L. E. (1978). *The unity of the senses.* New York: Academic Press.

Marshall, J. C., & Newcombe, F. (1973). Patterns of paralexia: A psycholinguistic approach. *Journal of Psycholinguistic Research* 2:175–199.

Martin, A., Wiggs, C. L., Ungerleider, L. G., & Haxby, J. V. (1996). Neural correlates of category-specific knowledge. *Nature* 379:649–652.

Martinet, C., Valdois, S., & Fayol, M. (2004). Lexical orthographic knowledge develops from the beginning of literacy acquisition. *Cognition* 91(2):B11–22.

Matsuzawa, T. (1985). Use of numbers by a chimpanzee. *Nature* 315(6014):57–59.

Maurer, D., Pathman, T., & Mondloch, C. J. (2006). The shape of boubas: Sound-shape correspondences in toddlers and adults. *Developmental Science* 9(3):316–322.

Maurer, U., Brem, S., Bucher, K., & Brandeis, D. (2005). Emerging neurophysiological specialization for letter strings. *Journal of Cognitive Neuroscience* 17(10):1532–1552.

Maurer, U., Brem, S., Kranz, F., Bucher, K., Benz, R., Halder, P., Steinhausen, H. C., & Brandeis, D. (2006). Coarse neural tuning for print peaks when children learn to read. *Neuroimage* 33(2):749–758.

Maurer, U., Bucher, K., Brem, S., & Brandeis, D. (2003). Altered responses to tone and phoneme mismatch in kindergartners at familial dyslexia risk. *NeuroReport* 14(17):2245–2250.

Mayall, K., Humphreys, G. W., Mechelli, A., Olson, A., & Price, C. J. (2001). The effects of case mixing on word recognition: evidence from a PET study. *Journal of Cognitive Neuroscience* 13(6):844–853.

Mayall, K., Humphreys, G. W., & Olson, A. (1997). Disruption to word or letter processing? The origins of case-mixing effects. *Journal of Experimental Psychology: Learning, Memory, and Cognition* 23(5):1275–1286.

Mazoyer, B. M., Dehaene, S., Tzourio, N., Frak, V., Syrota, A., Murayama, N., Levrier, O., Salamon, G., Cohen, L., & Mehler, J. (1993). The cortical representation of speech. *Journal of Cognitive Neuroscience* 5:467–479.

Mechelli, A., Gorno-Tempini, M. L., & Price, C. J. (2003). Neuroimaging studies of word and pseudoword reading: consistencies, inconsistencies, and limitations. *Journal of Cognitive Neuroscience* 15(2):260–271.

Mehler, J., Jusczyk, P., Lambertz, G., Halsted, N., Bertoncini, J., & Amiel-Tison, C. (1988). A precusor of language acquisition in young infants. *Cognition* 29: 143–178.

Mello, N. K. (1965). Interhemispheric reversal of mirror-image oblique lines following monocular training in pigeons. *Science* 148:252–254.

Mello, N. K. (1966). Interocular generalization: A study of mirror-image reversal following monocular discrimination training in the pigeon. *Journal of the Experimental Analysis of Behavior* 9(1):11–16.

———. (1967). Inter-hemispheric comparison of visual stimuli in the pigeon. *Nature* 214(84):144–145.

Meng, H., Smith, S. D., Hager, K., Held, M., Liu, J., Olson, R. K., Pennington, B. F., DeFries, J. C., Gelernter, J., O'Reilly-Pol, T., Somlo, S., Skudlarski, P., Shaywitz, S. E., Shaywitz, B. A., Marchione, K., Wang, Y., Paramasivam, M., LoTurco, J. J., Page, G. P., & Gruen, J. R. (2005). DCDC2 is associated with reading disability and modulates neuronal development in the brain. *Proceedings of the National Academy of Sciences* 102(47):17053–17058.

Merzenich, M. M., Jenkins, W. M., Johnston, P., Schreiner, C., Miller, S. L., & Tallal, P. (1996). Temporal processing deficits of language-learning impaired children ameliorated by training. *Science* 271(5245):77–81.

Miozzo, M., & Caramazza, A. (1998). Varieties of pure alexia: The case of failure to access graphemic representations. *Cognitive Neuropsychology* 15:203–238.

Mishkin, M., & Pribram, K. H. (1954). Visual discrimination performance following partial ablations of the temporal lobe. I. Ventral vs. lateral. *Journal of Comparative and Physiological Psychology* 47(1):14–20.

Mithen, S. (1996). *The prehistory of the mind: The cognitive origins of art, religion and science.* London: Thames & Hudson.

Miyashita, Y. (1988). Neuronal correlate of visual associative long-term memory in the primate temporal cortex. *Nature* 335(6193):817–820.

Molko, N., Cohen, L., Mangin, J. F., Chochon, F., Lehéricy, S., Le Bihan, D., & Dehaene, S. (2002). Visualizing the neural bases of a disconnection syndrome with diffusion tensor imaging. *Journal of Cognitive Neuroscience* 14:629–636.

Mondloch, C. J., & Maurer, D. (2004). Do small white balls squeak? Pitch-object correspondences in young children. *Cognitive, Affective, and Behavioral Neuroscience* 4(2):133–136.

Montant, M., & Behrmann, M. (2000). Pure alexia. *Neurocase* 6:265–294.

Morais, J., Bertelson, P., Cary, L., & Alegria, J. (1986). Literacy training and speech segmentation. *Cognition* 24:45–64.

Morais, J., Cary, L., Alegria, J., & Bertelson, P. (1979). Does awareness of speech as a sequence of phones arise spontaneously? *Cognition* 7:323–331.

Morrison, R. E., & Rayner, K. (1981). Saccade size in reading depends upon character spaces and not visual angle. *Perception and Psychophysics* 30(4):395–396.

Mozer, M. C. (1987). Early parallel processing in reading: A connectionist approach. In Coltheart, M. (Ed.), *Attention and performance XII: The psychology of reading* (pp. 83–104). Hillsdale, NJ: Erlbaum.

Naatanen, R., Lehtokoski, A., Lennes, M., Cheour, M., Huotilainen, M., Iivonen, A., Vainio, M., Alku, P., Ilmoniemi, R. J., Luuk, A., Allik, J., Sinkkonen, J., & Alho, K. (1997). Language-specific phoneme representations revealed by electric and magnetic brain responses. *Nature* 385(6615):432–434.

Naccache, L., & Dehaene, S. (2001). The priming method: Imaging unconscious repetition priming reveals an abstract representation of number in the parietal lobes. *Cerebral Cortex* 11(10):966–974.

Naccache, L., Gaillard, R., Adam, C., Hasboun, D., Clémenceau, S., Baulac, M., S., D., & L., C. (2005). A direct intracranial record of emotions evoked by subliminal words. *Proceedings of the National Academy of Sciences* 102:7713–7717.

Nakamura, K., Dehaene, S., Jobert, A., Le Bihan, D., & Kouider, S. (2005). Subliminal convergence of Kanji and Kana words: Further evidence for functional parcellation of the posterior temporal cortex in visual word perception. *Journal of Cognitive Neuroscience* 17(6):954–968.

Nakamura, K., Honda, M., Okada, T., Hanakawa, T., Toma, K., Fukuyama, H., Konishi, J., & Shibasaki, H. (2000). Participation of the left posterior inferior temporal cortex in writing and mental recall of Kanji orthography: A functional MRI study. *Brain* 123(Pt 5):954–967.

Nation, K., Allen, R., & Hulme, C. (2001). The limitations of orthographic analogy in early reading development: Performance on the clue-word task depends on phonological priming and elementary decoding skill, not the use of orthographic analogy. *Journal of Experimental Child Psychology* 80(1):75–94.

National Institute of Child Health and Human Development. (2000). *Report of the National Reading Panel. Teaching children to read: An evidence-based assessment of the scientific research literature on reading and its implications for reading instruction (NIH Publication No. 00–4769).* Washington, D.C.: U.S. Government Printing Office.

Nazir, T. A., Ben-Boutayab, N., Decoppet, N., Deutsch, A., & Frost, R. (2004). Reading habits, perceptual learning, and recognition of printed words. *Brain and Language* 88(3):294–311.

Nicolson, R. I., Fawcett, A. J., & Dean, P. (2001). Developmental dyslexia: The cerebellar deficit hypothesis. *Trends in Neurosciences* 24(9):508–511.

Nieder, A., Diester, I., & Tudusciuc, O. (2006). Temporal and spatial enumeration processes in the primate parietal cortex. *Science* 313(5792):1431–1435.

Nieder, A., & Miller, E. K. (2004). A parieto-frontal network for visual numerical information in the monkey. *Proceedings of the National Academy of Sciences* 101(19):7457–7462.

Nimchinsky, E. A., Gilissen, E., Allman, J. M., Perl, D. P., Erwin, J. M., & Hof, P. R. (1999). A neuronal morphologic type unique to humans and great apes. *Proceedings of the National Academy of Sciences* 96(9):5268–5273.

Niogi, S. N., & McCandliss, B. D. (2006). Left lateralized white matter microstructure accounts for individual differences in reading ability and disability. *Neuropsychologia* 44(11):2178–2188.

Nobre, A. C., Allison, T., & McCarthy, G. (1994). Word recognition in the human inferior temporal lobe. *Nature* 372(6503):260–263.

Nunez, R. E., & Lakoff, G. (2000). *Where mathematics comes from: How the embodied mind brings mathematics into being.* New York: Basic Books.

Olavarria, J. F., & Hiroi, R. (2003). Retinal influences specify cortico-cortical maps by postnatal day six in rats and mice. *Journal of Comparative Neurology* 459(2):156–172.

O'Regan, J. K. (1990). Eye movements and reading. *Reviews of Oculomotor Research* 4:395–453.

Orton, S. T. (1925). "Word-blindness" in school children. *Archives of Neurology and Psychiatry* 14:581–615.

Orton, S. T. (1937). *Reading, writing, and speech problems in children.* New York: Norton.

Paap, K. R., Newsome, S. L., & Noel, R. W. (1984). Word shape's in poor shape for the race to the lexicon. *Journal of Experimental Psychology: Human Perception and Performance* 10(3):413–428.

Pacton, S., Perruchet, P., Fayol, M., & Cleeremans, A. (2001). Implicit learning out of the lab: The case of orthographic regularities. *Journal of Experimental Psychology: General* 130(3):401–426.

Pammer, K., Hansen, P. C., Kringelbach, M. L., Holliday, I., Barnes, G., Hillebrand, A., Singh, K. D., & Cornelissen, P. L. (2004). Visual word recognition: The first half second. *Neuroimage* 22(4):1819–1825.

Paracchini, S., Thomas, A., Castro, S., Lai, C., Paramasivam, M., Wang, Y., Keating, B. J., Taylor, J. M., Hacking, D. F., Scerri, T., Francks, C., Richardson, A. J., Wade-Martins, R., Stein, J. F., Knight, J. C., Copp, A. J., Loturco, J., & Monaco, A. P. (2006). The chromosome 6p22 haplotype associated with dyslexia reduces the expression of KIAA0319, a novel gene involved in neuronal migration. *Human Molecular Genetics* 15(10):1659–1666.

Parviainen, T., Helenius, P., Poskiparta, E., Niemi, P., & Salmelin, R. (2006). Cortical sequence of word perception in beginning readers. *Journal of Neuroscience* 26(22):6052–6061.

Pascalis, O., de Haan, M., & Nelson, C. A. (2002). Is face processing species-specific during the first year of life? *Science* 296(5571):1321–1323.

Pascalis, O., & de Schonen, S. (1994). Recognition memory in 3- to 4-day-old human neonates. *NeuroReport* 5(14):1721–1724.

Pascalis, O., Scott, L. S., Kelly, D. J., Shannon, R. W., Nicholson, E., Coleman, M., & Nelson, C. A. (2005). Plasticity of face processing in infancy. *Proceedings of the National Academy of Sciences* 102(14):5297–5300.

Patterson, K., & Kay, J. (1982). Letter-by-letter reading: Psychological descriptions of a neurological syndrome. *Quarterly Journal of Experimental Psychology A* 34:411–441.

Paulesu, E., Demonet, J. F., Fazio, F., McCrory, E., Chanoine, V., Brunswick, N., Cappa, S. F., Cossu, G., Habib, M., Frith, C. D., & Frith, U. (2001). Dyslexia: Cultural diversity and biological unity. *Science* 291(5511):2165–2167.

Paulesu, E., Frith, C. D., & Frackowiak, R. S. J. (1993). The neural correlates of the verbal component of working memory. *Nature* 362:342–345.

Paulesu, E., Frith, U., Snowling, M., Gallagher, A., Morton, J., Frackowiak, R., & Frith, C. D. (1996). Is developmental dyslexia a disconnection syndrome? Evidence from PET scanning. *Brain* 119:143–157.

Paulesu, E., McCrory, E., Fazio, F., Menoncello, L., Brunswick, N., Cappa, S. F., Cotelli, M., Cossu, G., Corte, F., Lorusso, M., Pesenti, S., Gallagher, A., Perani, D., Price, C., Frith, C. D., & Frith, U. (2000). A cultural effect on brain function. *Nature Neuroscience* 3(1):91–96.

Pelli, D. G., Farell, B., & Moore, D. C. (2003). The remarkable inefficiency of word recognition. *Nature* 423(6941):752–756.

Pena, M., Maki, A., Kovacic, D., Dehaene-Lambertz, G., Koizumi, H., Bouquet, F., & Mehler, J. (2003). Sounds and silence: An optical topography study of language recognition at birth. *Proceedings of the National Academy of Sciences* 100(20):11702–11705.

Perea, M., & Lupker, S. J. (2003). Does jugde activate COURT? Transposed-letter similarity effects in masked associative priming. *Memory and Cognition* 31(6):829–841.

Peressotti, F., & Grainger, J. (1999). The role of letter identity and letter position in orthographic priming. *Perception and Psychophysics* 61(4):691–706.

Perfetti, C., & Bell, L. (1991). Phonemic activation during the first 40 ms of word identification: Evidence from backward masking and masked priming. *Journal of Memory and Language* 30:473–485.

Perrett, D. I., Mistlin, A. J., & Chitty, A. J. (1989). Visual neurones responsive to faces. *Trends in Neuroscience* 10:358–364.

Perry, C., Ziegler, J., & Zorzi, M. (2007). Nested incremental modeling in the development of computational theories: The CDP+ model of reading aloud. *Psychological Review* 114(2):273–315.

Petersen, S. E., Fox, P. T., Posner, M. I., Mintun, M., & Raichle, M. E. (1988). Positron emission tomographic studies of the cortical anatomy of single-word processing. *Nature* 331(6157):585–589.

————.(1989). Positron emission tomographic studies of the processing of single words. *Journal of Cognitive Neuroscience* 1:153–170.

Petersen, S. E., Fox, P. T., Snyder, A. Z., & Raichle, M. E. (1990). Activation of extrastriate and frontal cortical areas by visual words and word-like stimuli. *Science* 249:1041–1044.

Petersson, K. M., Silva, C., Castro-Caldas, A., Ingvar, M., & Reis, A. (2007). Literacy: A cultural influence on functional left-right differences in the inferior parietal cortex. *European Journal of Neuroscience* 26(3):791–799.

Pflugshaupt, T., Nyffeler, T., von Wartburg, R., Wurtz, P., Luthi, M., Hubl, D., Gutbrod, K., Juengling, F. D., Hess, C. W., & Muri, R. M. (2007). When left becomes right and vice versa: Mirrored vision after cerebral hypoxia. *Neuropsychologia* 45(9):2078–2091.

Pica, P., Lemer, C., Izard, V., & Dehaene, S. (2004). Exact and approximate arithmetic in an Amazonian indigene group. *Science* 306(5695):499–503.

Pinel, P., Dehaene, S., Riviere, D., & Le Bihan, D. (2001). Modulation of parietal activation by semantic distance in a number comparison task. *Neuroimage* 14(5):1013–1026.

Pinker, S. (2002). *The blank slate: The modern denial of human nature.* London: Penguin.

Plaut, D. C., McClelland, J. L., Seidenberg, M. S., & Patterson, K. (1996). Understanding normal and impaired word reading: computational principles in quasi-regular domains. *Psychological Review* 103(1):56–115.

Polk, T. A., & Farah, M. J. (2002). Functional MRI evidence for an abstract, not perceptual, word-form area. *Journal of Experimental Psychology: General* 131(1):65–72.

Polk, T. A., Stallcup, M., Aguirre, G. K., Alsop, D. C., D'Esposito, M., Detre, J. A., & Farah, M. J. (2002). Neural specialization for letter recognition. *Journal of Cognitive Neuroscience* 14(2):145–159.

Pollatsek, A., Bolozky, S., Well, A. D., & Rayner, K. (1981). Asymmetries in the perceptual span for Israeli readers. *Brain and Language* 14(1):174–180.

Posner, M. I., & McCandliss, B. D. (1999). Brain circuitry during reading. In Klein, R. M., & McMullen, P. A. (Eds.), *Converging methods for understanding reading and dyslexia* (pp. 305–337). Cambridge, MA: MIT Press.

Posner, M. I., Petersen, S. E., Fox, P. T., & Raichle, M. E. (1988). Localization of cognitive operations in the human brain. *Science* 240:1627–1631.

Pouratian, N., Bookheimer, S. Y., Rubino, G., Martin, N. A., & Toga, A. W. (2003). Category-specific naming deficit identified by intraoperative stimulation mapping and postoperative neuropsychological testing: Case report. *Journal of Neurosurgery* 99(1):170–176.

Price, C. (1998). The functional anatomy of word comprehension and production. *Trends in Cognitive Sciences* 2:281–288.

Price, C. J., & Devlin, J. T. (2003). The myth of the visual word form area. *Neuroimage* 19:473–481.

Price, C. J., Gorno-Tempini, M. L., Graham, K. S., Biggio, N., Mechelli, A., Patterson, K., & Noppeney, U. (2003). Normal and pathological reading: Converging data from lesion and imaging studies. *Neuroimage* 20(Suppl 1):S30–41.

Price, C. J., Moore, C. J., Humphreys, G. W., & Wise, R. J. S. (1997). Segregating semantic from phonological processes during reading. *Journal of Cognitive Neuroscience* 9:727–733.

Price, C. J., Wise, R. J. S., & Frackowiak, R. S. J. (1996). Demonstrating the implicit processing of visually presented words and pseudowords. *Cerebral Cortex* 6:62–70.

Priftis, K., Rusconi, E., Umilta, C., & Zorzi, M. (2003). Pure agnosia for mirror stimuli after right inferior parietal lesion. *Brain* 126(Pt 4):908–919.

Prinzmetal, W. (1990). Neon colors illuminate reading units. *Journal of Experimental Psychology: Human Perception and Performance* 16(3):584–597.

Prinzmetal, W., Treiman, R., & Rho, S. H. (1986). How to see a reading unit. *Journal of Memory and Language* 25:461–475.

Puce, A., Allison, T., Asgari, M., Gore, J. C., & McCarthy, G. (1996). Differential sensitivity of human visual cortex to faces, letterstrings, and textures: A functional magnetic resonance imaging study. *Journal of Neuroscience* 16:5205–5215.

Pugh, K. R., Shaywitz, B. A., Shaywitz, S. E., Constable, R. T., Skudlarski, P., Fulbright, R. K., Bronen, R. A., Shankweiler, D. P., Katz, L., Fletcher, J. M., & Gore, J. C. (1996). Cerebral organization of component processes in reading. *Brain* 119(Pt 4):1221–1238.

Pulvermuller, F. (2005). Brain mechanisms linking language and action. *Nature Reviews Neuroscience* 6(7):576–582.

Quartz, S. R., & Sejnowski, T. J. (1997). The neural basis of cognitive development: A constructivist manifesto. *Behavioral and Brain Sciences* 20(4):537–556; discussion 556–596.

Quiroga, R. Q., Reddy, L., Kreiman, G., Koch, C., & Fried, I. (2005). Invariant visual representation by single neurons in the human brain. *Nature* 435(7045):1102–1107.

Raichle, M. E., MacLeod, A. M., Snyder, A. Z., Powers, W. J., Gusnard, D. A., & Shulman, G. L. (2001). A default mode of brain function. *Proceedings of the National Academy of Sciences* 98(2):676–682.

Raij, T., Uutela, K., & Hari, R. (2000). Audiovisual integration of letters in the human brain. *Neuron* 28(2):617–625.

Ramachandran, V. S. (2005). *The artful brain.* New York: Fourth Estate.

Ramachandran, V. S., & Hubbard, E. M. (2001a). Psychophysical investigations into the neural basis of synaesthesia. *Proceedings of the Royal Society B: Biological Sciences* 268(1470):979–983.

Ramachandran, V. S., & Hubbard, E. M. (2001b). Synaesthesia—a window into perception, thought and language. *Journal of Consciousness Studies* 8:3–34.

Ramus, F. (2003). Developmental dyslexia: specific phonological deficit or general sensorimotor dysfunction? *Current Opinion in Neurobiology* 13(2):212–218.

Ramus, F. (2004). Neurobiology of dyslexia: A reinterpretation of the data. *Trends in Neurosciences* 27(12):720–726.

Ramus, F., Pidgeon, E., & Frith, U. (2003). The relationship between motor control and phonology in dyslexic children. *Journal of Child Psychology and Psychiatry* 44(5):712–722.

Ramus, F., Rosen, S., Dakin, S. C., Day, B. L., Castellote, J. M., White, S., & Frith, U. (2003). Theories of developmental dyslexia: Insights from a multiple case study of dyslexic adults. *Brain* 126(Pt 4):841–865.

Rastle, K., Davis, M. H., Marslen-Wilson, W. D., & Tyler, L. K. (2000). Morphological and semantic effects in visual word recognition: A time-course study. *Language and Cognitive Processes* 15:507–537.

Rayner, K. (1998). Eye movements in reading and information processing: 20 years of research. *Psychological Bulletin* 124(3):372–422.

Rayner, K., & Bertera, J. H. (1979). Reading without a fovea. *Science* 206(4417):468–469.

Rayner, K., Foorman, B. R., Perfetti, C. A., Pesetsky, D., & Seidenberg, M. S. (2001). How psychological science informs the teaching of reading. *Psychological Science* 2:31–74.

Rayner, K., Inhoff, A. W., Morrison, R. E., Slowiaczek, M. L., & Bertera, J. H. (1981). Masking of foveal and parafoveal vision during eye fixations in reading. *Journal of Experimental Psychology: Human Perception and Performance* 7(1):167–179.

Rayner, K., McConkie, G. W., & Zola, D. (1980). Integrating information across eye movements. *Cognitive Psychology* 12(2):206–226.

Rayner, K., & Pollatsek, A. (1989). *The psychology of reading*. Englewood Cliffs, NJ: Prentice Hall.

Rayner, K., Well, A. D., & Pollatsek, A. (1980). Asymmetry of the effective visual field in reading. *Perception and Psychophysics* 27(6):537–544.

Read, C., Zhang, Y. F., Nie, H. Y., & Ding, B. Q. (1986). The ability to manipulate speech sounds depends on knowing alphabetic writing. *Cognition* 24(1–2):31–44.

Reicher, G. M. (1969). Perceptual recognition as a function of meaningfulness of stimulus material. *Journal of Experimental Psychology* 81:274–280.

Rey, A., Jacobs, A. M., Schmidt-Weigand, F., & Ziegler, J. C. (1998). A phoneme effect in visual word recognition. *Cognition* 68(3):B71–80.

Rey, A., Ziegler, J. C., & Jacobs, A. M. (2000). Graphemes are perceptual reading units. *Cognition* 75(1):B1–12.

Richardson, U., Leppanen, P. H., Leiwo, M., & Lyytinen, H. (2003). Speech perception of infants with high familial risk for dyslexia differ at the age of 6 months. *Developmental Neuropsychology* 23(3):385–397.

Riddoch, M. J., & Humphreys, G. W. (1988). Description of a left-right coding deficit in a case of constructional apraxia. *Cognitive Neuropsychology* 5:289–315.

Riesenhuber, M., & Poggio, T. (1999). Hierarchical models of object recognition in cortex. *Nature Neuroscience* 2:1019–1025.

Rissman, J., Eliassen, J. C., & Blumstein, S. E. (2003). An event-related fMRI investigation of implicit semantic priming. *Journal of Cognitive Neuroscience* 15(8):1160–1175.

Robertson, L. C., & Lamb, M. R. (1991). Neuropsychological contributions to theories of part/whole organization. *Cognitive Psychology* 23(2):299–330.

Robinson, A. J., & Pascalis, O. (2004). Development of flexible visual recognition memory in human infants. *Developmental Science* 7(5):527–533.

Rodd, J. M., Davis, M. H., & Johnsrude, I. S. (2005). The neural mechanisms of speech comprehension: fMRI studies of semantic ambiguity. *Cerebral Cortex* 15(8):1261–1269.

Rodman, H. R., Scalaidhe, S. P. O., & Gross, C. G. (1993). Response properties of neurons in temporal cortical visual areas of infant monkeys. *Journal of Neurophysiology* 70:1115–1136.

Rollenhagen, J. E., & Olson, C. R. (2000). Mirror-image confusion in single neurons of the macaque inferotemporal cortex. *Science* 287(5457):1506–1508.

Rolls, E. T. (2000). Functions of the primate temporal lobe cortical visual areas in invariant visual object and face recognition. *Neuron* 27(2):205–218.

Rossion, B., Kung, C. C., & Tarr, M. J. (2004). Visual expertise with nonface objects leads to competition with the early perceptual processing of faces in the human occipitotemporal cortex. *Proceedings of the National Academy of Sciences* 101(40):14521–14526.

Rubenstein, H., Lewis, S. S., & Rubenstein, M. (1971). Evidence for phonemic coding in visual word recognition. *Journal of Verbal Learning and Verbal Behavior* 10:645–657.

Rubin, G. S., & Turano, K. (1992). Reading without saccadic eye movements. *Vision Research* 32(5):895–902.

Rumelhart, D. E., & McClelland, J. L. (1982). An interactive activation model of context effects in letter perception: Part 2. The contextual enhancement effect and some tests and extensions of the model. *Psychological Review* 89(1):60–94.

Saffran, J. R., Aslin, R. N., & Newport, E. L. (1996). Statistical learning by 8-month-old infants. *Science* 274(5294):1926–1928.

Sakai, K., & Miyashita, Y. (1991). Neural organization for the long-term memory of paired associates. *Nature* 354(6349):152–155.

Sakurai, Y., Ichikawa, Y., & Mannen, T. (2001). Pure alexia from a posterior occipital lesion. *Neurology* 56(6):778–781.

Sakurai, Y., Momose, T., Iwata, M., Sudo, Y., Ohtomo, K., & Kanazawa, I. (2000). Different cortical activity in reading of Kanji words, Kana words and Kana nonwords. *Brain Research Cognitive Brain Research* 9(1):111–115.

Sakurai, Y., Takeuchi, S., Takada, T., Horiuchi, E., Nakase, H., & Sakuta, M. (2000). Alexia caused by a fusiform or posterior inferior temporal lesion. *Journal of the Neurological Sciences* 178(1):42–51.

Salmelin, R., Service, E., Kiesila, P., Uutela, K., & Salonen, O. (1996). Impaired visual word processing in dyslexia revealed with magnetoencephalography. *Annals of Neurology* 40(2):157–162.

Sary, G., Vogels, R., & Orban, G. A. (1993). Cue-invariant shape selectivity of macaque inferior temporal neurons. *Science* 260(5110):995–997.

Sasaki, Y., Vanduffel, W., Knutsen, T., Tyler, C., & Tootell, R. (2005). Symmetry activates extrastriate visual cortex in human and nonhuman primates. *Proceedings of the National Academy of Sciences* 102(8):3159–3163.

Sasanuma, S. (1975). Kana and Kanji processing in Japanese aphasics. *Brain and Language* 2(3):369–383.

Schmandt-Besserat, D. (1996). *How writing came about*. Austin: University of Texas Press.

Schoenemann, P. T., Sheehan, M. J., & Glotzer, L. D. (2005). Prefrontal white matter volume is disproportionately larger in humans than in other primates. *Nature Neuroscience* 8(2):242–252.

Schoonbaert, S., & Grainger, J. (2004). Letter position coding in printed word perception: Effects of repeated and transposed letters. *Language and Cognitive Processes* 19:333–367.

Schwartz, E. L., Desimone, R., Albright, T. D., & Gross, C. G. (1983). Shape recognition and inferior temporal neurons. *Proceedings of the National Academy of Sciences* 80(18):5776–5778.

Segui, J., & Grainger, J. (1990). Priming word recognition with orthographic neighbors: Effects of relative prime-target frequency. *Journal of Experimental Psychology: Human Perception and Performance* 16(1):65–76.

Seidenberg, M. S., & McClelland, J. L. (1989). A distributed, developmental model of word recognition and naming. *Psychological Review* 96(4):523–568.

Seidenberg, M. S., Petersen, A., MacDonald, M. C., & Plaut, D. C. (1996). Pseudohomophone effects and models of word recognition. *Journal of Experimental Psychology: Learning, Memory, and Cognition* 22:48–62.

Seidenberg, M. S., Tanenhaus, M. K., Leiman, J. M., & Bienkowski, M. (1982). Automatic access of the meanings of ambiguous words in context: Some limitations of knowledge-based processing. *Cognitive Psychology* 14:489–537.

Selfridge, O. G. (1959). Pandemonium: A paradigm for learning. In Blake, D. V., & Uttley, A. M. (Eds.), *Proceedings of the Symposium on Mechanisation of Thought Processes* (pp. 511–529). London: H. M. Stationery Office.

Sere, B., Marendaz, C., & Herault, J. (2000). Nonhomogeneous resolution of images of natural scenes. *Perception* 29(12):1403–1412.

Seymour, P. H., Aro, M., & Erskine, J. M. (2003). Foundation literacy acquisition in European orthographies. *British Journal of Psychology* 94(Pt 2):143–174.

Shallice, T. (1988). *From neuropsychology to mental structure.* Cambridge, UK: Cambridge University Press.

Share, D. L. (1995). Phonological recoding and self-teaching: Sine qua non of reading acquisition. *Cognition* 55(2):151–218; discussion 219–226.

Share, D. L. (1999). Phonological recoding and orthographic learning: A direct test of the self-teaching hypothesis. *Journal of Experimental Child Psychology* 72(2):95–129.

Shaywitz, B. A., Shaywitz, S. E., Pugh, K. R., Mencl, W. E., Fulbright, R. K., Skudlarski, P., Constable, R. T., Marchione, K. E., Fletcher, J. M., Lyon, G. R., & Gore, J. C. (2002). Disruption of posterior brain systems for reading in children with developmental dyslexia. *Biological Psychiatry* 52(2):101–110.

Shaywitz, S. (2003). *Overcoming dyslexia.* New York: Random House.

Shaywitz, S. E., Escobar, M. D., Shaywitz, B. A., Fletcher, J. M., & Makuch, R. (1992). Evidence that dyslexia may represent the lower tail of a normal distribution of reading ability. *New England Journal of Medicine* 326(3):145–150.

Shaywitz, S. E., Shaywitz, B. A., Pugh, K. R., Fulbright, R. K., Constable, R. T., Mencl, W. E., Shankweiler, D. P., Liberman, A. M., Skudlarski, P., Fletcher, J. M., Katz, L., Marchione, K. E., Lacadie, C., Gatenby, C., & Gore, J. C. (1998). Functional disruption in the organization of the brain for reading in dyslexia. *Proceedings of the National Academy of Sciences* 95(5):2636–2641.

Shelton, J. R., Fouch, E., & Caramazza, A. (1998). The selective sparing of body part knowledge: A case study. *Neurocase* 4:339–351.

Shuwairi, S. M., Albert, M. K., & Johnson, S. P. (2007). Discrimination of possible and impossible objects in infancy. *Psychological Science* 18(4):303–307.

Sigman, M., & Gilbert, C. D. (2000). Learning to find a shape. *Nature Neuroscience* 3(3):264–269.

Sigman, M., Pan, H., Yang, Y., Stern, E., Silbersweig, D., & Gilbert, C. D. (2005). Topdown reorganization of activity in the visual pathway after learning a shape identification task. *Neuron* 46(5):823–835.

Silani, G., Frith, U., Demonet, J. F., Fazio, F., Perani, D., Price, C., Frith, C. D., & Paulesu, E. (2005). Brain abnormalities underlying altered activation in dyslexia: A voxel based morphometry study. *Brain* 128(Pt 10):2453–2461.

Simon, O., Mangin, J. F., Cohen, L., Le Bihan, D., & Dehaene, S. (2002). Topographical layout of hand, eye, calculation, and language-related areas in the human parietal lobe. *Neuron* 33(3):475–487.

Simos, P. G., Breier, J. I., Fletcher, J. M., Bergman, E., & Papanicolaou, A. C. (2000). Cerebral mechanisms involved in word reading in dyslexic children: A magnetic source imaging approach. *Cerebral Cortex* 10(8):809–816.

Simos, P. G., Breier, J. I., Fletcher, J. M., Foorman, B. R., Castillo, E. M., & Papanicolaou, A. C. (2002). Brain mechanisms for reading words and pseudowords: An integrated approach. *Cerebral Cortex* 12(3):297–305.

Simos, P. G., Breier, J. I., Fletcher, J. M., Foorman, B. R., Mouzaki, A., & Papanicolaou, A. C. (2001). Age-related changes in regional brain activation during phonological decoding and printed word recognition. *Developmental Neuropsychology* 19(2):191–210.

Simos, P. G., Fletcher, J. M., Bergman, E., Breier, J. I., Foorman, B. R., Castillo, E. M., Davis, R. N., Fitzgerald, M., & Papanicolaou, A. C. (2002). Dyslexia-specific brain activation profile becomes normal following successful remedial training. *Neurology* 58(8):1203–1213.

Siok, W. T., Perfetti, C. A., Jin, Z., & Tan, L. H. (2004). Biological abnormality of impaired reading is constrained by culture. *Nature* 431(7004):71–76.

Smith, L. B. (2003). Learning to recognize objects. *Psychological Science* 14(3):244–250.

Son, J. Y., Smith, L. B., & Goldstone, R. L. (2008). Simplicity and generalization: Short-cutting abstraction in children's object categorizations. *Cognition* 108(3):626–638.

Southgate, V., Csibra, G., Kaufman, J., & Johnson, M. H. (2008). Distinct processing of objects and faces in the infant brain. *Journal of Cognitive Neuroscience* 20(4):741–749.

Sperber, D. (1974). Contre certains a priori anthropologiques. In Morin, E. & Piatelli-Palmarini, M. (Eds.), *L'unité de l'homme: Invariants biologiques et universaux culturels* (pp. 491–512). Paris: Le Seuil.

———. (1996). *Explaining culture: A naturalistic approach.* London: Blackwell.

Sperber, D., & Hirschfeld, L. A. (2004). The cognitive foundations of cultural stability and diversity. *Trends in Cognitive Sciences* 8:40–46.

Sperling, J. M., Prvulovic, D., Linden, D. E., Singer, W., & Stirn, A. (2006). Neuronal correlates of colour-graphemic synaesthesia: an fMRI study. *Cortex* 42(2):295–303.

Spoehr, K. T., & Smith, E. E. (1975). The role of orthographic and phonotactic rules in perceiving letter patterns. *Journal of Experimental Psychology: Human Perception and Performance* 104(1):21–34.

Sprenger-Charolles, L., & Siegel, L. (1997). A longitudinal study of the effects of syllabic structure on the development of reading and spelling skills in French. *Applied Psycholinguistics* 18:485–505.

Sprenger-Charolles, L., Siegel, L. S., & Bonnet, P. (1998). Reading and spelling acquisition in French: The role of phonological mediation and orthographic factors. *Journal of Experimental Child Psychology* 68(2):134–165.

Stein, J. (2001). The magnocellular theory of developmental dyslexia. *Dyslexia* 7(1):12–36.

Streifler, M., & Hofman, S. (1976). Sinistrad mirror writing and reading after brain concussion in a bi-systemic (oriento-occidental) polyglot. *Cortex* 12:356–364.

Stuart, M. (1990). Processing strategies in a phoneme deletion task. *Quarterly Journal of Experimental Psychology A* 42:305–327.

Swinney, D. A., Onifer, W., Prather, P., & Hirshkowitz, M. (1979). Semantic facilitation across sensory modalities in the processing of individual words and sentences. *Mem Cognit* 7(3):159–165.

Taft, M. (1994). Interactive activation as a framework for understanding morphological processing. *Language and Cognitive Processes* 9:271–294.

Tagamets, M. A., Novick, J. M., Chalmers, M. L., & Friedman, R. B. (2000). A parametric approach to orthographic processing in the brain: An fMRI study. *Journal of Cognitive Neuroscience* 12(2):281–297.

Tallal, P., & Gaab, N. (2006). Dynamic auditory processing, musical experience and language development. *Trends in Neurosciences* 29(7):382–390.

Tamura, H., & Tanaka, K. (2001). Visual response properties of cells in the ventral and dorsal parts of the macaque inferotemporal cortex. *Cerebral Cortex* 11(5):384–399.

Tan, L. H., Liu, H. L., Perfetti, C. A., Spinks, J. A., Fox, P. T., & Gao, J. H. (2001). The neural system underlying Chinese logograph reading. *Neuroimage* 13(5):836–846.

Tan, L. H., Spinks, J. A., Gao, J. H., Liu, H. L., Perfetti, C. A., Xiong, J., Stofer, K. A., Pu, Y., Liu, Y., & Fox, P. T. (2000). Brain activation in the processing of Chinese characters and words: A functional MRI study. *Human Brain Mapping* 10(1):16–27.

Tanaka, K. (1996). Inferotemporal cortex and object vision. *Annual Review of Neuroscience* 19:109–139.

———. (2003). Columns for complex visual object features in the inferotemporal cortex: Clustering of cells with similar but slightly different stimulus selectivities. *Cerebral Cortex* 13(1):90–99.

Tanaka, M., Tomonaga, M., & Matsuzawa, T. (2003). Finger drawing by infant chimpanzees (Pan troglodytes). *Animal Cognition* 6(4):245–251.

Tarkiainen, A., Cornelissen, P. L., & Salmelin, R. (2002). Dynamics of visual feature analysis and object-level processing in face versus letter-string perception. *Brain* 125(Pt 5):1125–1136.

Tarkiainen, A., Helenius, P., Hansen, P. C., Cornelissen, P. L., & Salmelin, R. (1999). Dynamics of letter string perception in the human occipitotemporal cortex. *Brain* 122(Pt 11):2119–2132.

Tarr, M. J., & Gauthier, I. (2000). FFA: A flexible fusiform area for subordinate-level visual processing automatized by expertise. *Nature Neuroscience* 3(8):764–769.

Tarr, M. J., & Pinker, S. (1989). Mental rotation and orientation-dependence in shape recognition. *Cognitive Psychology* 21(2):233–282.

Temple, E., Deutsch, G. K., Poldrack, R. A., Miller, S. L., Tallal, P., Merzenich, M. M., & Gabrieli, J. D. (2003). Neural deficits in children with dyslexia ameliorated by behavioral remediation: evidence from functional MRI. *Proceedings of the National Academy of Sciences* 100(5):2860–2865.

Temple, E., Poldrack, R. A., Protopapas, A., Nagarajan, S., Salz, T., Tallal, P., Merzenich, M. M., & Gabrieli, J. D. (2000). Disruption of the neural response to rapid acoustic stimuli in dyslexia: Evidence from functional MRI. *Proceedings of the National Academy of Sciences* 97(25):13907–13912.

Temple, E., Poldrack, R. A., Salidis, J., Deutsch, G. K., Tallal, P., Merzenich, M. M., & Gabrieli, J. D. (2001). Disrupted neural responses to phonological and orthographic processing in dyslexic children: an fMRI study. *NeuroReport* 12(2):299–307.

Terepocki, M., Kruk, R. S., & Willows, D. M. (2002). The incidence and nature of letter orientation errors in reading disability. *Journal of Learning Disabilities* 35(3):214–233.

Thompson-Schill, S. L., D'Esposito, M., & Kan, I. P. (1999). Effects of repetition and competition on activity in left prefrontal cortex during word generation. *Neuron* 23(3):513–522.

Thorpe, S., Fize, D., & Marlot, C. (1996). Speed of processing in the human visual system. *Nature* 381(6582):520–522.

Tokunaga, H., Nishikawa, T., Ikejiri, Y., Nakagawa, Y., Yasuno, F., Hashikawa, K., Nishimura, T., Sugita, Y., & Takeda, M. (1999). Different neural substrates for Kanji and Kana writing: a PET study. *NeuroReport* 10(16):3315–3319.

Tomasello, M. (2000a). *The cultural origins of human cognition.* Cambridge, MA: Harvard University Press.

———. (2000b). Culture and cognitive development. *Current Directions in Psychological Science* 9:37–40.

Tomasello, M., Carpenter, M., Call, J., Behne, T., & Moll, H. (2005). Understanding and sharing intentions: The origins of cultural cognition. *Behavioral and Brain Sciences* 28(5):675–691; discussion 691–735.

Tomasello, M., Strosberg, R., & Akhtar, N. (1996). Eighteen-month-old children learn words in non-ostensive contexts. *Journal of Child Language* 23(1):157–176.

Torgesen, J. K. (2005). Recent discoveries on remedial interventions for children with dyslexia. In Snowling, M. J. & Hulme, C. (Eds.), *The science of reading: A handbook* (pp. 521–537). Oxford: Blackwell.

Tsao, D. Y., Freiwald, W. A., Tootell, R. B., & Livingstone, M. S. (2006). A cortical region consisting entirely of face-selective cells. *Science* 311(5761):670–674.

Tsunoda, K., Yamane, Y., Nishizaki, M., & Tanifuji, M. (2001). Complex objects are represented in macaque inferotemporal cortex by the combination of feature columns. *Nature Neuroscience* 4(8):832–838.

Turing, A. M. (1952). The chemical basis of morphogenesis. *Philosophical Transactions of the Royal Society of London B: Biological Sciences* 237:37–72.

Turkeltaub, P. E., Flowers, D. L., Verbalis, A., Miranda, M., Gareau, L., & Eden, G. F. (2004). The neural basis of hyperlexic reading: an FMRI case study. *Neuron* 41(1):11–25.

Turkeltaub, P. E., Gareau, L., Flowers, D. L., Zeffiro, T. A., & Eden, G. F. (2003). Development of neural mechanisms for reading. *Nature Neuroscience* 6(7):767–773.

Turnbull, O. H. (1997). A double dissociation between knowledge of object identity and object orientation. *Neuropsychologia* 35:567–570.

Turnbull, O. H., Beschin, N., & Della Sala, S. (1997). Agnosia for object orientation: Implications for theories of object recognition. *Neuropsychologia* 35(2):153–163.

Turnbull, O. H., & McCarthy, R. A. (1996). Failure to discriminate between mirror-image objects: A case of viewpoint-independent object recognition? *Neurocase* 2:63–72.

Tzourio-Mazoyer, N., De Schonen, S., Crivello, F., Reutter, B., Aujard, Y., & Mazoyer, B. (2002). Neural correlates of woman face processing by 2-month-old infants. *Neuroimage* 15(2):454–461.

Ungerleider, L. G., & Mishkin, M. (1982). Two cortical visual systems. In Ingle, D. J., Goodalem M. A., & Mansfield, R. J. (Eds.), *Analysis of visual behavior* (pp. 549–586). Cambridge, MA: MIT Press.

Valdois, S., Bosse, M. L., & Tainturier, M. J. (2004). The cognitive deficits responsible for developmental dyslexia: Review of evidence for a selective visual attentional disorder. *Dyslexia* 10(4):339–363.

van Atteveldt, N., Formisano, E., Goebel, R., & Blomert, L. (2004). Integration of letters and speech sounds in the human brain. *Neuron* 43(2):271–282.

Van Essen, D. C., Lewis, J. W., Drury, H. A., Hadjikhani, N., Tootell, R. B., Bakircioglu, M., & Miller, M. I. (2001). Mapping visual cortex in monkeys and humans using surface-based atlases. *Vision Research* 41(10–11):1359–1378.

Van Orden, G. C., Johnston, J. C., & Hale, B. L. (1988). Word identification in reading proceeds from spelling to sound to meaning. *Journal of Experimental Psychology: Learning, Memory, and Cognition* 14(3):371–386.

Vandenberghe, R., Nobre, A. C., & Price, C. J. (2002). The response of left temporal cortex to sentences. *Journal of Cognitive Neuroscience* 14(4):550–560.

Vandenberghe, R., Price, C., Wise, R., Josephs, O., & Frackowiak, R. S. (1996). Functional anatomy of a common semantic system for words and pictures. *Nature* 383(6597):254–256.

VanRullen, R., & Thorpe, S. J. (2002). Surfing a spike wave down the ventral stream. *Vision Research* 42(23):2593–2615.

Vellutino, F. R., Fletcher, J. M., Snowling, M. J., & Scanlon, D. M. (2004). Specific reading disability (dyslexia): what have we learned in the past four decades? *Journal of Child Psychology and Psychiatry* 45(1):2–40.

Vinckenbosch, E., Robichon, F., & Eliez, S. (2005). Gray matter alteration in dyslexia: Converging evidence from volumetric and voxel-by-voxel MRI analyses. *Neuropsychologia* 43(3):324–331.

Vinckier, F., Dehaene, S., Jobert, A., Dubus, J. P., Sigman, M., & Cohen, L. (2007). Hierarchical coding of letter strings in the ventral stream: Dissecting the inner organization of the visual word-form system. *Neuron* 55(1):143–156.

Vinckier, F., Naccache, L., Papeix, C., Forget, J., Hahn-Barma, V., Dehaene, S., & Cohen, L. (2006). "What" and "where" in word reading: Ventral coding of written words revealed by parietal atrophy. *Journal of Cognitive Neuroscience* 18(12):1998–2012.

Vogels, R., & Biederman, I. (2002). Effects of illumination intensity and direction on object coding in macaque inferior temporal cortex. *Cerebral Cortex* 12(7):756–766.

Vogels, R., Biederman, I., Bar, M., & Lorincz, A. (2001). Inferior temporal neurons show greater sensitivity to nonaccidental than to metric shape differences. *Journal of Cognitive Neuroscience* 13(4):444–453.

Vuilleumier, P., Henson, R. N., Driver, J., & Dolan, R. J. (2002). Multiple levels of visual object constancy revealed by event-related fMRI of repetition priming. *Nature Neuroscience* 5(5):491–499.

Wade, J. B., & Hart, R. P. (1991). Mirror phenomena in language and nonverbal activities—a case report. *Journal of Clinical and Experimental Neuropsychology* 13(2):299–308.

Wallin, N. L., Merker, B., & Brown, S. (2000). *The origins of music.* Cambridge, MA: MIT Press.

Walsh, V., & Butler, S. R. (1996). The effects of visual cortex lesions on the perception of rotated shapes. *Behavioural Brain Research* 76(1–2):127–142.

Wang, S. H., & Baillargeon, R. (2008). Detecting impossible changes in infancy: a three-system account. *Trends in Cognitive Sciences* 12(1):17–23.

Warrington, E. K., & Davidoff, J. (2000). Failure at object identification improves mirror image matching. *Neuropsychologia* 38(9):1229–1234.

Warrington, E. K., & Shallice, T. (1980). Word-form dyslexia. *Brain* 103:99–112.

———. (1984). Category-specific semantic impairments. *Brain* 107:829–854.

Washburn, D. A., & Rumbaugh, D. M. (1991). Ordinal judgments of numerical symbols by macaques (*Macaca mulatta*). *Psychological Science* 2:190–193.

Weekes, B. S. (1997). Differential effects of number of letters on word and nonword naming latency. *Quarterly Journal of Experimental Psychology A* 50:439–456.

Weiskrantz, L., & Saunders, R. C. (1984). Impairments of visual object transforms in monkeys. *Brain* 107(Pt 4):1033–1072.

Werker, J. F., & Tees, R. C. (1984). Cross-language speech perception: Evidence for perceptual reorganization during the first year of life. *Infant Behavior and Development* 7:49–63.

White, S., Milne, E., Rosen, S., Hansen, P., Swettenham, J., Frith, U., & Ramus, F. (2006). The role of sensorimotor impairments in dyslexia: A multiple case study of dyslexic children. *Developmental Science* 9(3):237–255; discussion 265–239.

Whiten, A., Goodall, J., McGrew, W. C., Nishida, T., Reynolds, V., Sugiyama, Y., Tutin, C. E., Wrangham, R. W., & Boesch, C. (1999). Cultures in chimpanzees. *Nature* 399(6737):682–685.

Whitney, C. (2001). How the brain encodes the order of letters in a printed word: The SERIOL model and selective literature review. *Psychonomic Bulletin and Review* 8(2):221–243.

Wilson, A. J., Dehaene, S., Pinel, P., Revkin, S. K., Cohen, L., & Cohen, D. (2006). Principles underlying the design of "The Number Race," an adaptive computer game for remediation of dyscalculia. *Behavioral and Brain Functions* 2(1):19.

Wilson, E. O. (1998). *Consilience: The unity of knowledge.* New York: Knopf.

Wolff, P. H., & Melngailis, I. (1996). Reversing letters and reading transformed text in dyslexia: A reassessment. *Reading and Writing* 8:341–355.

Wong, A. C., Gauthier, I., Woroch, B., DeBuse, C., & Curran, T. (2005). An early electrophysiological response associated with expertise in letter perception. *Cognitive, Affective, and Behavioral Neuroscience* 5(3):306–318.

Wydell, T. N., & Butterworth, B. (1999). A case study of an English-Japanese bilingual with monolingual dyslexia. *Cognition* 70(3):273–305.

Xu, B., Grafman, J., Gaillard, W. D., Ishii, K., Vega-Bermudez, F., Pietrini, P., Reeves-Tyer, P., Di Camillo, P., & Theodore, W. (2001). Conjoint and extended neural networks for the computation of speech codes: The neural basis of selective impairment in reading words and pseudowords. *Cerebral Cortex* 11(3):267–277.

Yoncheva, Y. N., Blau, V. C., Maurer, U., & McCandliss, B. D. (2006). *Strategic focus during learning impacts the neural basis of expertise in reading.* Poster presented at the Association for Psychological Science Convention, New York, May 25–28.

Zali, A., & Berthier, A. (Eds.). (1997). *L'aventure des écritures: Naissances.* Paris: Bibliothèque nationale de France.

Zeki, S. (2000). *Inner vision: An exploration of art and the brain.* New York: Oxford University Press.

Ziegler, J. C., & Goswami, U. (2005). Reading acquisition, developmental dyslexia, and skilled reading across languages: a psycholinguistic grain size theory. *Psychological Bulletin* 131(1):3–29.

Zoccolotti, P., De Luca, M., Di Pace, E., Gasperini, F., Judica, A., & Spinelli, D. (2005). Word length effect in early reading and in developmental dyslexia. *Brain and Language* 93(3):369–373.

Zorzi, M., Houghton, G., & Butterworth, B. (1998). Two routes or one in reading aloud? A connectionist dual-process model. *Journal of Experimental Psychology: Human Perception and Performance* 24:1131–1161.

INDEX

Italic page numbers refer to illustrations.

Index

Figure Credits

Figures 1.1, 1.2, 1.3, 1.4, 1.5, 2.2, 2.4, 2.6, 2.15, 2.20, 3.1, 3.8, 3.9, 4.1, 4.2, 5.2, 5.3, 7.1 (bottom), 7.2, 7.3, 7.5 (top), 8.2 (top) and the anatomical figure at the front of the book were created by the author.

Figure 2.1, composed from elements in Déjerine, J. (1892). Contribution à l'étude anatomo-pathologique et clinique des différentes variétés de cécité verbale. *Mémoires de la Société de Biologie* 4: 61–90; and Figures 4 and 7 from Cohen, L., Martinaud, O., Lemer, C., Lehéricy, S., Samson, Y., Obadia, M., Slachevsky, A., & Dehaene, S. (2003). Visual word recognition in the left and right hemispheres: Anatomical and functional correlates of peripheral alexias. *Cerebral Cortex* 13: 1313–1333. Reproduced by permission of Oxford University Press.

Figure 2.3, Courtesy of Michael Posner and Marc Raichle.

Figure 2.4, composed from elements in Dehaene, S., Le Clec, 'H. G., Poline, J. B., Le Bihan, D., & Cohen, L. (2002). The visual word form area: a prelexical representation of visual words in the fusiform gyrus. *Neuroreport* 13(3): 321–325. With permission of Wolters Kluwer/Lippincott, Williams & Wilkins.

Figure 2.7, adapted from Figure 5, page 5210 of Puce, A., Allison, T., Asgari, M., Gore, J. C., & McCarthy, G. (1996). Differential sensitivity of human visual cortex to faces, letter strings, and textures: a functional magnetic resonance imaging study. *Journal of Neuroscience* 16: 5205–5215. With permission of the Society for Neuroscience.

Figure 2.8, adapted from Figures 6 and 7 of Tarkiainen, A., Cornelissen, P. L., & Salmelin, R. (2002). Dynamics of visual feature analysis and object-level processing in face versus letter-string perception. *Brain* 125(Pt 5): 1125–1136. With permission of Oxford University Press.

Figure 2.9, adapted from elements of Figures 1, 4 and 6, of Allison, T., Puce, A., Spencer, D. D., & McCarthy, G. (1999). Electrophysiological studies of human face perception. I: Potentials generated in occipitotemporal cortex by face and non-face stimuli. *Cerebral Cortex* 9(5): 415–430. With permission of Oxford University Press.

Figure 2.10, adapted from Figures 5 and 6 of Cohen, L., Dehaene, S., Naccache, L., Lehéricy, S., Dehaene-Lambertz, G., Hénaff, M. A., & Michel, F. (2000). The visual word form area: Spatial and temporal characterization of an initial stage of reading in normal subjects and posterior split-brain patients. *Brain* 123: 291–307. With permission of Oxford University Press.

Figure 2.11, adapted from Figures 1, 4 and 5 in Cohen, L., Dehaene, S., Naccache, L., Lehéricy, S., Dehaene-Lambertz, G., Hénaff, M. A., & Michel, F. (2000). The visual word form area: Spatial and temporal characterization of an initial stage of reading in normal subjects and posterior split-brain patients. *Brain* 123: 291–307. With permission of Oxford University Press; and from Figures 1 and 3 in Molko, N., Cohen, L., Mangin, J. F., Chochon, F., Lehéricy, S., Le Bihan, D., & Dehaene, S. (2002). Visualizing the neural bases of a disconnection syndrome with diffusion tensor imaging. *Journal of Cognitive Neuroscience* 14: 629–636. With permission of MIT Press.

Figure 2.12, adapted from Figures 4 and 5 in Dehaene, S., Naccache, L., Cohen, L., Bihan, D. L., Mangin, J. F., Poline, J. B., & Riviere, D. (2001). Cerebral mechanisms of word masking and unconscious repetition priming. *Nature Neuroscience* 4(7): 752–758. With permission of the Nature Group.

Figure 2.13, adapted from Figure 1 in Dehaene, S., Jobert, A., Naccache, L., Ciuciu, P., Poline, J. B., Le Bihan, D., & Cohen, L. (2004). Letter binding and invariant recognition of masked words: behavioral and neuroimaging evidence. *Psychological Science* 15(5): 307–313.

Figure 2.14, adapted from Figure 4 in Cohen, L., Lehéricy, S., Chochon, F., Lemer, C., Rivaud, S., & Dehaene, S. (2002). Language-specific tuning of visual cortex? Functional properties of the Visual Word Form Area. *Brain* 125(Pt 5): 1054–1069. With permission of Oxford University Press; and from Figure 4 in Tarkiainen, A., Helenius, P., Hansen, P. C., Cornelissen, P. L., & Salmelin, R. (1999). Dynamics of letter string perception in the human occipitotemporal cortex. *Brain* 122(Pt 11): 2119–2132. With permission of Oxford University Press.

Figure 2.16, adapted from Figures 5 and 8 in Catani, M., Jones, D. K., Donato, R., & Ffytche, D. H. (2003). Occipitotemporal connections in the human brain. *Brain* 126(Pt 9): 2093–2107. With permission of Oxford University Press.

Figure 2.17, adapted from elements of Figure 1 in Marinkovic, K., Dhond, R. P., Dale, A. M., Glessner, M., Carr, V., & Halgren, E. (2003). Spatiotemporal dynamics of modality-specific and supramodal word processing. *Neuron* 38(3): 487–497. With permission of Elsevier.

Figure 2.18, adapted from elements of Figure 5 in Jobard, G., Crivello, F., & Tzourio-Mazoyer, N. (2003). Evaluation of the dual route theory of reading: a metanalysis of 35 neuroimaging studies. *Neuroimage* 20(2): 693–712. With permission of Elsevier.

Figure 2.19, adapted from Figure 1 in van Atteveldt, N., Formisano, E., Goebel, R., & Blomert, L. (2004). Integration of letters and speech sounds in the human brain. *Neuron* 43(2): 271–282. With permission of Elsevier.

Figure 3.2, adapted from Figure 2 and 3 in Tamura, H., & Tanaka, K. (2001). Visual response properties of cells in the ventral and dorsal parts of the macaque inferotemporal cortex. *Cerebral Cortex* 11(5): 384–399. With permission of Oxford University Press.

Figure 3.3, adapted from Figure 6, page 95, of Tanaka, K. (2003). Columns for complex visual object features in the inferotemporal cortex: clustering of cells with similar but slightly different stimulus selectivities. *Cerebral Cortex* 13(1): 90–99. With permission of Oxford University Press.

Figure 3.4, adapted from Figures 1 and 3 in Booth, M., & Rolls, E. (1998). View-invariant representations of familiar objects by neurons in the inferior temporal visual cortex. *Cerebral Cortex* 8(6): 510–523. With permission of Oxford University Press.

Figure 3.5, adapted from Figures 1 and 5 in Rolls, E. T. (2000). Functions of the primate temporal lobe cortical visual areas in invariant visual object and face recognition. *Neuron* 27(2): 205–218. With permission of Elsevier.

Figure 3.6, adapted from Figures 1 and 7 in Tanaka, K. (2003). Columns for complex visual object features in the inferotemporal cortex: clustering of cells with similar but slightly different stimulus selectivities. *Cerebral Cortex* 13(1): 90–99. With permission of Oxford University Press.

Figure 3.7, adapted from Biederman, I. (1987). Recognition-by-components : A theory of human image understanding. *Psychological Review* 94: 115–147. With permission of the American Psychological Association.

Figure 3.10, adapted from Figures 2 and 4 in Hasson, U., Levy, I., Behrmann, M., Hendler, T., & Malach, R. (2002). Eccentricity bias as an organizing principle for human high-order object areas. *Neuron* 34(3): 479–490. With permission from Elsevier.

Figure 3.11 from Figures 1 and 2 in Cohen, L., Lehéricy, S., Henry, C., Bourgeois, M., Larroque, C., Sainte-Rose, C., Dehaene, S., & Hertz-Pannier, L. (2004). Learning to read without a left occipital lobe: right-hemispheric shift of visual word form area. *Annals of Neurology* 56(6): 890–894.

Figure 4.3, from Web site www.wam.umd.edu/~rfradkin/alphapage.html, with permission of Robert Fradkin.

Figure 5.1, courtesy of Ed Hubbard.

Figure 6.1, (top left) courtesy of Eraldo Paulesu; (top right) adapted from Figure 3 panel D in Paulesu, E., Demonet, J. F., Fazio, F., McCrory, E., Chanoine, V., Brunswick, N., Cappa, S. F., Cossu, G., Habib, M., Frith, C. D., & Frith, U. (2001). Dyslexia: cultural diversity and biological unity. *Science* 291(5511): 2165–2167. Reprinted with permission of the American Association for the Advancement of Science (AAAS). (middle) adapted from Figure 3 in Silani, G., Frith, U., Demonet, J. F., Fazio, F., Perani, D., Price, C., Frith, C. D., & Paulesu, E. (2005). Brain abnormalities underlying altered activation in dyslexia: a voxel based morphometry study. *Brain* 128(Pt 10): 2453–2461. With permission of Oxford University Press; (bottom left and bottom right) courtesy of Albert Galaburda.

Figure 6.2, adapted from figure 1a in Klingberg, T., Hedehus, M., Temple, E., Salz, T., Gabrieli, J. D., Moseley, M. E., & Poldrack, R. A. (2000). Microstructure of temporo-parietal white matter as a basis for reading ability: evidence from diffusion tensor magnetic resonance imaging [see comments]. *Neuron* 25(2): 493–500. With permission of Elsevier; and from Figures 3a and 4a in Beaulieu, C., Plewes, C., Paulson, L. A., Roy, D., Snook, L., Concha, L., & Phillips, L. (2005). Imaging brain connectivity in children with diverse reading ability. *Neuroimage* 25(4): 1266–1271. With permission of Elsevier.

Figure 6.3, adapted from Figure 1 in Temple, E., Deutsch, G. K., Poldrack, R. A., Miller, S. L., Tallal, P., Merzenich, M. M., & Gabrieli, J. D. (2003). Neural deficits in children with dyslexia ameliorated by behavioral remediation: evidence from functional MRI. *Proceedings of the National Academy of Sciences, USA*, 100(5): 2860–2865. Copyright 2003 National Academy of Sciences, USA.

Figure 7.1 (top) Adapted from Walsh, V., & Butler, S. R. (1996). The effects of visual cortex lesions on the perception of rotated shapes. *Behavioral Brain Research* 76(1–2): 127–142. With permission of Elsevier.

Figure 7.4, adapted from Logothetis, N. K., & Pauls, J. (1995). Psychophysical and physiological evidence for viewer-centered object representations in the primate. *Cerebral Cortex* 5(3): 270–288. By permission of Oxford University Press.

Figure 7.5 (bottom), adapted from Figure 3 in Houzel, J. C., Carvalho, M. L., & Lent, R. (2002). Interhemispheric connections between primary visual areas: beyond the midline rule. *Brazilian Journal of Medical and Biological Research* 35(12): 1441–1453, with permission from the publisher.

Figure 7.6, adapted from figures 6 and 7 in Turnbull, O. H., Beschin, N., & Della Sala, S. (1997). Agnosia for object orientation: implications for theories of object recognition. *Neuropsychologia* 35(2): 153–163. With permission of Elsevier.

Figure 7.7, adapted from Figure 1 in Gottfried, J. A., Sancar, F., & Chatterjee, A. (2003). Acquired mirror writing and reading: evidence for reflected graphemic representations. *Neuropsychologia*, 41(1): 96–107. With permission of Elsevier.

Figure 7.8, adapted from Table 2 and Figure 1 in McCloskey, M., & Rapp, B. (2000). A visually based developmental reading deficit. *Journal of Memory and Language* 43: 157–181. With permission of Elsevier; and from McCloskey, M., Rapp, B., Yantis, S., Rubin, G., Bacon, W., Dagnelie, G., Gordon, B., Aliminosa, D., Boatman, D. F., Badecker, W., Johnson, D. N., Tusa, R. J., & Palmer, E. (1995). A developmental deficit in localizing objects from vision. *Psychological Science* 6: 112–117.

Figure 8.1 (left) adapted from Tanaka, M., Tomonaga, M., & Matsuzawa, T. (2003). Finger drawing by infant chimpanzees (Pan troglodytes). *Animal Cognition* 6(4): 245–251. With kind permission from Springer Science+Business Media. (right) Image reproduced with permission from web site http://www.great-apes.com, © Canadian Ape Alliance.

Figure 8.2 (bottom), adapted from Figure 3 in Goldman-Rakic, P. S. (1988). Topography of cognition: Parallel distributed networks in primate association cortex. *Annual Review of Neuroscience* 11: 137–156; and from Figure 1 in Catani, M., & Ffytche, D. H. (2005). The rises and falls of disconnection syndromes. *Brain* 128(Pt 10): 2224–2239. With permission of Oxford University Press.